A HISTORY OF COMPARATIVE ANATOMY
FROM ARISTOTLE TO THE EIGHTEENTH CENTURY

A HISTORY OF

COMPARATIVE ANATOMY

FROM ARISTOTLE
TO THE EIGHTEENTH CENTURY

BY

F. J. COLE

D.SC. (OXON), F.R.S.
EMERITUS PROFESSOR OF ZOOLOGY
UNIVERSITY OF READING

DOVER PUBLICATIONS, INC.
NEW YORK

Published in Canada by General Publishing Company, Ltd., 30 Lesmill Road, Don Mills, Toronto, Ontario.
Published in the United Kingdom by Constable and Company, Ltd., 10 Orange Street, London WC 2.

This Dover edition, first published in 1975, is an unabridged and unaltered republication of the work first published in 1949. It is reprinted by special arrangement with the original publisher, The Macmillan Press, Ltd., 4 Little Essex Street, London WC2R 3LF, England.

International Standard Book Number: 0-486-60224-9
Library of Congress Catalog Card Number: 75-12174

Manufactured in the United States of America
Dover Publications, Inc.
180 Varick Street
New York, N.Y. 10014

PREFACE

THIS is not the book I had designed to write. The original plan leaned to the merits and concerns of the little masters of comparative anatomy, whose memoirs have been almost completely forgotten, and appear to be unknown even to those who write on the history of zoology. Much of this forsaken *corpus anatomicum* is accurately recorded in Dryander's *Catalogus Bibliothecæ Historico-Naturalis Josephi Banks* — that *opus aureum* as Haller would have described it had he been privileged to be familiar with its pages. After protracted and laborious searches, relieved from time to time, it is true, by some rare and curious discovery, I have succeeded in assembling in my own library a large proportion of these minor works. But a book such as I had proposed to myself was nothing less than an exhaustive history of zoological *discovery*, and in times when considerations of brevity must not be ignored its prospects of birth and survival were not impressive. Moreover any attempt at abridgement could but result in a discursive analytical bibliography, lacking in relief and essential historical unity. Since therefore necessity regulated the choice, I decided to abandon for the time the purpose and industry of years, and to undertake the less formidable task of re-studying the masters themselves, together with the developments to which their works had given rise. The present volume covers such a field up to the beginning of the eighteenth century. Some overlap into the first years of that century is inevitable, and indeed in Chapter VIII on the anatomical museum the limit had to be discarded altogether.

It must be conceded that the results of my labours bear out the damping reflection of the poet that " science moves, but slowly slowly, creeping on from point to point ", nor is our admiration stirred by the attitude of the early anatomists towards the accumulations of factual detail which they were industriously piling up. The fundamental inference to be drawn from it all was there right under their eyes, but they saw only its shadow. The tradition of special creation was so firmly established in the minds of men that its validity was never questioned or even suspected. Darwin himself, working some two centuries later, hesitated to attack this venerable and petrified dogma, and approached the task with misgivings which the event amply justified. Hence the historian of these pioneer days can build up a narrative only as imposing and continuous as the material he has discovered justifies, and he must perforce forgo the happy ending as a dream that eluded the grasp of the medieval anatomist.

Readers will notice that biographical detail has been reduced to a minimum, and relegated to an appendix. This aspect of the history of science has been traversed so often that it seemed more profitable to confine the text of the book to material which had hitherto partly or wholly escaped attention, and of *that* there seems to be no end. I regret that the masterly treatise on Fabricius by Prof. Adelmann, although dated 1942, reached me too late to be used in the present work.

The history of medicine has at all times drawn the attention of learned members of the Faculty, but the history of science is a study of more recent growth. Men of science in the past have not reacted to the attractions and significance of wisdom in the making, and they have consequently failed to honour the very workers who made possible their own researches. This neglect, happily, is passing away, and the change is due in the main to the personality and inspired example of George Sarton and Charles Singer. We owe much to their vision and constructive leadership, and long may the discipline which they have imposed permeate the studies of their successors.

In preparing the manuscript for the press I am under a deep obligation to three former students and present friends. Mr. L. J. F. Brimble, joint editor of *Nature*, has revised the final draft, and given me the benefit of his knowledge of book-making and typographical ritual; Dr. K. M. White, honorary research fellow in the University of Reading, typed and checked the whole of the letterpress; and Dr. N. B. Eales, lecturer in zoology in the University of Reading, read the typescript and proofs to the advantage of accuracy and the clarification of obscure passages. The print reproduced as Fig. 130 was courteously supplied by the Director of the Wellcome Historical Medical Museum, and I have also made brief use of three of my own papers published by the Liverpool University Press, the Quekett Microscopical Club and *Nature*. The photographic illustrations are the work of Mr. F. C. Padley, and they testify to the experience and skill which he has brought to bear upon them. My successor at the University of Reading, Prof. C. H. O'Donoghue, very kindly placed the excellent photographic equipment and material of the Zoology Department at my disposal, which resulted in a considerable saving of time and effort.

The dates in this work are, as far as possible, New Style, and conform to modern usage.

To the library of the Royal Society most students of the history of science gravitate sooner or later. The Society wisely and generously makes its library accessible to all serious workers, and to many of them it becomes their spiritual home. Such it has been in my own case for more than forty years, and the debt is hereby warmly acknowledged.

Kingwood, Henley-on-Thames　　　　　　　　　　F. J. COLE
　　May 1943

CONTENTS

Our purpose is to set forth the things which are, as they are.
                                                FREDERICK THE SECOND, *c.* 1248

Some ther be that do defye
All that is newr, and euer do crye
The olde is better, awaye with the new
Becaufe it is falfe, and the olde is true :
Let them this boke reade and beholde
For it preferreth the learnyng moft olde.
                                                WYLIAM TURNER, 1537

Giulio Casserio, 1552–1616   [*Frontispiece*

INTRODUCTION

I

FIRST STATEMENTS OF THE COMPARATIVE METHOD

THE study of the history of any branch of science is one which modern tendencies of scientific thought and research do little to encourage. And this because the current literature to be mastered is so oppressive in bulk and complexity that it swamps the work of the old masters, and also because science is too intimately concerned with the promise of the future to adventure into an obsolete and neglected past. The scientist may admire, but cannot accept, the paradox that the beliefs and knowledge of antiquity have been superseded only by the more rational ignorance of to-day. It is thus the necessity of the man of science, if it is also his misfortune, to focus his efforts on the prospect that lies in front. There, the untrodden fields are so numerous and inviting, and the temptation to add his own small contribution to the sum of human knowledge is in itself so laudable, that he prefers the yoke of the pioneer, and turns his back on that fading but deeply instructive record of wisdom, error and defeat on which, nevertheless, he needs must build. He believes in the mission of his own generation, and in the adequacy of the knowledge and methods he finds at his hand. What he fails to perceive is that the same mental limitations which frustrated the observer of the seventeenth century may wreck his own investigations. Wherefore let us bear in mind that this ancient knowledge still has its uses, if only as a warning of the dangers of research indifferently planned and weakly followed up. It behoves us therefore to examine the works of the early anatomists, to learn from them what we can, and, above all, to take heed lest we sow as they had sown and reap a similar harvest.

Now the study of the history of science, whilst it cannot exclude error, completes the learning of the modern schools and assists in keeping the worker within his powers. It transports him into a quaint and pleasant atmosphere. It stirs his imagination. He can foregather with men, surely as great as any now, and with this added advantage. He can see them as they really are ; his judgment, if condescending, is free from prejudice, and the wit

3

and genius of these old heathens, as Bentley described them, is neither magnified nor obscured by the closeness of the observer. When Robert Boyle approaches him and says : " I have demonstrated the Spring of the Air ", he replies : " Excellent, admitted, but *why* did you advise me to cure my complaints by frying a live toad on a shovel and hanging its ashes round my neck ? " In this mingling of a great discovery with the backwash of romantic medievalism the man of genius is discovered as mortal. In one of his more lucid moments Boyle warns us that " we are not near so competent judges of wisdom as we are of justice and veracity " — a profound truth which he proceeded to exemplify by swallowing his own medicines. Thus history teaches us to be cautious in honouring the drafts of authority, and to accept a statement, not because it is made by Galen and deeply rooted in the learning of the period, but because it is *sound*.

There is, however, another reason why the history of biology should be cultivated. It introduces the element of philosophical interest, perhaps not the least important of the differences which separate the craftsman from the creative artist. Philemon Holland's translation of Pliny's *History of the World* (1601) and the translation by Martin Lluelyn [1] of Harvey's work on *Generation* (1653), are an education in the formidable combination of literature and science. The student who can be persuaded to abate the strong and occasionally contentious waters of modern biology with some such diluents can look forward to grateful and enduring memories of the time thus spent. We run a serious risk in neglecting the humane side of scientific literature, and the academic world has incurred the contemptuous reproach of Edward Gibbon, who tells us that after he left the University his interest in books *began to revive*. He himself practised the wisdom of extending his mental horizon in every direction, and he attended a course of lectures on anatomy by Dr. William Hunter, the eloquent brother of the more famous John, with the result that we may " sometimes track him in our own snow ". No one who studies the literature of zoology of the first half of the nineteenth century can fail to be impressed by the revolution which has since descended on, and restricted, the outlook of the scientific worker. In modern times

[1] So spelt on his grave in High Wycombe church. There are, however, several variants.

Fig. 1.—Vignette from Edward Forbes' *British Starfishes*, 1841

the transformation of the naturalist into the morphologist, accompanied, as was inevitable, by the shifting of interest from the field to the laboratory, and the substitution of mechanical and analytical for observational methods of research, has destroyed that element of genial humanity and art which makes the works of the older biologists a pleasure to handle. Edward Forbes is dead, and is scarcely even a memory to the present generation. No longer are we under the spell of his heartening example. Nor can we imagine him living happily in a despirited and robotized world. What modern author would have the imagination, or the courage, to submit among his illustrations a delightful sketch of a gnome, adorned with horns, wings and a tail, attempting to rescue a starfish from the deadly grip of an oyster — an inversion of Nature, it is true, since it is the oyster that would be in need of protection from the wily attentions of the starfish. Even the sombre Vesalius does not abhor the role of comedian, throwing his dissections into humorous and suggestive attitudes and staging them in attractive but superfluous surroundings. Such illustrations, however, are not necessarily, or even generally, facetious. More often they are relevant to the subject, but fasten on the artistic and personal aspects of it, albeit with complete good-humour and originality. Examples are, on the anatomical side, Cowper's *Myotomia* (1724), Cheselden's *Osteographia* (1723) and Roesel's *Frogs* (1758); and on the zoological side, Réaumur's *Insects* (1734–42), Trembley's *Polypes* (1744) and Forbes' *Starfishes* (1841).

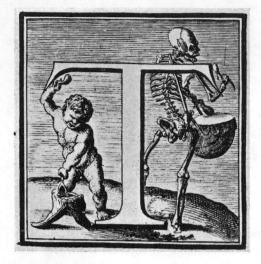

FIG. 2.—Historiated capital from Cowper's *Myotomia*, 1724, figuring the auditory ossicles. Death beating the drum with two mallei. Youth with a malleus forging the stapes on the incus

To insist that there can be no comparative anatomy without evolution is to be bound by the limitations of an academic mind. Biological principles are themselves subject to the inexorable law of evolution, and they must of necessity pass through stages in which the current phase can be recognized only with difficulty. The elucidation of an organ in one animal by comparison with the corresponding feature in another was in fact practised in the earliest days of comparative anatomy. The *explanation* of these homologies is another story, although a closely related one. The birth of a new idea is slow and uncertain, and we must not expect, in searching for the roots of knowledge, to find the developments which distinguish a later growth in the freedom of the atmosphere. In the early days of anatomy the zootomist, who merely dissected an animal, should not be arbitrarily separated from the comparative anatomist, who resolved the relations of its parts. To do so would be to apply a modern standard to an ancient work, and to deny that comparative anatomy had arisen by any process of natural growth with which we are acquainted. Before the period of evolution community of structure might be established between two organs without the inference being drawn that they were *historically* related. Such community was often regarded as evidence of design in the sublime scheme of the Creation, in which case organisms were as independent of each other as are the variations on a theme in a musical composition, each of which arises out of the original subject, with which only is it organically related. Before the coming of evolution the variations were known, but the theme, although dimly perceived, had not taken shape. Darwin disclosed the theme, and revealed to us the composition as a whole.

It is now relevant to enquire when and in what form the conception arose that animal anatomy was worth exploring for its own sake, and especially for the light it might throw on the structure of man and of animals themselves. It is indeed remarkable that the most acute and significant contribution to the comparative method should have come from almost the first naturalist who devoted any attention to it, but in spite of this early lead no attempt was made for long afterwards to found any philosophic system of anatomy with comparison as its guiding principle. If the most important pronouncements on the purpose and expansion of anatomy be arranged chronologically, it will be possible to trace the history of the morphological idea, and to observe how the dull progress of monographic anatomy or zootomy was periodically illuminated by shrewd anticipations of the comparative method as we know it to-day.

1555. In this year Belon published his classic figures illustrating the comparative anatomy of the skeleton of man and bird. These figures, unaltered, were re-issued by Belon himself in 1557,[1] by Aldrovandus in 1642 (a poor, inaccurate copy), and by many subsequent authors. The earlier anatomists mistook the heel of the bird for the knee *turned round*, and were therefore compelled to attribute *two* long bones to the thigh. In correcting this ancient error Belon indirectly assisted in formulating the " law " of the uniformity of composition of animal structure, to which he would certainly have subscribed himself, and which the transcendentalists exploited for much more than it was worth two centuries later. Crié (1882) regards Belon as the " father of comparative anatomy ", but he had predecessors. Leonardo correctly compared the bones of the leg in the horse and man, and Belon himself is

---

[1] Printed from the original wood blocks, but with the lettering set up from type of a different face. In the writer's copy of the 1555 issue the upper mandible of the bird is lettered *A*, instead of the *AB* of all other copies examined. The fact, however, that there is a faint blind impression of the letter *B* shows that this apparent and important variation is due to defective inking.

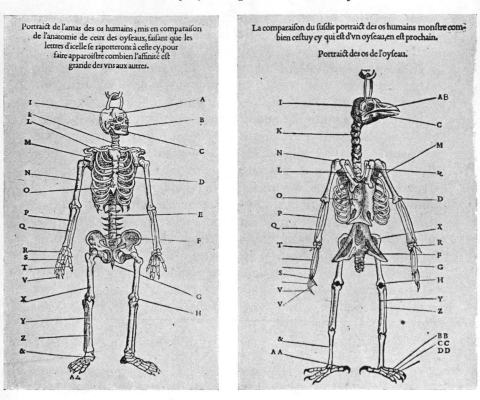

FIG. 3.—Belon, 1555. Comparison of the skeleton of man and bird. Note that homologies are indicated by the use of corresponding reference letters

guilty of the astonishing renunciation of attributing his own achievement to Aristotle, who, he says, not only recognized the similarity of plan in the bones of the wing and leg of a bird and the limbs of a quadruped and man, but also promised to carry out similar comparisons in other animals. There is nothing in Aristotle to justify this generous but unnecessary act of abnegation on the part of Belon. On the contrary, Aristotle completely misinterpreted the bones of the leg in birds, regarding the femur as the ischium, the tibio-tarsus as the femur, and the tarso-metatarsus as the shin bone. These homologies were indeed rectified as early as the thirteenth century by Frederick II, Emperor of Germany (1194–1250), in a work on falconry first published in 1596.[1] Belon, who does not mention the identity of the bird

[1] Composed *c.* 1248. Cf. also the Biographical Notes.

figured by him, tells us that he had dissected two hundred diverse species of birds, and adds " it is therefore not surprising that I am able to describe and figure the bones of birds so accurately ". We may well share his satisfaction, especially when we remember that he had no knowledge of the development of the skeleton, from which alone is it possible to homologize accurately the bones of the wing and leg in birds. Nevertheless the following parts were correctly identified by him : scapula, humerus, radius and ulna ; [1] the two free carpals (recognized as wrist bones or carpals) ; the first digit or thumb and two other digits in the wing ; femur, tibia and fibula ; [2] position of the knee and tarsal joints ; and the elongated metatarsal region or ankle. His mistakes are by comparison venial. The coracoid is naturally described as the clavicle, and the clavicle itself is said to be a bone peculiar to birds. He is puzzled by the elongated carpo-metacarpus, which is left unlettered, although the reference lines are engraved. The bones lettered as metacarpals, if they *are* bones, have no existence in any bird. The proximal phalanx of the second digit of the wing is unlettered and represented as *double*. In some birds this bone is perforated, which would produce the effect figured by Belon, who in fact seems to regard it as a single bone. In his summary he holds that there are in the wing of a bird the same bones as in the arm of man and the fore-limb of a quadruped. A comparison of his two figures, he says, shows how great is the affinity between the skeleton of man and bird, and how much more closely they agree than had hitherto been suspected. Had Belon gone badly astray in his interpretations, as others had done before him, it would have been excusable. It must be allowed that this famous *coup de maître* was the first successful attempt to solve in detail a difficult problem in comparative anatomy, and if the result it produced on his own mind was teleological, it will be admitted that such a reaction was governed by the time in which he lived. In the 1557 version of the figures Belon adds the following two verses, which may be freely translated as claiming that all birds are alike, and that their structural conformity with man is one of the subtleties of Nature :

[1] In the human skeleton the confusion of radius and ulna in the figure must be a printer's error, since the text on this point is sound.

[2] The tibia and fibula are mislettered in the figure of the human skeleton—again a printer's error.

La fection d'vn oyfeau feulement
De tous oyfeaux l'interieur demonftre :
Bien que les vns foient de petite monftre,
Autres trop grands, autres moyennement.

Des os humains la vraye pourtraiture
Soit des oyfeaux mife en comparaifon,
Et lon verra que non pas fans raifon
En fes effaits fe iouë la nature.

1563. Eustachius observes that if you would contemplate the varied and admirable workmanship of Nature it is necessary to examine the anatomy of brutes, since much knowledge is to be obtained thereby which is not at first revealed in the structure of man.

1623. Lord Bacon was the first to use the expression *comparative anatomy*, but not in the modern sense. He complains that whilst pure (human or monographic) anatomy has been diligently studied, comparative anatomy, or the comparison of members *of the same species* [individual variation], has been neglected. He recommends the careful study of many different dissections.

1628. William Harvey drew freely on the animal kingdom in his experimental researches on the circulation of the blood, and mentions some fifty species which he had personally examined. It is remarkable therefore that he refers only once to the comparative method ; but it is in language which, though concise, shows that he fully grasped its great importance. It will be noticed that his conception of the method is the modern one and not Lord Bacon's. This is what he says : " Had anatomists been as conversant with the dissection of the lower animals as they were with that of the human body, the matters which have hitherto kept them in a perplexity of doubt would, in my opinion, have been freed from every kind of difficulty ". In his second letter to Riolan (1649) he refers to some criticisms of his comparative studies in the following words : " There are some who say that I have shown a vainglorious love of vivisections, and who deride the introduction of frogs, serpents, flies and other lower animals upon the scene, as a piece of puerile levity, not even refraining from opprobrious epithets ". The time was evidently not ready for the acceptance of the new approach to the study of animal structure.

1645. Severino introduces the term *zootomia*. It rapidly passed into general use, but has been almost abandoned in modern times. He derives it from ζῷον and ἄτομον, and it means animal dissection carried to the point of indivisibility — animals resolved into their smallest parts or atoms. The derivation usually given is τομή, a cutting up of animals, which is incorrect. Severino uses zootomia as a plural term to include animal dissections generally, and zootomis is a single animal dissection. Also an anatomy lesson is " an anatomy ".

1649. John Bulwer speaks of his correspondence with " Doctor *Wright* Junior [? Dr. Robert Wright] . . . He shewing me the hint of his grand undertaking, which was *Anatomia Comparata*, that great Defect in Anatomy noted by my L. Bacon in his Book *De Augmentis Scientiarum*." [1] It is clear that Bulwer's conception of comparative anatomy is no advance on Lord Bacon's, and therefore he cannot be regarded as the first to use the term in a morphological sense.

1654. Whitlock employs the word zootomia in the title-page of a work which has no connection with anatomy, but is a satire on the manners, customs and follies of the seventeenth century. He says nothing on the derivation of the term, and it appears only on the engraved and printed titles.

1661. Malpighi in his famous letter on the structure of the lungs remarks : " This is what I have been able to discover by observations made with the object of linking up anatomical practice — ad usum anatomicum comparâsse ". He is here not suggesting comparative anatomy as a scientific term, but the idea of converting anatomical knowledge into a science by the comparative method seems to be present in his mind.

1663. Boyle defines *andratomy* as " the Dissection of Mans Body " and *zootomy* as " the Dissections of the Bodies of other

---

[1] Dr. Hubert J. Norman was the first to direct attention to this passage.

Animals ". " The Naturalist by his Zootomy ", he says, " may be very serviceable to the Physitian in his Anatomical Inquiries." " The diligence of Zootomists may much contribute to illustrate the Doctrine of *Andratomy*, and both inform Physitians of the true use of the parts of a humane Body, and help to decide divers Anatomical Controversies. For as in general 'tis scarce possible to learn the true Nature of any Creature, from the consideration of the single Creature itself : so particularly of divers parts of a humane Body 'tis very difficult to learn the true use, without consulting the Bodies of other Animals."

1664. De Graaf in his classic work on the pancreas remarks : " We consider that it is worth while to examine many animals of different species because what is often more condensed or more concealed in one species Nature displays more clearly and openly in another ". This opinion was adopted and repeated in almost identical language by Swammerdam *c.* 1673.

Willis (1664) supports " Anatomia Comparata " because it contributes to the understanding of the uses of the parts. Hence, he says, we must study first of all the brains of other animals, such as birds and fishes. He recognizes an analogy between the brains of fishes and birds, and of four-footed beasts and man, chiefly on the ground that the former were brought forth on the fifth day of the Creation and the latter on the sixth.

1666. Malpighi's views on comparative anatomy have strengthened since 1661. He now considers that the anatomy of his own time is superior to that of the ancients, because the lower forms of life are investigated and compared with the higher. " Since in the higher red-blooded animals ", he says, " it often happens that their [basic] simplicity is involved in many obscurities, it is hence necessary to commence our work with the examination of the lower animals."

1667. W. Needham heads Chapter VII of his work on the foetus with the expression : " Embryotomia comparata sive directio cultri ". What this means is open to speculation, especially as the term *embryotomia* has been applied to the dissection of the foetus, the extraction of the foetus from the uterus by operation, and even to the separation of the foetus from the placenta by severing the umbilical cord, but the first sentence of the chapter seems to make it clear that Needham has in mind the coining of a term analogous to *anatomia comparata*. He says : " I do not write for the initiated but for beginners, for the benefit of whom I shall describe historically what happens in various animals ".

1668. Charleton speaks of " Anatomia comparativa " in the restricted sense of *animal* anatomy in his introduction to Ent's zootomical contribution to the *Onomasticon Zoicon*. Ent himself makes no attempt at comparison in his descriptions of the anatomy of *Lophius*, *Galeus* and *Rana*.

1670. An anonymous French writer in a communication dated 1670, but not published until 1733, observes : " Nothing is more advantageous for Anatomy than the comparison of parts in different animals. Often a structure will be invisible in one species and quite obvious in another, and where you find two different mechanisms which should be the same the one which is the more developed and more clearly adapted to a particular purpose serves to explain the working and use of the other which is the more obscure."

c. 1670–3. First published, 1738. Swammerdam refers twice to the comparative method in the following terms : " The manner in which the blood circulates in the adult frog is of no small importance. It most highly recommends compared anatomy, since without it one cannot arrive at a true knowledge of the use of the viscera." " The structure of one animal throws a light on, and explains, that of another, for compared anatomy is the most liberal and accurate guide, and parts which are dark and incomprehensible in one animal may often be understood very clearly in another." The latter quotation, as already pointed out, is based on de Graaf of 1664.

1672. First published, 1757. W. Croone says : " This [the amnion] is the second of those membranes which in embryotomia comparata my most worthy and learned friend D. D. Needham has enumerated ". Here no comparison in the modern sense is understood, but only the embryology of animals as distinct from that of man.

Grew warns us to " neglect not the comparative Anatomy ; for as some things are better seen in one estate, so in one Vegetable, than another ". This is the first appearance of the term, having a modern connotation, in English.

In his index Hoboken introduces the question — " Anatomia comparata cui bono ? " In the text he answers it as follows : " Comparative anatomy is in most cases certainly of great assistance, because it throws considerable light on truths which would otherwise remain obscure ".

Willis has a somewhat ambiguous reference to " Anatomia multiplex ac comparata ", which may mean that dissections should be often repeated and compared with each other [for variations], without involving the true comparative method. Valentini (1720), however, gives Willis the benefit of the doubt, especially as he distinguishes the dissection of beasts as zootomy.

Wepfer expects that " Anatomia Comparata " will throw some light on the problems of hermaphroditism.

1674. T. Bartholin is interested in the pathological aspect of animal anatomy, and is " of opinion that Zootomia, or the dissection of *sick* beasts, should by no means be neglected ", on the ground that animal diseases may resemble those of man.

1675. The first work in which the phrase comparative anatomy occurs on the title-page is *The Comparative Anatomy of Trunks* [of trees], by Nehemiah Grew, read before the Royal Society in 1674–5. It was translated into Latin in 1680 under the title *Comparativa Anatomia Truncorum*, but in the second edition of 1682, issued during Grew's lifetime and under his supervision, the word *comparative* is omitted, for reasons which are not stated. An anonymous reviewer of this work in 1676 refers to the " Absolute and Comparative Anatomy of Roots ", but he does not explain this expression, nor was it used by Grew himself in the treatise under review.

1676. In lectures read before the Royal Society in 1676, but not published until 1681, Grew produces another interesting title-page which reads *The Comparative Anatomy of Stomachs and Guts begun*, but this time he does more than that, and demonstrates that he fully grasps the implications of the comparative method. In comparing the compound stomach of ruminants with the simple stomach of man he reaches the following accurate conclusion : " The Fourth *Venter* is called *Abomasus* : by *Butchers*, the *Read*. The only analogous one to that in a Man." [1] He therefore recognizes that the paunch, honeycomb and psalterium are new chambers not represented in the stomach of other mammals.

1677. Charleton refers to Lord Bacon's views on the usefulness of " Anatomia Comparativa ", but does not explain what he himself understands by this term. This passage is wanting in the first edition of 1668.

Glisson defines anatomia comparativa as a comparison of man with any other kind of animal, and also of diverse individuals of the same species (in Lord Bacon's sense).

1679. An anonymous reviewer of Lorenzini's monograph on the *Torpedo* (1678) says that upon all occasions the author confirms his assertions " by Arguments, Experiments, and new comparative Anatomy ". The title-page of the English translation of Lorenzini's work (1705) reads " The Curious and Accurate Observations of Mr. *Stephen Lorenzini* of Florence, on the Dissections of the Cramp-Fish : containing The *Comparative Anatomy* of that and some other Fish, with *Experiments* ", but the title-page of the original Italian edition includes no reference to comparative anatomy.

1680. Charleton divides anatomy into *simple* and *comparative*. The former relates to man only, and the latter is subdivided into (1) a comparison of the whole body with all its most remarkable parts, that is, the mutual relationship of the whole and its parts ; (2) a comparison of the same parts in human bodies of different ages, sexes and countries — this is Lord Bacon's comparative anatomy ; and (3) a comparison of the parts of quadrupeds, birds, fishes, insects and worms with the same or like parts in man, so that common characters and properties may be observed. Charleton

[1] Cf. Severino, who had already suggested this homology in 1645.

states that comparative anatomy, recent as the name is, was practised by the Ancients, but by this he means *animal* anatomy and not comparative anatomy *sensu stricto*. He points out that Galen was well versed in comparative anatomy, and criticizes Lord Bacon for declaring that it was not sufficiently studied. Charleton admits that more might have been made of it, but claims that it had not been neglected. His own conception of comparative anatomy, however, is teleological, and is restricted to the search for a common structural type to which all animals more or less conform. He applies the word *Zootomie* to the dissection of the bodies of brutes of various kinds.

The first bibliography of comparative anatomy was published at Heidelberg in 1680 by G. Franck von Franckenau under the title of *Bibliotheca parva Zootomica*.

In the same year Guenellon greatly recommends comparative anatomy, chiefly because it is possible to teach much human anatomy from the dissections of brutes owing to the great similarity between them — a method, he says, which has given occasion to so many beautiful discoveries.

Tyson, who also separates animal anatomy or zootomy from human anatomy, and has " a design of giving the Comparative Anatomy of the organ of hearing in various animals ", develops his views on the method in this passage : " In nothing have our endeavours been more successful than in making a *comparative* survey. Nature when more shy in one, hath more freely confest and shewn herself in another ; and a Fly sometimes hath given greater light towards the true knowledge of the structure and the uses of the Parts in Humane Bodies, than an often repeated dissection of the same might have done."

1682. An anonymous reviewer writes : " Many *industrious persons* of *all* Ages, especially of *this* we live in, have thought the comparative *Anatomy* of *Animals* a subject well worthy their cultivating " ; but zootomy only is here indicated.

Moulin takes a wider view. He says : " I could not spare time to make a comparative Anatomy of the Eyes of most Animals to be had in our Countries, which I take to be the best way of informing our selves in their structure, and in finding out the uses of their several parts ".

1683. Muralt's convictions, as expressed in the following translation, are in advance of his practice : " Anatomia comparata is the son of Medicine in the image of Ariadne, and with its assistance we can extricate ourselves from the entanglements of doubt, and discover an unimpeded approach to the sublime secrets of Nature ".

1685. Samuel Collins' large and imposing volumes represent the first attempt to produce a comprehensive treatise on comparative anatomy based on new material. The previous work of Blasius (1681) is largely a compilation, in which no attempt is made to co-ordinate the numerous quotations and abstracts of which it is composed. Collins goes further, but his outlook is restricted to an attempt to unravel the intricacies of man, and is therefore only partially comparative. He says : " And I humbly conceive the great use of comparative anatomy is to illustrate the structure, actions, and uses of man's body, which are sometimes more clear in that of other animals, than in ours : as I have discovered in frequent dissections to my great satisfaction, pleasure and admiration ".

1687. Pitfeild remarks that as regards " the Construction, Fabrick and Genuine Use of the Parts of Animals, and even of Man : A Knowledge no way better to be obtained than from the Comparative Anatomy of divers Animals ; that Texture of Parts being discoverable in one Animal, which Nature has conceal'd and made more obscure in another ".

1694. An anonymous reviewer of Lister's work of the same date speaks of animal anatomy and comparative anatomy as if they were synonymous terms. Lister himself does not. He is dealing with the anatomy of Gastropods, and in that limited field his method deserves the name he applies to it — " Anatomia comparativa ". He says that only internal anatomy can give us any true knowledge of these animals, nor are the uses of the parts discoverable unless that anatomy be *comparative*. In the following year (1695) he claims to have instituted the comparative anatomy of Gastropods, and emphasises the very great assistance rendered by the comparative method in interpreting the parts in cases where they are variously developed.

1695. In discussing the circulation of the blood in the brain Ridley provides us with a good example of the importance of comparison in anatomy. He refers to certain obscurities in man, due to the adoption of the erect attitude, which are cleared up by examining the same parts in brutes. This, he adds, " therefore shews the great usefulness of Comparative Anatomy ".

1697. An anonymous reviewer of Malpighi's posthumous works mentions that " Zootomy is either for compleating natural History . . . or for the better Attainment of the Cure of Diseases ". In this he is repeating Malpighi himself, whose own views on comparative anatomy are scarcely characteristic of so distinguished an observer of Nature. After speaking of the picturesque but imaginative discussion between Hippocrates and Democritus on the dissection of the lower animals, and on Galen's anatomical studies of various mammals, birds, reptiles and fishes, he quotes Giraldi (1693) as stating that zootomy is a useful occupation if only because vivisections are possible on animals. In fact certain obscurities in the structure and physiology of man are more readily comprehensible when examined by the comparative method. Malpighi refers to the dissection of animals as zootomia, and adds that it completes or rounds off natural history.

1699. Published 1702. Duverney commends comparative anatomy as a method which has contributed so much to clarify our knowledge of the structure and function of the parts of the body in man and animals.

In the same year Tyson strikes a new note, and introduces us to an anatomical " Scale of Beings ". He says : " To render this *Disquisition* more useful, I have made a *Comparative* Survey of this *Animal* [Chimpanzee], with a *Monkey*, an *Ape*, and a *Man*. By viewing the same Parts of all these together, we may the better observe *Nature's Gradation* in the Formation of *Animal* bodies, and the Transitions made from one to another ; than which, nothing can more conduce to the Attainment of the true Knowledge, both of the *Fabrick*, and *Uses* of the Parts. By following *Nature's* Clew in this wonderful *Labyrinth* of the *Creation*, we may be more easily admitted into her *Secret Recesses*, which Thread if we miss, we must needs err and be bewilder'd."

1701. C. Bartholin dwells on the importance of zootomy to the understanding of the human body. There are many parts, he says, the meaning of which would be as dark as night but for a knowledge of their analogues in brutes.

1705. Greenhill's opinions on the place of comparative anatomy in the system belong to an earlier period, and to an attitude of mind already discredited. After remarking that we must make most of our enquiries into human nature by *dissections*, he concludes : " and tho' Brutes may sometimes be useful in *Comparative Anatomy*, yet Man being the *Epitome* and Perfection of the *Macrocosm*, his Body shews a more wonderful Mechanism than all other Creatures can do ".

1707. Drake is still obsessed with the importance of the human body over all others. He says : " The Subject of [anatomy] is a *Humane Body* : And tho' *Brutes* and *other Animals* . . . fall under its Consideration, it is only in order to illustrate the former ; and is called *Comparative Anatomy* ".

1723. Stukeley brings his florid rhetoric to bear on an alien and unresponsive subject, but, as is often the case with rhetoricians, the thought struggling for utterance is difficult to recover from the coils of verbosity in which it is entangled. " The great uses of comparative anatomy ", he observes, " are so glaring and manifest, that nothing need be said to recommend it. . . . We need not insist upon the great light arising therefrom, towards explaining difficult, points in our own economy. . . . Comparative anatomy is the metaphorical oratory of nature, a divine sermon, where she explains her purposes by pleasing circumlocutions, and a redundancy of invention, that strikes us with inconceivable ravishment."

1728. Sir Hans Sloane approximates to the modern attitude in the following passage : " It would be an Object well worthy the Inquiries of ingenious Anatomists, to make a Sort of comparative Anatomy of Bones. . . . This would doubtless lead us into many Discoveries, and is otherwise one of those Things, which seem to be wanting to make Anatomy a Science still more perfect and compleat."

1733. J. J. Baier comments on the results to be obtained by the study of " *comparata* (quam dicimus) *anatomia* ". A comparison of man with quadrupeds which closely resemble him, he says, would throw much light on his anatomy and physiology.

1744. Monro *primus* is the author of the first *general* treatise on comparative anatomy [1] in which this term appears on the title-page, but the work is a transcript of his lectures at the University of Edinburgh prepared by one of his students and published anonymously, without his knowledge or consent. In the authorised edition of the essay, however, edited by his son and published in 1783, the following passage is reproduced without alteration : " The great Use of Comparative Anatomy is the Light it casts on several Functions in the human Oeconomy about which there have been so many Disputes among Anatomists : These will be in a great Measure cleared up by exhibiting the Structure of the same Parts in different Animals, and comparing the several Organs imployed in performing the same Action, which in the human Body is brought about by one more complex ". The doctrine is sound, but the practice indifferent. Monro's work is a series of isolated animal anatomies, which are quickened only by occasional attempts at comparison.

1753. Daubenton very properly points out, and not before it was necessary, that what had been called comparative anatomy was in most cases not comparative, but only monographic or systematic anatomy of individual animal types. Structural comparisons of one animal with others of the same or related groups scarcely existed, nor was it customary to work to any generalized and philosophical system. He admits that in the *Memoires* of the Parisian group comparisons on a limited scale were successfully accomplished, but Valentini's *Amphitheatrum Zootomicum*, the most celebrated treatise of its class, which might have given us such a system, failed to do so. Nevertheless it would have been possible to have compiled a defensible scheme from this work alone. Later, in 1786, Vicq-D'Azyr expresses himself to the same purpose, but still more vigorously. He says that when we examine the miscellanies of Blasius and Valentini, and note the masses of undigested

[1] Cf. the more modern general treatise of 1808–9 by G. Jacopi.

and incongruous facts which are there assembled, we can understand how, in the midst of all these riches, we experience a feeling
of fatigue and weariness.

1757.   Haller acknowledges that his great work on Physiology
is based largely on "comparatae anatomes utilitates".

1774.   Haller attributes the rise of comparative anatomy to
the fact that early in the seventeenth century the practice of human
anatomy had been subject to so much obstruction and expense
that the scalpels of medical men were diverted to the dissection of
animals, and especially of living animals, since on them alone were
physiological experiments possible.

c. 1780.   John Hunter's conception of comparative anatomy
accepts man as the central object, with whom other types are to be
compared.   Therefore the human body must be mastered first, for
if we do not understand the standard we cannot expect to understand the variations from it.   " When ", he says, " I compared
the human with the quadruped . . . it always put me in mind of
two machines of the same kind, one made by an artist, the other
only an imitation of it made by a novice."   This carries us no
further towards a true comparative anatomy, nor was the final
advance possible until the demonstration of the historical relationships of animals had placed the whole problem on an entirely new
footing.

1797.   The foregoing selection of opinions on the rational basis
of comparative anatomy leads us, with some surprise, to the
chastening reflection that the zootomist is implicit in the butcher.
Silas James, in his *Narrative of a Voyage to Arabia*, remarks that
" the cook lives in East Smithfield, where he exercises the trade of
Zootomy ".

1812.   Our last quotation, from T. Thomson, would almost
suggest that we finish where we started : " The term *anatomy*,
without any epithet, is usually applied to human anatomy ; while
the anatomy of the inferior animals, is called *comparative* anatomy ".

The old anatomists were not slow in reacting to the invasion
of their territory by the " viler creatures ", but for a long time the

change is evident in their attitude of mind rather than in their practice. The tendencies at work weakened their convictions without disturbing their habits. The publications which have been quoted are representative of the better known workers, but many unexpected and important omissions will be noted, such as T. Bartholin, Perrault, Redi, Leeuwenhoek, Steno and Monro *secundus*, who were apparently too fascinated with the details of the new anatomy to devote much attention to underlying principles. In addition there is a large body of purely descriptive minor zootomists whose work is unknown even to historians of zoology, but whose publications have for the most part been accurately recorded in that model bibliography — the *Catalogus Bibliothecæ Historico-Naturalis Josephi Banks*.

What material progress was made during these early days ? The answer is neither manifest nor free from perplexities. Progress depends on the influence of some energizing and integrative speculation, sound or unsound, which is sufficiently bold and provocative to attract supporters and opponents, eager to investigate its validity. Without such a stimulus the observer is for ever groping in the dark for the switch he cannot reach. We know now that the theoretical basis of comparative anatomy is the *genetical* history of animals, or in other words the principle of evolution. Was *any* advance possible before this became operative, and without it can we say that comparative anatomy existed as an established branch of science ? Immediately preceding evolution we have the transcendentalists, over-confident and patronizing, who demanded unquestioning acceptance of that fantastic misconception known as the " Unity of Composition ". According to these misguided enthusiasts there is only one animal type, which is manifest to us as a scattered multitude of variants or departures from the type. The purpose of the transcendentalist was the reconciliation of these departures with the central abstraction — the reduction of variety to uniformity, or, in the terms of the principle of Leibniz, the perception of uniformity in diversity. *Ab uno disce omnes*.

The alternative was the teleological doctrine of " Design in Nature ", which accepted the dogma of uniformity in diversity, but saw in it evidence of a universe constructed according to plan, and hence by a Divine Architect. It should be noted that the transcendentalist, the teleologist and the evolutionist made use of the same data, and in much the same way — they were all seeking after homologies, but their interpretations were different. If the writings of these builders of philosophical biology be compared, they will be found to exhibit remarkable points of agreement. For example, a few unimportant modifications would convert Goethe's "transcendental" speculations on the intermaxillary bone (1784) into an essay in evolution, to which the same adjective, but in its literal sense, could be applied. Therefore some restricted and even significant advances were possible before the coming of evolution, and the comparative anatomist might have made his contribution to the foundations of his science even when he could only faintly visualize the character of the building, itself transitory, which his successors would hopefully erect on them. It must, however, be recorded that this possibility was but imperfectly realized. The old anatomists frequently speak of the light which comparative anatomy would throw on the structure of man. Boyle, Glisson, Guenellon, Collins, Duverney, C. Bartholin, Drake, Stukeley, Sloane, Monro *primus* and Haller all do this, but few of them are sufficiently venturesome to put their opinions to the test of experience, nor do any of them, except perhaps Duverney, appear to suspect that comparative anatomy is entitled to exist apart from the problems associated with the human body.[1]

Again, many of them, such as Belon, Harvey, Malpighi, de Graaf, Swammerdam, Grew, Hoboken, Tyson, Moulin, Pitfeild, Lister and Ridley, dwell on the scientific implications of the comparative method with a vision almost prophetic, but Belon is almost the only one who gives us a critical example worked out in detail. To Severino, Needham, Charleton, Croone, Willis and Greenhill comparative anatomy has no theoretical bearings, and is a study concerned only with the descriptive anatomy of *animals* (zootomy). In the pre-evolutionary period a comprehensive system of comparative anatomy based on strictly morphological conceptions was scarcely to be expected. It is true that Etienne Geoffroy Saint-Hilaire, a nascent evolutionist, projected an ambitious work of this calibre, and even made an impressive beginning, but before it could be completed *Naturphilosophie* had wilted under the punishing criticism of Cuvier.

[1] This was, however, clearly recognized later by Geoffroy.

I

THE CONTRIBUTION OF GREECE

II

ARISTOTLE

THE history of modern zoological research is a record which is not necessarily continuous or progressive. There are times of stagnation, of failure to develop advances already begun, and even occasionally of reaction. No great discovery is the work of one man, or even of one generation, but may represent centuries of human endeavour. And when the final stage has been reached, rejection and lack of understanding may limit and delay its impact on the science of the period. This capricious ebb and flow, however, may be grouped into three well-defined eras : first, the period of Greek science, covering some 800 years, and forming a tradition which the Romans did not inherit ; second, the Dark Ages, followed by the scholastic period, which, though not as black as they have been painted, must still be regarded as an epoch when true learning was suspended for over 1300 years ; and third, the revival of original research, or the period of modern science, which has an unbroken history from the sixteenth century to our own times.

The most important, or perhaps we should say the best known, of the Greek biologists are Aristotle (384–322 B.C.), his pupil and successor Theophrastus the botanist (374–286 B.C.), and the encyclopaedic Galen (A.D. 130–200). It must, however, on no account be assumed that the outstanding figures of Aristotle and Galen exhaust the list of Greek zoologists and anatomists. For example, in the writings of the Hippocratic School, there is an attempt, the details of which have been lost, to compare the skeleton of man with that of other vertebrates. In the fifth century B.C., Anaxagoras dissected the head of a ram, Empedocles examined the structure of many animals and discovered the cochlea of the ear, and Democritus, on the somewhat doubtful authority of the elder Pliny, is supposed to have dissected the chameleon, and " verely made so great reckoning of this beast, that hee compiled one entire

24

booke expressely of it, and hath anatomized everie severall member thereof ".[1] Many of the works of the Greek biologists have perished beyond recovery. Indeed, but for chance references to them and their activities in the writings of their contemporaries, even their names would be unknown to us.

The zoological and anatomical works of Aristotle which have survived, and they are by no means all of them, cover a field so vast in extent that it is necessary, however briefly, to consider them as a whole. Greek science was preceded by centuries of activity in the civilizations of Babylon and Egypt, and it was from this great heritage in the hands of the Greek philosophers that the various sciences began to assume the form with which we are now familiar. The Greeks, however, were observers of Nature only in so far as such observations bore on some problem grateful to the human mind, or important to human life. In other words they did not observe Nature for its own sake, or because they loved it, but rather as an intellectual recreation or a practical exercise. Thus Aristotle's fauna embraces relatively few species, but the structure and habits of living things concern him vitally as a human being, and his success in elucidating them is indeed remarkable. It is significant that Greek art represents the superficial anatomy of the human form with astonishing faithfulness, but the surroundings in which man is placed receive only a formal recognition. Again, animals as being more akin to man than plants were studied first ; and when plants do begin to receive attention, their importance to human food and medicine is obviously the chief stimulus.

The relevant anatomical works of Aristotle, the founder of biological science, are the *History of Animals*, the *Parts of Animals*, and the *Generation of Animals* — written probably in that order. He himself refers to two other biological writings which have not survived. To what extent he was an original worker or compiler, or both, can only be conjectured from the internal evidence of the texts themselves. They do not appear to be finished works prepared expressly for circulation, but rather rough lecture notes containing errors so flagrant that it is difficult to believe they could have been made by Aristotle himself.[2] Many of these lapses are

---

[1] Holland's free translation.

[2] Huxley was astounded that Aristotle should have committed himself to absurdities which were contradicted by the plainest observation and excused by no theoretical prepossession.

due to the blind acceptance of statements by fishermen, huntsmen, travellers and other untrained and prejudiced witnesses. Aristotle adopted and perpetuated, but did not originate, them. It is greatly to be regretted that the diagrams to which he refers in many places have never been recovered. According to Platt the treatises of Aristotle " have suffered terribly in the process of transmission to us, and are full of grievous blunders committed by scribes ; whole passages have often fallen out and we can only guess what was in them ; other bits have been added by people too ignorant to avoid supplying nonsense for sense ".

Aristotle's science was accepted almost without question until the revival of learning, when his authority at last began to weaken. The accuracy and bearings of his works were now vigorously canvassed, and even their authenticity was debated. Others respected Aristotle himself, but derided the Aristotelians. The immediate result was that his writings generally declined in favour until he was almost forgotten, but this fate was not shared by his zoological works, which were more carefully studied and quoted than ever. His methods, however, presented fewer openings to his critics, and we must not forget that it was Aristotle who dictated the form of the modern scientific memoir — first, the statement of the problem, then, a discussion of the contributions to that problem made by predecessors, and finally, the author's own contribution to it. In his natural history works Aristotle's first concern is with observation ; he aims at simplicity and clearness of literary style, and shuns metaphysical discussion ; but his language in places can almost defy comprehension. It must also be admitted that he does not always acknowledge the sources of his information, and he occasionally fails to observe the excellent rules which he drew up for his own guidance in research. If he had done so, he could not, for example, have believed that the lion and the wolf had only one bone in their necks, nor would he have indulged in so many generalizations devoid of any observational basis or check of his own.

No definite or consistent classification of animals was drawn up by Aristotle ; but it is possible to construct one or more from his writings. However, his views on this subject are not based on a knowledge even of the whole of the comparatively few species, predominantly marine, known to him. Also his diagnoses are not always sufficiently precise, with the result that a species may have a choice of groups, and a group may include obviously unrelated species. In spite of this, it must be allowed that he was feeling his way towards a *natural* classification, and in some respects he was more sound than Linnaeus and other systematists of modern times. For example he separates the Crustacea from the Insecta, although he groups together as " Entoma " the Myriapods, Arachnids and Insects. His division of the animal kingdom into equivalents of the modern Invertebrata and Vertebrata, although his diagnostic feature here is the absence or presence of red blood, is fundamental, and his observations on the position of the Cetacea are equally acute. He holds that these animals are *not* fishes for the following reasons : they have hair — a mammalian feature ; they have lungs and breathe like a terrestrial animal, and there are no gills ; they suckle their young by means of mammae ; they are " internally viviparous ", that is, no egg is discernible ; they do not *constantly* take in water, but do so incidentally when feeding ; and finally the structure of Cetacean bone is that of ordinary or mammalian bone, whereas fish bone is different and is only *analogous* to bone. In addition, he points out that the young grow very rapidly and that the dolphin is full grown at ten years of age — a conclusion which has been confirmed in modern times. In 1551 Belon reversed Aristotle's conclusion that the Cetacea are mammals, although he found the placenta, and they were for a long time included among the fishes by many naturalists, in spite of, or perhaps rather in ignorance of, Aristotle's sagacious and accurate pronouncement on their affinities. His classification of birds based on the structure of the foot has had its counterpart in modern systems, and his separation of the cartilaginous from the bony fishes has also been amply justified. He errs in regarding *Lophius* as a cartilaginous fish, but its skeleton is actually more cartilaginous than that of bony fishes in general.

Aristotle separates the Cephalopods from the other Mollusca, and has closely and successfully investigated their structure. His Invertebrate groups, except the Ostrakoderma, are not unacceptable to the modern zoologist, but it is necessary to note that his "Holothouria " were probably not Echinoderms, and may not even have been animals. He is sounder than many of his successors in regarding sponges as animals, and his reasons for doing so partly anticipate

modern experimental proof.   Intermediate types engage much of
his attention.   The hermit-crab, he says, links up the Crustacea
and the Mollusca.   In itself it is a typical Crustacean, but by forcing
its body into a Molluscan shell it comes partly to resemble the
animal which normally inhabits that structure.

A Scala Naturae or progressive scale of beings was not formally
attempted by Aristotle, but the idea was undoubtedly present in
his mind.   Any such scheme of his necessarily omits the lower
forms of life, of which he had no knowledge.   He says it is difficult
to construct a series because the stages are so blended with each
other that it is almost impossible to identify the boundaries which
separate them.   On the other hand some gradations are incon-
testible ;   for example, animals which have red blood are higher in
the scale than those which are " bloodless ".   In plants there is a
continuous scale of ascent towards the animal, and in the sea there
are forms of life, such as the zoophytes and sponges, which have
relations with both animals and plants.   He attaches importance to
sensibility or responsiveness, especially the sense of touch, as a
criterion between animals and plants, and, as a less important
feature; attachment, or absence of motion, the latter being essentially characteristic of plants.   A species which is attached but yet
has some sensibility, such as an Ascidian, is intermediate between
animals and plants.   The medusae are plant-like in the imperfection
of their structure, but animal in their habits.   They use the asperity
of their bodies [the nematocysts] as a protection against their
enemies — also an animal character.

We may now consider some representative examples of
Aristotle's researches and discoveries.[1]   They form a series of
outstanding but disconnected observations and speculations, on
which a philosophy of biology might have been, but was not,
founded.   They suggest the notebook rather than the finished
treatise — the indispensable material awaiting integration into a
science.

Aristotle was the first to describe the hectocotylized arm of
Cephalopods, and to regard it as an external male genital organ
which is introduced into the funnel of the female.   These observations, however, were completely forgotten until 1852, when Siebold
directed attention to them in consequence of the interest aroused

[1] Cf. H. Balss, *Arch. Stor. Scienza*, 5 (1924).

by the work of Verany and H. Müller. Siebold remarks : these authors " will learn with astonishment that Aristotle may fairly contest with them the priority of their discovery of the relation of the male *Octopus* to the *Hectocotylus* arm ". Aristotle's species was *Octopus vulgaris*, and he states correctly that the sexual arm is more pointed than the others, and has two very large suckers. He is not certain whether this arm is merely an accessory copulatory organ, or whether it is instrumental in conveying the male semen into the body of the female ; nor was he aware that the arm subsequently became detached. In this matter he appears to be recording the testimony of fishermen rather than his own observations, and his statements amount to little more than that the hectocotylized arm is an organ of generation concerned somehow with fertilization. Modern zoology teaches us that in the Cephalopods one of the arms of the male is modified structurally as a sexual or copulatory organ, for the purpose of receiving the spermatophores expelled from the funnel, and transferring them into the bursa copulatrix of the female. This process reaches its extreme state of specialization in *Argonauta*, where the hectocotylized arm *becomes detached*, and, packed with spermatophores, finds its way eventually by its own movements into the mantle cavity of the female. It can swim, climb and attach itself. In other species, however, the arm may *not* be detached at all, or after copulation only, in which case the two sexes meet, and the tapering extremity of the arm has been stated to be inserted into the terminal portion of the oviduct of the female. There are indeed all gradations between the sexual ritual of *Argonauta* and species in which sexual dimorphism is scarcely apparent, the hectocotylized arm being only slightly modified and not breaking away at any stage. In *Argonauta* the large individual with the shell and expanded arms is the female. The male has no shell, and is so small that it was at first mistaken for an embryo, and the unfolded hectocotylized arm for the yolk sac. For a long time the examination of large numbers of individuals failed to establish the existence of males.

In 1827 delle Chiaic discovered an " organism " attached to a female *Argonauta*. For reasons which should not have influenced so distinguished a naturalist he concluded that it was a parasitic Nematode, and named it *Tricocephalus* [*sic*] [1] *acetabularis*. Two

---

[1] This Nematode genus was first seen in 1740 by Morgagni in man.

years later Cuvier received from Nice five other specimens taken from an octopus. He also regarded them as parasites, but rejected the Nematode identification in favour of the equally erroneous one that the " parasite " was a Trematode. He named his own " species " *Hectocotylus octopodis*, and, without any justification, renamed delle Chiaie's " species " *H. argonautae*. He describes the anatomy of the " parasite ", which was four to five inches long, and the fifty-two pairs of suckers, basing his generic name on the latter character. Stomach and intestines, genital organs and a single external opening are also distinguished. Cuvier commends his *Hectocotylus* to those philosophers who profess to detect in the parasitic worms the substance and reflection of their hosts. "Here", he says, " we have the body of a Polypus which has for its parasite a worm so like the arm of a Polypus that the illusion could not be greater. Let it be judged how many theories might be founded on this extraordinary resemblance. Never has the imagination been exercised on so curious a subject."

In 1841 O. G. Costa described another specimen as a " pretended parasite of *Argonauta* ". He considers it to be, however, a part of the argonaut — perhaps a peculiar type of spermatophore. Later Defilippi noted that the long arm of *Octopus carena* might drop off when touched, and that it was the *H. octopodis* of Cuvier. The other arms could be removed only by rupture.[1] Dujardin in 1845, after studying some museum specimens, correctly surmised that they might be " portions of some Cephalopod detached in order to subserve fecundation ". In 1845–6 Kölliker carefully re-examined a third species of the " parasite ". He commenced his investigation fatally prejudiced by the fact that the hectocotylized arm is capable of maintaining a separate existence, and continues to be active and exhibit voluntary movements for hours in that state. When, therefore, he found that the skin was provided with the contractile pigment cells characteristic of Cephalopods, that the muscle layers were arranged exactly as in that group, that the acetabula closely resembled those of *Argonauta*, that the spermatozoa had the same size and form as those of *Octopus vulgaris*, and finally that *all* the specimens were *males*, he concluded that the Hectocotyli were the long-sought *males* of Octopods,

---

[1] This important observation was reported by Verany in 1851, but apparently was not published by Defilippi himself.

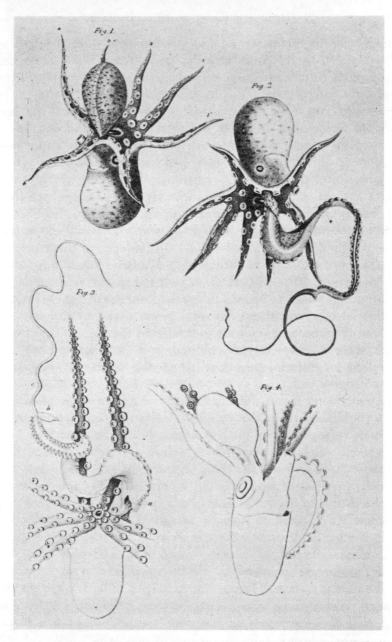

FIG. 4.—H. Müller, 1852. Male of *Argonauta argo*. 1, with the hectocotylized arm enclosed in the capsule ; 2, with the arm liberated. Vérany and Vogt, 1852. 3 and 4, similar stages of " *Octopus carena* ". *a*, aperture of the ruptured capsule ; *b*, remains of the sac which contained the (now extended) whip or penis, *c*

arrested in development and living parasitically on the female. The fact that dwarf parasitic males had already been recognized in various Crustacea removed the element of improbability from Kölliker's interpretation. Now assuming it to be correct, the "parasite" must be a *complete* organism, and hence he was tempted to look for, and find, a digestive system and caecum,[1] a heart with auricle and ventricle, branchial arteries and veins, respiratory gills, a nervous system and male genitalia. He admits the absence of sense organs.[2] Such are the pitfalls which entrap even the most eminent naturalists who attempt to argue back from preconception to fact. In one respect, however, Kölliker is entitled to our sympathy. If *Hectocotylus argonautae* is the true male, it should be developed as such in the eggs of the argonaut. In a series of papers dating from 1837, Madame Power, supported by Maravigna, affirmed that elongated, naked embryos similar to *H. argonautae* were to be seen in special bunches of eggs belonging to this species, and they even maintained that they had watched the young vermiform argonaut escape from the egg. These assertions were subsequently discredited, but not before they had exercised an unfortunate influence on the course of Kölliker's investigations.[3]

One chapter more, and this comedy of errors is complete. In 1836 (published 1839) Verany had collected in the Mediterranean a single octopod three inches long, which he described, but imperfectly diagnosed, as a new species — *Octopus carenae*. The hectocotylized arm was the third on the *right* side, and hence this species, as appeared later, was not an argonaut. In the figure the arm is shown in the bulbous unfolded condition, but its structure was not investigated, nor was it identified as a *Hectocotylus*. In 1851 Verany re-described the species as *O. carena*, and re-figures individuals *without* the hectocotylized arm, although they are stated in the text to possess it. In the meantime an interpretation of the hectocotylus had been attempted by Kölliker, Dujardin and others. Verany now tells us that between 1836 and 1850 he had collected five specimens of the new species, all of which had possessed the hectocotylized arm except one, which had evidently

[1] This is the seminal vesicle and its efferent duct.
[2] Kölliker is here describing the Hectocotylus of *Tremoctopus*.
[3] In 1853 Kölliker admitted the error, having, as he says, put too great a value on the statements of Maravigna and Power.

cast it. Even now he does not recognize that his *O. carena* must have been a true complete male, but believes that the hectocotylus may be a deciduous arm of a Cephalopod which carries the male organs, and that it is probably shed periodically as it matures. He is unable, however, to accept *H. argonautae* as belonging to that species, on account of its small size, and the fact that the adult argonaut has never been found to lack an arm. Seeing that the said adult, as then known, was always the much larger female, the latter point is deprived of any element of surprise. In an added plate Verany figures an extended hectocotylus of *O. carena*, but does not apparently recognize that the bulbous condition, which is also figured, is the same organ in an earlier state. In 1852 H. Müller brushed aside Verany's doubts and hesitations when he discovered and correctly identified the dwarf but perfect male of *Argonauta*, the *Hectocotylus argonautae* lying coiled up in the sac which occupied the position of the third arm of the *left* side. Müller also records in two cases a successful copulation between a hectocotylus and a mature female in *Tremoctopus*. In 1894 Racovitza observed congress in *Octopus vulgaris*. The hectocotylized arm, he alleges, with its contained spermatozoa, is introduced through the funnel into the mantle cavity of the female. The sexes finally separate and the copulatory arm breaks off, being left in the mantle cavity of the female adhering by its suckers.

It is instructive to reflect that, had Aristotle's researches on Cephalopods been known to these modern zoologists, how different would have been the story which has just been unfolded.

Aristotle's discovery of placental fish is another anticipation of contemporary work, this time, however, without the diverting consequences which distinguish the previous case. The mam-

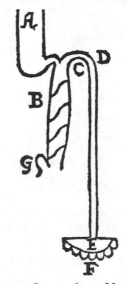

FIG. 5.—Steno, 1673. *Mustelus laevis.* Diagram illustrating the relations of the placenta to the foetal gut. *A*, stomach; *B*, spiral intestine; *C*, point where the tubular yolk-stalk opens into the gut; *D*, yolk-stalk; *E*, its connection with the placenta; *F*, placenta attached to the oviduct; *G*, rectal gland

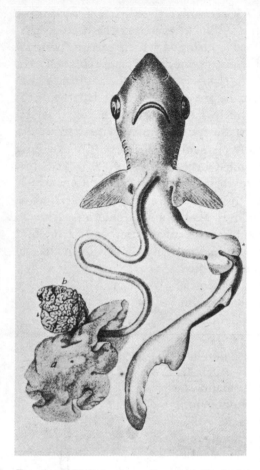

FIG. 6.—J. Müller, 1840. Foetus of *Carcharias glaucus* with the navel string and placenta, *b*. *a*, part of the yolk-sac

malian placenta was known to Aristotle, but he denied that an ovum took any part in mammalian development, classifying them as the *internally* viviparous animals. On the other hand in viviparous species which are not mammals there *is* an egg, which therefore develops within the body of the parent. These are the *externally* viviparous animals, in which no placenta and no organic connection between young and parent are to be found. Thus the Ovipara may produce a " perfect egg " which is laid and hatches outside the body, or an " imperfect egg " which is not laid and perforce develops within the body of the parent. Now Aristotle knew that fishes were oviparous, and that some cartilaginous fishes were *externally* viviparous, but he knew also that there was one shark-like animal [*Mustelus*] *which was internally viviparous like a mammal*, such as man and the horse. Here the wall of the yolk sac at one point becomes thickened and attached to a corresponding thickening of the wall of the uterus. Hence a [yolk-sac] placenta is formed similar to that of quadrupeds, and the embryo is nourished in the same way.

In modern times Belon (1553) observed the attachment of the embryo to the wall of the oviduct by a navel string in a cartilaginous fish, and Rondelet (1554) figured this connection in *Mustelus laevis*,

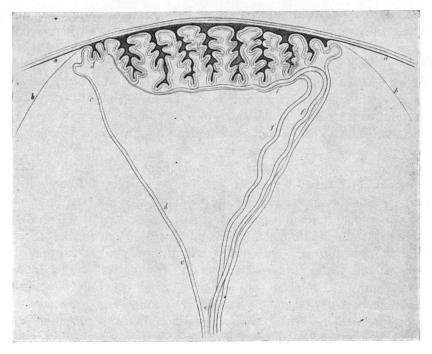

FIG. 7.—J. Müller, 1840. *Mustelus laevis*. Diagrammatic transverse section of the placenta. *a*, uterus ; *b*, shell membrane of the ovum which separates the foetal from the maternal placenta ; *c*, outer boundary of the foetal placenta ; *d*, inner boundary of same ; *e*, navel string ; *f*, foetal placental artery and vein

but it was the Danish anatomist Steno (1673) who was the first, after Aristotle, to claim that the connection was a functional placenta, by the activity of whose blood vessels soluble food matters were transferred from the parent to the cavity of the yolk-sac, and thence by its stalk to the gut of the embryo.[1] Finally J. Müller (1839–42) gives us the best modern description of this curious fish placenta. He worked out the structural details and vascular supply, and showed that the placenta consisted of interdigitating foetal and maternal portions as in mammals. He found the same type of placenta, but in a much more complex form, in *Carcharias* as well as in *Mustelus*. Müller also confirmed Rondelet in the surprising discovery that a placenta occurs in *M. laevis* but not in

---

[1] It should be noted that, with the formation of the placenta, the yolk-stalk is closed and the connection between placenta and embryo becomes purely vascular, as in mammals.

M. vulgaris! In the latter species the yolk-sac is quite free and unattached to the uterus, and there is no placenta. There is no *direct* connection between the foetal and maternal blood vessels, and in this respect also fish and mammalian placentas agree. When Müller wrote his first paper he was not aware that he had been anticipated by Steno, nor was either of them aware that the honour of the discovery belonged to Aristotle.[1] That observations of such skill and importance were to be found in Aristotle seems to have astonished Müller and his generation.

The compound stomach of ruminants such as the sheep, ox, deer and goat was well known to Aristotle, and he must have dissected this organ very carefully. He points out that horned animals which have no incisor teeth in the upper jaw, have, as a kind of compensation, a four-chambered stomach in which the food is broken down into a pulp. He gives a good description of the paunch, honeycomb, psalterium and reed with their characteristic linings. Ungulates and other animals which have incisor teeth in both jaws, such as the pig and rabbit, have, he says, a simple stomach. After Aristotle the compound stomach is not again mentioned in zoological literature until it was described by Coiter in 1572, Aldrovandus in 1613 and Fabricius in 1618 ; but it must have been known to butchers long before then.

Although Aristotle was impressed by the community of structure which characterizes man, apes and quadrupeds, there is nothing in his writings to suggest that it had given rise in his mind to any suspicion of historical relationship or mutability of species. This is surprising in view of the fact that morphological speculation was very attractive to him. For example, he points out that tusks and horns are mutually exclusive, and a carnivorous dentition is associated with absence of tusks and horns. On the other hand when upper incisors are wanting, horns are present, which he attempts to explain by assuming that the material which would otherwise have gone into the teeth has been diverted to the formation of the horns. Any exceptions to these conclusions could scarcely have been known to Aristotle, and in any event would have no greater significance than the inevitable exceptions to every general rule. Again, he was convinced that hands and talons, nails and hoofs, and feathers, scales and scutes are *analogous* structures,

[1] Cf. W. Haberling, *Arch. Gesch. Math. Naturwiss.*, **10** (1927).

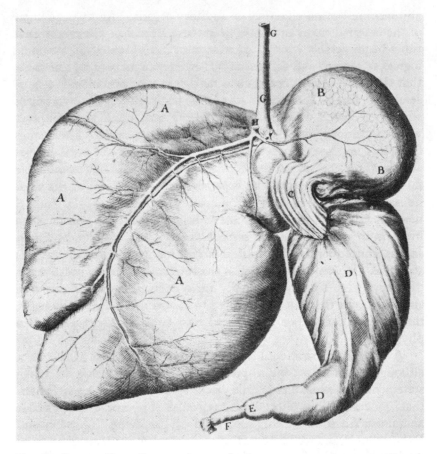

Fig. 8.—Peyer, 1685.　Compound stomach of young goat.　*A*, rumen; *B*, reticulum; *C*, omasum; *D*, abomasum; *E*, pylorus; *F*, duodenum; *G*, oesophagus; *H*, vein; *I*, artery

and he is only less happy in claiming that the arms and fore limbs of mammals, the wings of birds and the pectoral fins of fishes are comparable organs.　He narrowly missed an important generalization when he observed that changes occur during life in the consistency and thickness of the scales of fishes ; but he does not perceive in these changes the possibility of determining the age of the fish.

Errors of judgment, observation and faulty reasoning are to be found in Aristotle as in all other zoological works, but they are surprisingly few in number.　Fishes, he says, have a sense of smell and hearing but no olfactory or auditory organs. And yet he knew of the external nares and otoliths in these animals. He can even at times be perversely wrong. Having convinced himself on grounds, logical as regards the argument but erroneous as regards the facts, that the sense organs could have no direct connection with a non-sensory brain, he declined to accept the testimony of his own senses when he himself dissected a chameleon and found that the eyes were continuous with the brain [via the optic nerves]. He escapes from this impasse, however, by maintaining that the optic nerves are ducts for the transmission of some convenient but impalpable fluid. On the brain itself he is at his worst. The seat of the soul, he says, and the control of voluntary movement — in fact of nervous functions in general, are to be sought in the heart.[1] The brain is an organ of minor importance ; it has no blood supply, and is consequently the coldest structure in the body, and in-sensitive. It was disastrous that the relations of brain and liver to the heart were so fundamentally misconceived by both Aristotle and Galen, for their views gave rise to opposing schools of thought, the triumph of either of which was fatal to the progress of physiology.

A mistake which could not possibly have been Aristotle's, although he must accept responsibility for perpetuating it, is the statement that when a crocodile opens its mouth it is the *upper* jaw that moves. This fable was repeated by many respectable anatomists (including Leonardo, Vesalius, Rondelet, Columbus and Robert Boyle), as well as by Oliver Goldsmith, and survived until modern times. Aristotle's plausible attempt to explain it shows that he himself could not have examined the skull of a crocodile.

Aristotle failed to homologize correctly the long bones in the leg of a bird. That this could be done without a knowledge of their development was shown by Belon in 1555, who compared the leg bones of man and bird much as we do now.

Aristotle maintained that the testes were not an essential organ of reproduction and were actually wanting in some fishes and reptiles, although he observed them correctly in others. The male semen, he says, is secreted by the seminal ducts. It is surprising that he should have overlooked the testes in some of the lower

---

[1] Galen's comment on this view is that it can be held only by those who are ignorant of anatomy.

Vertebrates, but perhaps the fact that these organs have not always a compact oval form deceived him. This is all the more probable when we note that he is much sounder on the testes and their associated structures in mammals.

The blood system claimed much of Aristotle's attention. He describes the movements of the heart of a chameleon when exposed during dissection, and of a tortoise after removal from the body. These observations, however, do not necessarily involve vivisection, as has been supposed. The blood vessels, he says, are associated with, and originate in, the heart, which is the first part of the embryo to be formed and contains blood from its inception. He saw, but regarded as nerves, the chordae tendineae, and described the pericardium. He did not consider that the Invertebrates had a true heart or blood, by which he means *red* blood, and the structures he describes as the heart in those animals are certainly not that organ. He had no knowledge of the circulation, or even of a to-and-fro movement of the blood in relation to the heart, and he regards the heart as a kind of excrescence on the precaval and postcaval veins, which are supposed to pass *through* it. The aorta on the other hand merely arises from the heart. Aristotle's account of the distribution of the blood vessels is confused and inaccurate. The blood, he says, issues from the heart and is dissipated variously in the tissues. Hence it does not, and cannot, return to the heart, nor does he understand the constitution and distribution of the portal system, the dissection of which makes no great demands on the skill and vision of the observer. Nevertheless Aristotle's work on the vascular system was a marked advance on that of his predecessors, and his successors would have stumbled even more than they did but for his pioneer efforts.

The interpretation of Aristotle's views on the anatomy of the heart is still a subject for speculation, and his exact meaning must remain obscure. In 1681 Grew voiced the prevailing opinion in the following words : " The hearts of all great animals, saith Aristotle, have three ventricles ; of lesser, two ; of all, at least one. One would a little wonder, how so observing a man, should discover so many mistakes, in so few words." There is scarcely a passage in Aristotle's anatomical works which has been so minutely scrutinized by commentators as this. In succession Galen, Vesalius, Columbus, Caesalpinus, Harvey, Haller, Huxley, Ogle,

Payne, D'Arcy Thompson, Platt and others have attempted interpretations which should be consonant with the text without arraigning the accuracy of its author. The fact that the text is in a " chaotic state ", and the presumption that parts of it are wanting, sufficiently explain the difficulties. It has been suggested that Aristotle's left chamber is the left auricle, the intermediate chamber is the left ventricle, and the right chamber is the right ventricle. The right auricle was ignored because in the suffocated animals with which he worked it would appear to be a part of the caval system. This solution does not explain Aristotle's assertion that large animals have more cavities in their hearts than small ones, unless he is here drawing a distinction between the hearts of the higher and lower *vertebrates*, and not of mammals, as the text seems to imply. His description of the heart is also interesting as containing a dubious reference to the ductus arteriosus.

Although Aristotle had no sound conception of the process of respiration he noted that the ramifications of the bronchial tubes in the lungs were accompanied everywhere by blood vessels, and he suggested that air passed from the former into the pulmonary artery and vein, through which vessels it reached the right and left chambers of the heart. At the same time he did not believe that there was any palpable or *direct* connection between air and blood vessels in the lung.

Aristotle's works are packed with observations of which all are interesting, some are important, and most have been forgotten. The following are a few remarkable examples. Aristotle's lantern is the scraped *test* of the Echinoid (with, however, the five teeth), and not the structure so named in modern literature. The gastric mill and its teeth of lobsters and crabs is well described. He undoubtedly saw the palatal organ of the carp, which he says is so fleshy that it might be mistaken for a tongue. This interesting sense organ, long regarded as a tongue by fishermen, was first re-described by Rondelet in 1554, and again in 1667, presumably by Swammerdam. Its gustatory nature was first maintained by E. H. Weber in 1827.

The comparative anatomy of the gut of fishes is examined, and the muscular gizzard-like stomach of the grey mullet is noted. The number and occurrence of the pyloric caeca arouse his interest, and he discusses the possibility of rumination in the parrot wrasse.

The tunny (*Pelamys*), he says, has a long gall-bladder extending in a zigzag course along the whole of its intestine, and in the dragonet (*Callionymus*) the gall-bladder is relatively larger than in any other fish. He points out that in snakes there is only one lung and one lobe to the liver, which is true for the species examined by him, but he fails to associate these peculiarities with the assumption of an elongated body shape in the Ophidia. He gives a very good description of the gut of birds, including the crop, in which no digestion is stated to occur, the proventriculus, which he regarded as a storage chamber only, the gizzard (including the structure of its wall), and the caeca, the number of which is not stated. Mammalian anatomy claimed a large share of Aristotle's attention. He knew of the nictitating membrane of the eye, the os cordis of the horse and ox, and the os penis in several Carnivora. His dissections of the elephant are quoted with approval by Camper. He is an early, if not the first, describer of the pancreas, and he mentions a duct passing from the ear to the mouth which must have been the Eustachian tube. The latter discovery, wrongly attributed to Alcmaeon, was confirmed in the sixteenth century by Ingrassias, Vesalius and Eustachius. The comparative anatomy of the mammalian gall-bladder is also explored, and its occurrence and position described. He lists several animals, including the elephant and the horse, which are without a gall-bladder, and even notes its occasional absence in man. A gall-bladder, however, is present in some of the species in which he was unable to find it. Aristotle failed to observe the mammalian ovary.

Nothing is known of Aristotle's methods except that he appears to have practised dissection, especially on Vertebrates and some higher Invertebrates. The wealth of anatomical data included in his works, however, is not necessarily the product of his own labours. According to Lones, Aristotle dissected carefully twelve mammals, nine birds, four reptiles, two amphibia, ten fishes, a simple Ascidian, seven molluscs, three arthropods and a sea urchin. He did *not* dissect the human body, but examined a human foetus without going into detail. He doubtless dissected many other animals, but either less carefully or without making full use of his notes in his writings. Lones states that there are anatomical details involving the dissection of some 110 animals, but in at least some of these cases Aristotle obtained his information at second hand. It is remotely possible that this applies to most of them, but his manner of dealing with practical questions is that of a man who has handled the material himself.

III
GALEN

Galen, who was an Asiatic Greek, lived during the period of the Greek decline. He differs from Aristotle in that his interests were addressed rather to the physiological and morphological aspects of biology, and, although he owed a part of his training to the Peripatetic School, the speculative and philosophical approaches to an objective world did not specially attract him. He was in fact an observer, and above all an experimentalist, whereas the Aristotelians, as he says, prefer disputation to dissection. But neither does he favour pure anatomy, to which he objects that it errs in concentrating on the parts, whereas a living organism is not a summation of parts but an indivisible unity. He was, indeed, the first to establish a system of physiology based on experiments on living animals, and in this for many centuries he had no rival. Erasistratus, however, was in some respects a predecessor.

Again, like Aristotle, Galen propounds no theory of organic evolution. Nevertheless it is not specifically excluded, since the occurrence of climatic adaptations and also slight departures from the normal, *which become considerable with time*, are foreshadowed. It may here be noted that Galen's experimental and anatomical researches have in the past been but imperfectly studied, and in consequence his defects and merits alike have been misjudged. It is only in more recent times that a new generation of commentators has laboriously re-examined the Corpus Galenicum, and demonstrated the truly great part it has played in the evolution of biological learning. It is necessary to bear this in mind when assessing the status of the *Fabrica* of Vesalius, which has Galen's anatomy for its basis, and may even be regarded as a corrected and expanded version of it. The tendency to see nothing in Galen but his errors reveals a lack of knowledge and understanding, and is just as wrong as was the servile faith which for centuries proclaimed his infallibility.

Galen's writings are largely a record of his own researches, which were considerable, but probably also summarize fully, without, however, quoting, the anatomical tradition of his time. Prendergast states that Galen took nothing on trust, and " described nothing that he had not personally investigated ". Further, many of his major works have come down to us more or less intact. The following points have been selected as examples of the scope of his anatomical and physiological work. The *osteology* is good and complete, the bones and sutures of the skull being defined much as they would be now. The vertebrae, said to be twenty-four in number, are grouped into cervical, dorsal, lumbar, sacral and coccygeal series, and the thoracic and limb bones are accurately described. If the *myology* is not so complete, although some three hundred muscles are included, it is more original, and many muscles are described for the first time. The eye muscles, including the choanoid, the cutaneous muscles of the eyelids and lips, and the muscles of the larynx, hyoid, tongue, pharynx, lower jaw, trunk and limbs, are all carefully explored. Even his terms are often retained in modern usage, and we have, for example, to thank him for naming the broad cutaneous muscle of the neck the platysma myoides. In dealing with the *vascular system* Galen was much less successful than his predecessor Erasistratus, whom, notwithstanding, he severely criticizes. Galen was, however, responsible for one fundamental discovery — that in the living animal the left ventricle and arteries generally contained blood and not air. In this he was correcting a belief which had survived for three hundred years. His own errors, to mention only those which could not fail to vitiate any theory of the circulation which he might devise, were in assuming that the liver was the source of the veins, and that the ventricular septum of the heart was perforated. As regards the former point Galen decided against Aristotle, who correctly regarded the veins as related to the heart, but in opposition to this view the Galenists adduced the portal vein which was associated with the liver only. Galen's second lapse is readily understandable. He knew that the blood *must* pass from the right ventricle to the left, and a perforated septum provided the simplest solution to this problem. He knew also that the septum was pitted, and he was unable to convince himself that these pits could be blind and lead nowhere. Microscopic confirmation was naturally out of the question. The passage of the blood, therefore, from the right ventricle of the heart to the left through invisible pores in the septum seemed an irresistible inference.[1] Unhappily it developed from an inference into a dogma, the blind acceptance of which for more than 1300 years delayed the discovery of the circulation of the blood.   In other respects the heart was a stumbling-block to Galen.   He regards its substance as something *sui generis* and denies that it is muscular because it is too hard, and, unlike somatic muscle, is never static.   Nor will he concede any importance to the auricles, which seem to him to be nothing more than dilatations of the great veins.   The heart, therefore, has only two cavities — the cavities of the ventricles.   Its function is to produce " innate heat " by slow combustion, the temperature being determined and controlled by the lungs in respiration.   As a set-off to these failures he gives a good account of the valves of the heart, and describes the distribution of the blood by the arteries to all parts of the body with some accuracy.   He also saw the lacteal vessels, although in this he had been anticipated by Herophilus and Erasistratus, and he noted that the addition of air to the dark venous blood in the lungs made it light in colour ; but having failed to discover any blood in the pulmonary vein he concluded with Erasistratus that that vessel transmitted air only.

According to Galen, who in this was extending the system of his forerunner Erasistratus, the three central organs of the body are the liver, heart and brain, and associated with each of these is a subtle, impalpable effluvium, having a special function in each case.   The liver produces the *natural spirits*, which are found in all living organisms, and constitute the essence of growth.   The heart is the laboratory of the *vital spirits*, which are necessary for movement and muscular activity, whilst the brain lodges the *animal spirits*, derived from the rete mirabile.   The animal spirits account for all nervous activities and are contained in the ventricles of the brain, whence they are transmitted through the nerves to the body.   He is not certain whether the animal spirits are a fluid or what we should now term a stimulus.   These three imaginary substances play a considerable part in Galen's long-accepted but misguided views on the functions of the vascular system.

Galen's contributions to *neurology* represent perhaps his

[1] Even as recently as 1654 T. Bartholin was describing a perforated interventricular septum in the pig.

greatest work. Not only was he the founder of the physiology of the nervous system, but also he was so far in advance of his time that there was no one to dispute his pre-eminence until 1811, when Charles Bell confirmed the law to which his name has been given. On the anatomical side Galen demonstrated that the nerves originate in the brain and cord and not in the heart, as Aristotle had perversely maintained. As regards the structure of the brain he had as predecessors Erasistratus and Herophilus. He defines correctly the main divisions of the brain, including the choroid plexuses and pineal body, and describes the ventricles and their connections, a part of the iter, and the foramen of Monro. He distinguished only seven cranial nerves, which are, however, described with great accuracy. They represent the second, third, fifth (Galen's third and fourth pairs), seventh + eighth, ninth + tenth + eleventh and the twelfth of the modern anatomist. The olfactory is not considered to be a separate cranial nerve, and the fourth and sixth nerves are not mentioned. The sympathetic is only partly recognized, and is regarded as a branch of his fourth pair of cranial nerves. He gives close attention to the comparative anatomy and physiology of the recurrent laryngeal nerve (not, however, discovered by him), especially in the ox, dog and bear. His physiological observations include a comparison between the brain and cord, the former being the organ of thought, sensation and volition operating through the nerves of sensation, and the latter functioning as a conducting organ associated with the nerves of motion. Sensory and motor nerves are distinguished by differences in texture, and not by the specific nature of the stimuli transmitted by them. The control of muscle response by brain and cord, the production of voice by the larynx, the physiology of the respiratory muscles and of the phrenic nerve and diaphragm, are among the problems which were thoroughly investigated by experimental methods having a strangely modern outlook.

In the *thorax* and *abdomen* Galen recognized the relations of the pleurae and peritoneum to the thoracic and abdominal viscera, and of the pericardium to the heart — a subtle point in these early days. The viscera themselves are in general accurately and minutely described, and he noted that the liver had a parenchyma in addition to its blood vessels. The genitalia are not so successfully treated. Galen also wrote an important treatise on anatomical *methods* —

the first work to be produced on practical anatomy. He lived in a brutalized and corrupt age, and this is reflected in his ruthless experimental work, in which the operator not only ignores, but also is apparently oblivious of, the sufferings of his victims. Apart from this, the genius of Galen was superior to the circumstances and failings of his time.

What was the material available to Galen in his anatomical researches ? Haller was of opinion that he had never dissected the human body. Although in places Galen admits that he had anatomized only monkeys and other lower animals except insects,[1] these passages for various reasons were completely overlooked. On the other hand in his best-known works the reader is almost invited to assume that the author had dissected the *human* body, and this was universally believed until the sixteenth century, when a prolonged and acrimonious controversy was waged around this point. On one side it was claimed that the text of Galen himself decided the issue as against the human interpretation, to which it was replied that the text was corrupt, and it was even suggested by Sylvius that the anatomy of man had *changed* since the time of Galen.[2] For example, we are told, it might well be that the human femur was curved when Galen described it, but had since straightened out in response to the adoption of cylindrical nether garments. Sylvius even went so far as to defend the statement of Galen that the human sternum consisted of seven bones by suggesting that " in ancient times the robust chests of heroes might very well have had more bones than our degenerate day can boast ".

The following quotations from Galen's anatomy show that in these respects at least he was not describing the anatomy of man. The lower jaw, he says, is in two halves ; there is a separate premaxilla ; there are 7 distinct segments in the sternum [6 fusing to 3 in man] ; the transverse processes of the lumbar vertebrae are directed forwards [transversely in man] ; the sacrum and coccyx have three pieces each [5 in the sacrum and usually 4 in the coccyx in man] ; an os cordis is present [occurs in horse and ox but not in man[3]] ; the two carotids arise from a *single* stem [the

---

[1] *De anatomicis administrationibus*, Lib. VI, cap. i.

[2] This was bluntly stigmatized by Boerhaave and Albinus as a " ridiculous and worse than puerile lie ".

[3] T. Bartholin in 1654 described an os cordis in man " similar to those found in deer ". The ossicle appears for the first time in the *printed* book in 1498. Galen and

left carotid arises separately from the aorta, and the right carotid from the innominate in man] ; a rete mirabile is conspicuous on the base of the brain [in ruminants but not in man]. This last structure was discovered by Herophilus, but Galen gives us the first detailed description of it. According to Roth, Vesalius in his earlier days had conformed to the Galenic tradition, and " in his anatomy lessons of the years 1537-9 he had not ventured, as he himself confesses, on any criticism of Galen's osteology. In order to demonstrate to his pupils the Plexus mirabilis Galeni, *which he could not discover in man*, he had perforce to fall back upon the head of an ox or sheep, fearful lest it be said of him that he was incapable of finding the plexus." [1]

It seems certain that Galen worked very largely with tailed and tailless apes, the dog, horse and ruminants, and transferred his results by analogy to the human body. He was not, however, the only anatomist to do this. Leonardo and Willis were others, and even Vesalius, who scornfully attacked Galen for practising this very deception, is not guiltless of it himself. It should be added that Galen also dissected an elephant, goat, pig, bear, lion, cat, wolf, weasel, lynx, rat, and various birds, reptiles and fishes, but for obvious reasons Invertebrates were outside the range of his interests.

With the death of Galen, Greek science came to an end, and the tradition of biological learning in Europe was completely broken until the eleventh and twelfth centuries, when the translation of Greek–Arabic scientific texts into Latin brought about a partial revival, and inaugurated the reign of dogma known as the Scholastic Period, which was characterized by a pious and unreasoning acceptance of Aristotle and Galen. For many centuries mankind was either indifferent to the secrets of Nature, or hoped to penetrate them by vain speculation and literary research. It is remarkable that the revival of learning when it did come, with all its vast possibilities for the future of science, should have received its first and greatest impetus from one of the simplest of mechanical contrivances — the invention of printing.

Columbus mention a large os cordis in the elephant, where, however, it does not normally occur. The " os de corde cerui " was a popular medieval drug.
  [1] Berenger of Carpi (1521) appears to have been the first to deny that the rete occurs in man. " This net ", he says, " I never saw." Servetus refers to it in 1553.

II

ZOOTOMY DOWN TO THE SIXTEENTH CENTURY

IV
THE SCHOLASTIC PERIOD

THE fact that Greek biology is separated from modern science by a stagnant period of some 1300 years is perhaps the most astonishing phenomenon in the history of the academic world. Admitted that much has been claimed for the scholarship of the Dark and Middle Ages, it cannot be denied that it was an age in which scientific *research* was considered to be superfluous, and in some centres was even sternly forbidden in the statutes of universities. Greek science dominated the minds of men, and was elevated to a position of authority as if it exhausted the possibilities of knowledge. During the scholastic period, problems which could have been solved by a few simple observations or experiments were abandoned to industrious but servile commentators and textual critics, who disdained experiments, and succeeded only in meriting the satire of Pope on the ignorance of the learned. Thus the anatomist was left with no choice but to study and expound the works of Aristotle and Galen.

It is therefore not surprising that between the eclipse of classical learning and the invention of printing in the middle of the fifteenth century anatomical research was largely in abeyance, and the period is represented only by such studies as the twelfth-century " Anatomia Porci ", erroneously ascribed since 1531 to a teacher at Salernum named Copho,[1] of whom nothing is known, and the avian anatomy of Frederick II. The earliest recovered manuscript of the former was probably written *c.* 1150, and there are eleven printed versions dating from 1502 to 1852. It is nothing more than a description of the public dissection of a pig, and its only merit is the light it throws on the depressed state of biological science at the time it was written.

We note that " Copho " accuses Galen of basing his anatomy

---

[1] Cf. Le Roy Crummer, *Ann. Med. Hist.* (1927); K. Sudhoff, *Arch. Gesch. Math. Naturwiss.*, **10** (1927); R. Creutz, *Arch. Gesch. Med.*, **31** (1938); Redeker, *Die Anatomia . . . Chophonis . . .*, Leipzig (1917), and Corner (1927).

on the dissection of brutes, and hence anticipates the controversy which was to rage so fiercely in the sixteenth century. " Copho " himself prefers the pig as a subject for dissection, for the reason that there is no animal so like man *internally*,[1] as he politely adds. He gives some details of the anatomy of the cervical region, such as the epiglottis, larynx, recurrent laryngeal nerve and thyroid gland. He then proceeds briefly to describe the contents of the abdomen and thorax, including the pleura, pericardium and diaphragm. The lung, he says, is hollow, as can be shown by inflating it with a quill, and he repeats the old error that air passes from the lung to the heart. The larger arteries and veins are mentioned, and the postcaval after entering the " inferior auricle " of the heart is said to be *continued as the aorta*, from which arise the arteries " proceeding to the members ". This curious mistake is common to all anatomical texts of the period. The abdominal portions of the aorta and postcaval are believed to be formed by the union of all the arteries and veins of the head, which reads as if " Copho " had not been aware that the aorta was continuous throughout its entire length, although this fact is clearly recognized in other and more detailed twelfth-century anatomies. " Copho " summarizes the structure of the genitalia of the female, and his views on generation are of the well-known traditional type — the uterus is " nature's field which is cultivated that it may bear fruit, and in which, when the seed is sown, it remains as on good ground . . . and sends out roots by which it is attached to the uterus ". Finally he scarcely more than mentions the brain and its membranes and the eye. But " Copho " was evidently a physician, and to him anatomy is only a means to promote *medical* science.

V

LEONARDO DA VINCI

Standing between the Scholastic Period and the Revival of Learning is the majestic and lonely figure of Leonardo da Vinci — one of the greatest universal geniuses the human race has produced. He was born in 1452 and died in 1519. It is often said that in all

---

[1] A widespread belief which survived to modern times. Tyson (1683) dismisses it as a " vulgar Error ". In twelfth-century Salerno the type selected to illustrate the structure of man was usually the pig.

phases of intellectual activity, excepting only commerce and politics, he was a master.[1] His interest in comparative anatomy is the indirect result of his activities as a painter and sculptor. Following the example of the Greek sculptors he made a profound study of the superficial anatomy of the human body, and the anatomy of expression. This led him to investigate the *general* structure of man, and, when he became mystified, to turn to comparative anatomy for solutions of his difficulties.[2] In this way he began to study anatomy not merely as a means to an end but as an end in itself, worthy of his concentrated interest and attention. His first concern, however, was always with human anatomy, and he mentions having dissected more than thirty human bodies. As a comparative anatomist his researches produced a wealth of detail and stimulating inference which, had they been published at the time, must have exercised a profound influence on the future of animal biology. He combined, as no one else has ever done, the detailed objectivity of the anatomist with the vision and aesthetic sense of the artist. Nevertheless his biological activities were not appreciated by his contemporaries, and we learn that he was denounced by the Pope for his practice of anatomy. William Hunter, who was familiar with the Leonardo manuscripts in the King's Library at Windsor, announced in 1782 that he had obtained permission to publish them, but his intention was frustrated by death. All these anatomical manuscripts, together with others by Leonardo in France and Italy, have now been transcribed and published, and some of them have been carefully analysed by modern anatomists.[3]

Leonardo's methods are only partly known. He was the first after Aristotle and Galen to revive the comparative method, and was tardily followed in this by Belon and Coiter. He devised numerous plans for the expansion of comparative research which apparently were never developed. Scholasticism made no appeal to him. All science that ends in words, he says, has death rather than life. He was the first to make wax casts of the ventricles of the brain, and to employ (optical) serial sections. The animals which he mentions as having been the subjects of his anatomical

---

[1] A pardonable exaggeration.
[2] Michelangelo also studied human and comparative anatomy.
[3] Cf. particularly J. P. McMurrich, 1930.

and physiological work are : *Gordius*, moths, flies, fish, frog, crocodile, birds (mechanics of flight, anatomy of wing and foot, eye and nictitating membrane of the owl, and anatomy and movement of the tongue of the woodpecker), horse, ox, sheep, bear, lion, dog, cat, bat and monkey. He also favoured the tradition initiated by Aristotle that all animal biologists must at some time examine the development of the chick and mammal. Thus he gives a good description with figures of the cotyledonary placenta of the cow. He notes that each cotyledon consists of foetal and maternal portions which separate at birth, and he saw the interlocking of foetal villi and maternal crypts. As regards the relations of the two bloods he correctly implies in places that they remain quite separate, but in others he states the opposite view — that they are continuous, being influenced by his erroneous belief that the heart does not beat during foetal life. In *c.* 1490 he compares the limbs of man and the horse, and shows that the

latter moves on the tips of its phalanges. There is also a good description of the facial muscles and nerves of the horse, in which animal, he says, they are naturally displayed and can be most conveniently dissected. He is credited, wrongly according to McMurrich (if we except Leonardo's knowledge of the muscles of the limbs), with having written a whole treatise on the anatomy of the horse in preparation for the model of his colossal equestrian monument to Francesco Sforza (1493), but if he did so it has not survived. Neither has the model. His important

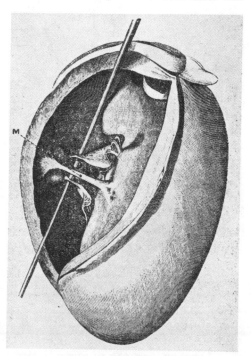

Fig. 9.—Rolleston, 1873. Right ventricle of sheep laid open. The probe passes under the moderator band, *M*

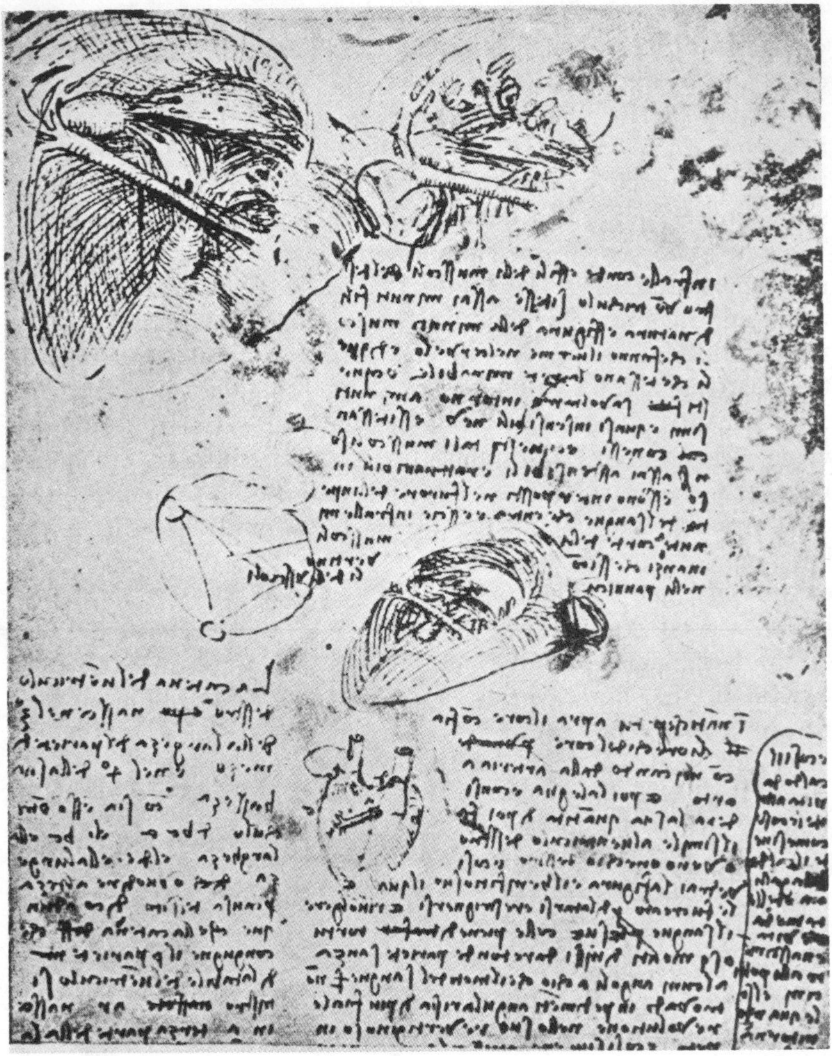

FIG. 10.—Leonardo's sketches of moderator band of sheep.   Upper figure on left
shows interior of right ventricle with the tricuspid valve, papillary muscles and
moderator band.   *Copyright of H.M. The King*

work on the structure of the heart is based largely on the hearts of
animals — principally cattle. He describes, for example, the os cordis of the ox, and the moderator band in the right ventricle of the
sheep, which is concerned with the regulation of the ventricular
musculature. This muscle was re-discovered by Lower in 1669 and

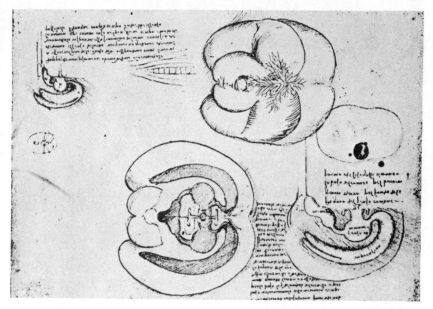

FIG. 11.—Leonardo's figure of rete mirabile of ox, and casts in wax of cerebral ventricles. *Copyright of H.M. The King*

by King in 1837. Nobody before Leonardo had explored and depicted the heart as he had done. He insists, contrary to Galen, that the heart is a *muscle* — in fact the most powerful of all muscles, and, like other muscles, it is nourished by arteries and veins. And yet his views on the circulation of the blood are almost pure Galenism, and there is no evidence for the statement, often made, that he came near to the discovery of the circulation. For example, he believed that the auricles filled up on the contraction of the ventricles, and he accepted, and even illustrated, Galen's imaginary perforations of the inter-ventricular septum, which facts must be held to rule him out as a forerunner of Harvey.

From the numerous observations in comparative anatomy described and figured in the Leonardo manuscripts the following typical examples may be selected. His experiments on the spinal cord of the frog induced him to draw a conclusion which did not naturally follow from them, namely, that the cord is the centre of life. He also failed to observe the two roots of the spinal nerves, which were first seen by Coiter in 1572 ; but he was more successful in his account of the aorta and rete mirabile of the ox, the latter of

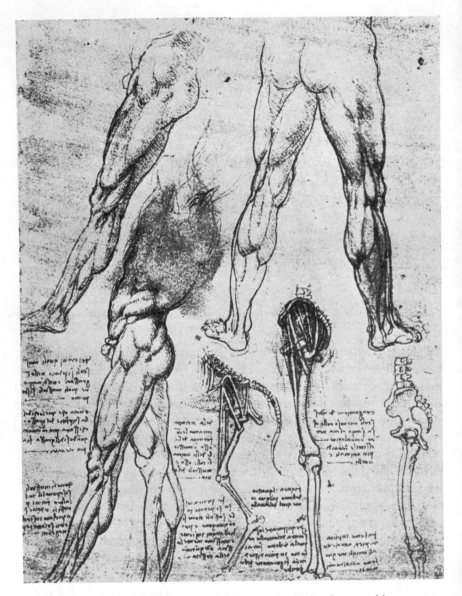

FIG. 12.—Leonardo's comparison of the posterior limbs of man and horse.
Copyright of H.M. The King

which is figured, although not very successfully. Of the carnivores, the dog, lion and bear were dissected, Leonardo devoting particular attention to the structure of the larynx and hyoid, the brain, nose and turbinal bones, and the foot. He is particularly interested in the anatomy and movements of the limbs in the higher mammals, and his treatment of the bones and muscles of the arm and leg of the anthropoid betrays the subtle hand of the creative artist. The mechanistics of the biceps brachii are demonstrated, and he points out that the power of this muscle depends on the position of its insertion with respect to the hand.

VI
VESALIUS[1]

It is scarcely necessary to emphasize that before the invention of printing with movable metal types the diffusion of learning was gradual and restricted. The manuscript copy, to be generally available, involved the development of a stereotyped form of calligraphy on hard wearing and costly material. The production of black-letter script in any quantity was excessively slow and laborious, and, to make matters worse, it soon became a highly skilled and beautiful craft, which was practised *for its own sake*, with scant regard for the significance of the matter which was being reproduced. Thus output was still further reduced, and the possibility that works of real importance would be brought into even this limited circulation was remote. The effect which the absence of adequate means of publication must have had on the advancement of science may be deduced from the number and distribution of early manuscripts in English libraries at the present time. The biological works of Aristotle were more extensively copied than those of any other scientific writer during the Middle Ages, and yet, for the three hundred years preceding the advent of the printing press, only eighty-two Aristotelian manuscripts have survived, three-quarters of which are in the libraries of the old Universities of Oxford and Cambridge. Moreover, the early printers were at first dominated by the scholastic tradition, and did little to encourage the work of *original* research, with the result that the revival of scientific learning was not one of the immediate results of the foremost and most beneficent invention of modern times.

[1] Cf. R. Schmutzer, *Ergebn. Anat.*, **32** (1938).

The first of the great biological works to be published in the Renaissance period was the *Fabrica* of Vesalius (1543). Vesalius, however, important as he is as the father of modern anatomy, was not, except incidentally, a *comparative* anatomist, nor can he be said to have realized the importance of comparison as a philosophical approach to human anatomy. Also, magnificent and imposing as his book is, the strangest of all facts about it, as Richardson says, is that its author " could have disclosed so much and not discovered more. The whole mechanism of the circulation, for example, is laid bare by his hand, and yet he knew nothing of the circulation. Heart, arteries, veins, valves — he dissected them all, but to him they were dead." It is therefore the less surprising that the only structures associated with the name of Vesalius should be the ossa vesaliana and the foramen vesalii — neither of which is a normal constituent of the human body. His mind was analytical and unimaginative rather than constructive. His service to anatomy was in establishing the anatomical *method* — in demonstrating once and for all that progress in science is possible only by *research*, and that authority, however firmly and comfortably established, must yield to original investigation. He weakened scholasticism in biology, but in doing so aroused the most implacable and unscrupulous opposition, even among those who were competent to appreciate his work and should therefore have welcomed it. The year 1543 is perhaps the most memorable in the annals of modern science. It witnessed the publication of works by Copernicus and Vesalius, the former of which laid the foundation of a true knowledge of the Universe or Macrocosm, and the latter that of the Living World or Microcosm.

The debt of Vesalius to Galen is too manifest to be questioned. Vesalian physiology is a revised version of Galen's experimental system, and to that extent only is it his own. Apart from the figures, which are obviously and magnificently original, the *Fabrica* as a whole is based on Galen, its outlook and qualities being conditioned by the standards of the Greek anatomist. According to Spencer, " Vesalius claimed that he had corrected 200 of the errors which Galen had imported from animal anatomy ; it is needless to say that he left a great number uncorrected ". These omissions in the work of so dynamic and skilful an anatomist may certainly be ascribed in great part to the difficulty of obtaining bodies for dissection in condition suitable for accurate research. The investigation of the respiratory movements of chest, lungs and diaphragm by the vivisection of dogs and pigs is clearly inspired by Galen. Vesalius complains that the ecclesiastical caucus would not countenance the vivisection of the brain, but he was able to show that when the recurrent laryngeal nerves were ligatured or cut the voice was silenced. The animals were kept alive during these operations by inflating the lungs through a tube of rush or reed, which, he adds, also revives the action of the heart.

To Vesalius the human body was, as Singer remarks, " a fabric, a piece of workmanship by the Great Craftsman ". His treatment of anatomy is not morphological but descriptive and topographical. This presumably explains his failure to be more interested in *comparative* anatomy, which has the more speculative appeal. He introduces, or rather stabilizes, but few new terms, of which well-known examples are atlas (spelt athlas) for the seventh cervical vertebra ; alveoli, suggested by the honeycomb of bees ; corpus callosum, a translation of Galen's term for this structure ; incus, the small anvil ; and mitral valve, the two segments of which are compared with a bishop's cap or mitre. He figures the malleus and incus, but omits the more curious stapes.[1] Nor does he see the two roots of the spinal nerves, or discover the true relations of the sympathetic, which is regarded, not unnaturally, as a branch of the vagus. He sometimes repeats the errors of others, thereby showing that his own work is partly a compilation. For example, he subscribes to the ancient and incomprehensible fiction that the gape of the crocodile is the result of the movement of the *upper* jaw — a belief so firmly established in the minds of anatomists that in 1681 Duverney was constrained to devote a special demonstration to its refutation, and

[1] This ossicle had not been discovered when the first edition of the *Fabrica* was published in 1543.

Fig. 13.—Vesalius, 1543. Auditory ossicles of man. Stapes described by Ingrassias in 1546

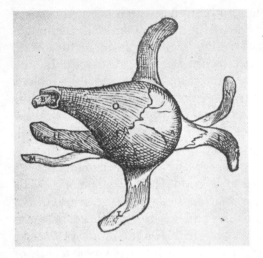

FIG. 14.—Vesalius, 1543. Seventh or choanoid eye muscle, *O*; *B*, optic nerve. Subject not stated, but obviously the ox

Grew in the same year rejected it as "ridiculous".

Vesalius cannot be acquitted of adopting a practice, which he himself reprehends, whereby the reader is not advised that certain descriptions are taken from comparative anatomy and not from the human subject. We have seen that this applies almost to the whole of Galen's anatomy, but Vesalius is only an occasional offender. His figure of the hyoid is evidently taken from the dog, and so also are his figures and description of the kidney, as was originally pointed out by Eustachius, and more recently by Monro *primus*. The figure of the recurrent laryngeal nerve again is not human, nor should a human foetus be grafted on to the zonary placenta of a dog, but the most interesting case is the description and figures of a seventh eye muscle, first seen by Galen. Vesalius found this muscle originally in the dog, where it

FIG. 15.—Vesalius, 1543. Historiated capital illustrating dissection of dog

FIG. 16.—Vesalius, 1543. Historiated capital illustrating dissection of pig

consists of four separate portions, but he was not aware that it was absent in man and the higher mammals.[1] It reaches its highest development in some ruminants, where it forms a continuous un-broken cone surrounding the entry of the optic nerve into the orbit, as figured in the *Fabrica*. The muscle has received a variety of names, such as choanoides, suspensor oculi, retractor oculi or bulbi and posterior rectus.

The lack of interest displayed by Vesalius in comparative anatomy as a subject to be studied for its own sake is some-what surprising when it is remembered that as a boy he dissected the internal organs of mice, moles, cats and dogs, and later studied the bones of quadrupeds and birds and the general structure of monkeys. It

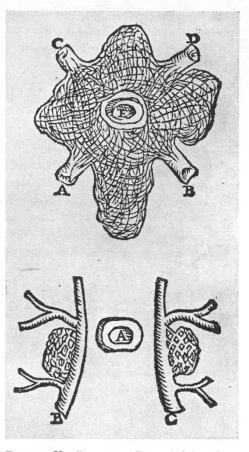

FIG. 17.—Vesalius, 1543. Rete mirabile on base of brain as seen in the sheep and ox. In the upper figure based on Galen, *E* is the pituitary body, and *A B*, *C D* are respectively the vessels entering and leaving the plexus

was the latter work which convinced him that Galen had dis-sected monkeys and not man. Vesalius takes his comparative anatomy largely from the dog, which was his chief anatomical and physiological subject. Two of the historiated capitals of the *Fabrica* illustrate vivisections of the dog and pig, the former being repeated twelve times. He figures the rete mirabile of the brain as typified in the sheep and ox, and compares the skull and other parts

---

[1] Fallopius (1561) was the first to demonstrate that the choanoid muscle does not occur in man. Eustachius (1552) correctly figures six.

of the skeleton of dog and man, showing that there is no separate premaxilla in man. His statement that the brains of monkey, dog, cat, horse and other four-footed animals, as well as those of birds and fish, correspond generally in every part with the brain of man expresses, it is true, the fundamental unity of plan of the vertebrate brain, but is not defensible in the sense understood by Vesalius.

VII

BELON

In comparative anatomy the revival of research into animal structure may be dated from the important work by Belon published in 1551 on the dissection of certain Cetacea and other marine animals. It is inevitable that these early studies should be deficient in morphological interest, and have but an indirect relation to the inception of the comparative method. Plain descriptive anatomy had to come first, and even when comparisons were initiated their true significance was for centuries only dimly perceived. Belon appears to have been a learned man and a stylist who, according to his biographer Delaunay, was one of the creators of French scientific prose. During his numerous travels he visited Oxford in 1550 and gave a demonstration to the lecturers in the Faculty of Medicine on the anatomy and physiology of the perch, *Acerina cernua*, during which he showed that the heart continued to beat for two days after it had been laid bare. In spite, however, of his extensive faunistic

Fig. 18.—Title-page of Belon's work on the anatomy of certain marine " Fishes "

experience his classification is casual and unenlightened, and he cannot be named among the philosophic systematists of the pioneer age. He says that the status of the bats was a problem on which he pondered for a long time, which makes even less excusable the fact that, after correctly enumerating all their mammalian characters, he rejects the evidence he himself had collected and puts them among the nocturnal birds of prey. With the same disregard for reason the Cetacea are relegated to the fishes, the hippopotamus figures with the Cetacea, and he even fails to include *Argonauta* among the Cephalopoda, as if he had not observed that its eight arms were like those of *Octopus*. His classification of fishes is only a copy of Pliny's unnatural assemblage of aquatic animals, and comprises some mammals and reptiles, the true fishes and many invertebrates. In these schemes anatomy is completely disregarded, but he attaches considerable importance to habitat. Thus an animal which moves unsupported through the air is a bird, or if it lives in water it is a fish, even if it exhibits mammalian characters in all its organs. His fishes are the fishes not of naturalists but of cooks and lexicographers.

As an anatomist, however, Belon concedes few grounds to the critic. He dissected and compared three Cetacean types — *Delphinus*, *Phocaena* and *Tursiops*, but there are no anatomical figures. He notes that the milk glands are two in number and mammalian in character, and mentions the occurrence of bristles in the upper lip of a foetal porpoise. The gut and its appendages are well described, and he saw the divided spleen, but could not find a caecum. The liver, he says, is split into lobes in young animals but undivided in the adult, as in man. There is no gall-bladder. He failed to recognize the compound nature of the stomach, but his " pylorus " is the long tubular pyloric chamber of that viscus. The respiratory organs attracted his close attention. He discovered and understood the function of the intranarial epiglottis, and realized that these animals, although aquatic, have lungs of the human type and are air breathers, but that the intake and exit are through the nostril and not the mouth. The heart, he says, has two auricles and two ventricles, and " in every respect is similar to that of man ". He does not distinguish arteries from veins, but the main blood vessels are described, especially the portal, azygos and postcaval veins. The skeleton as a whole, apart from the absence of the hind limbs, is stated to conform to the human plan, which, he says, can plainly be recognized in the sutures of the skull and in the condition of the tympano-periotic elements. The sternum is more human than in quadrupeds, and the fore limbs, though short, have the same bones as in man. There are five digits. The brain has all the parts and the ventricles of the human brain, and he noted the retrogressive behaviour of the olfactory organ of the dolphin during development. The genitalia belong to the mammalian type, and he observed the difference, as regards the relation of foetus to parent, between the viviparous mammal and the viviparous fish, both, however, agreeing in being formed from eggs in the uterus. He gives a crude figure of the female genitalia and diffuse placenta of the porpoise. Having established all this, and demonstrated that the Cetacea were essentially mammalian in all their characters, it says much for the native perversity of the human mind that Belon should still have assigned the Cetacea to the fishes.[1]

Belon also records a number of valuable observations on the anatomy of fish. He saw the highly muscular stomach of the mullet, and says this is the only fish which has a gizzard like a bird. He was the first after Aristotle to describe the pyloric caeca, and he refers to their variable development in many fishes. He discusses further the comparative anatomy of the fish gut, liver and biliary apparatus, and made extensive enquiries into the structure of the different types of poison apparatus found in fishes.

VIII

RONDELET

It is a singular coincidence that the first three eminent disciples of Ichthyology after Aristotle should have published their works almost simultaneously — Belon in 1553, Rondelet in 1554–5 and Salviani (issued in parts) in 1554–7. All were medical men. Rondelet is in many respects the most remarkable of the three, and his works are illustrated by original figures so carefully executed that it is possible to identify the genus and often the species which was being examined. The artist is unknown. Rondelet studied

---

[1] In 1554 Rondelet likewise underlines the differences between Cetacea and fishes without drawing the consequent distinction.

GVLIELMI

RONDELETII

DOCTORIS MEDICI,

ET MEDICINAE IN SCHOLA

MONSPELIENSI·PRO-

FESSORIS RE-

GII

Vniuersæ aquatilium Historiæ pars altera,
cum veris ipsorum Imaginibus.

His accesserunt Indices necessarij.

LVGDVNI,
Apud Matthiam Bonhomme.

M. D. LV.

Cum Priuilegio Regis ad duodecim annos.

Fig. 19.—Title-page of Part II of Rondelet's work on the fauna of the
Mediterranean

medicine at Montpellier, where at a later date he founded the first anatomical theatre the University was to possess. He entered as a medical student in 1529, and a year later Rabelais, some fifteen years his senior, also embarked on the career of medicine in the same University. Rabelais has been undeservedly reproached by several writers with disloyalty to his old friend. In a fictitious and extravagant narrative of a medical consultation,[1] the central figure is " Dr. Rondibilis " — an obvious play on the name and shape of our spherical naturalist. A perusal of Rabelais' sketch, however, shows it to be broadly humorous but not contumelious, and, although Rondelet is treated with some levity, he by no means appears in an unfavourable or ridiculous light, but rather as a kindly disingenuous man of good sense and worldly wisdom.[2]

Rondelet lived at a time when the scholastic approach to knowledge was giving way to the modern. It is therefore not surprising that he found it difficult completely to throw off the influence of the ancients and to trust to his own genius and fortune. To him Aristotle, Oppian, Athenaeus, Pliny and Galen are not merely material for the historian, but current authorities to be studied and quoted by the working naturalist. In particular Aristotle's works are reverently and meticulously studied. Nevertheless Rondelet is not the complete Aristotelian. He is too original an anatomist to accept statements contrary to his own observations, although disposed to do so when a personal check could not be applied. Thus he still believed in the spontaneous generation of carp in mountain lakes, and the production of eels from the corruption of dead horses. He just failed to understand that his part was not that of the follower and interpreter of Aristotle, but *himself* to found the science of marine biology. Further, we may regret that a trained anatomist, as he was, should not have perceived that diffuseness was a peril, and one easy to avoid. His dissections are much too numerous and superficial. Fewer attempted but more precisely carried out would, in the hands of a man of his skill, have yielded results of the first importance. He tells us very little of his methods, but describes a curious experiment on the fish *Lophius*. The stomach and the whole of the

---

[1] *Pantagruel*, Liv. III, caps. 31–5.

[2] An early biographer says of him : " Hilaris erat et facetus ". It should be pointed out that the identity of Rondibilis and Rondelet rests on conjecture.

viscera were drawn out through the mouth, and the body wall was stretched until it became transparent. The shell could then be converted into a lantern by putting a candle inside. The result of this novel method of studying anatomy was, he says, an horrific spectacle, in which the animal appeared in all its foul and loathsome aspects. No wonder the Italians had named it the Devil of the Sea! Even its flesh, he adds, as if determined that nothing good should be said of the monster, has a bad taste, a bad smell and is sore to digest.

After reviewing the nature and scope of his labours and the criticisms which might be passed on them, Rondelet is encouraged to end on a note of modest elation. " I have ", he says, " with

Fig. 20.—Portrait of Rondelet from Part I of the *De Piscibus*

great labour and at great cost produced a work in which you will find many good things of profit and satisfaction to studious men, and worthy of their commendation."

Before proceeding to examine the contents of Rondelet's great work, it is necessary to consider the meaning attached by him and his contemporaries to the term Pisces. He distinguishes anatomically, as regards the respiratory, alimentary, vascular and genital systems, and also as regards the manner of generation, between the gill-breathing fishes and the lung-breathing aquatic mammals. He perceives also that the latter not only differ structurally from the fishes, but resemble the terrestrial mammals. For example, he points out that the aquatic mammals have genitalia of the typical terrestrial pattern, quite different from the analogous parts in fishes, which latter he is of the opinion have no true testes. Nevertheless he includes *all* aquatic animals, whether vertebrates or invertebrates, among his " Fishes ". We must, however, not lose sight of the possibility that the early naturalists may employ the word Pisces, not in the strict taxonomic sense, but as an omnibus term applicable to *any* aquatic animal. On the other hand it must be admitted that an appropriate expression embracing aquatic life in general was available and used at the time, and the works of Belon (1553) and Gesner (1558) are in fact entitled *De Aquatilibus*. Aldrovandus (1613) includes the true fishes only in his " Pisces ", and the aquatic mammals are " Cete ", and even Rondelet is in two minds as to the meaning to be assigned to Pisces, since in places he distinguishes as fishes only the scaly and cartilaginous fishes. He was also fully aware that his authority Aristotle had given sound reasons for separating the Cetacea from the fishes proper. In the contemporary dictionary of Calepinus, " Piscis " is defined as an animal which lives in the water *and is typically covered with scales*, whilst " Aquatilis " is " *whatsoever* liveth in the water ". There is therefore little excuse for the misuse of the term Pisces by Belon and Rondelet, and it must certainly be regarded as a philosophical lapse that they should have so clearly mastered the facts and yet allowed the inference to escape them.

Rondelet's researches in comparative anatomy are directed chiefly to the fishes, but there are also some observations on other groups, especially the mammals. External characters only are figured. He knew the marine and river forms of the lamprey, but regarded it as belonging, with the important exception of its head, to the same type of fish structure as the eel. He is also misled by superficial resemblances into comparing in detail the dorsal nasal opening of the lamprey with the blow-hole of Cetacea, and believes that it conducts air and water into the pharynx, from which it follows that, as in the Cetacea, the animal would be suffocated if held under water. He notes that in the lamprey a segmented vertebral column is absent, being represented by a continuous " chorda ", and that the heart is partly enclosed in a cartilaginous " pericardium ", but these features had already been described by Belon, and the notochord was known to cooks from the earliest times.[1] Rondelet has much to tell us of the habits of fishes, and occasionally these descriptions have a morphological bearing. He confirms Aristotle in his account of the fishing frog (*Lophius*), which, he says, has two filamentous appendages in the neighbourhood of its mouth " with which it fishes as with a line ". In *Uranoscopus* he was familiar with the long, delicate filament below and before the tongue. On account of its resemblance to a small worm the filament acts as a lure and attracts small fishes which are thereupon seized and devoured. This, he says, has not been observed before. Rondelet failed badly in his attempt to resolve the asymmetry of the Pleuronectidae. He concludes that they are flattened dorso-ventrally like the *Torpedo* and the skate, and therefore the dorsal and anal fins become lateral paired fins, and the lateral line marks the sagittal plane. He should have seen that this interpretation is manifestly inconsistent with the disposition of the paired fins, gills, heart and viscera. Had he mastered this baffling problem it would have been an achievement to be ranked with Belon's successful comparison of the skeleton of man and bird.

The tongue of fishes, according to Rondelet, is a *taste* organ and takes no part in sound production or mastication, as it does in mammals. But many fishes, he says, have no tongue. He saw the palatal organ of the carp, and looked upon it as a *substitute* for the tongue. This well-known organ, he adds, is accepted as a true tongue by many learned men, although it had been correctly described by Aristotle, who held that it could be regarded as a tongue only by those who were ignorant of its relations. Rondelet

[1] *Athenaeus*, 7, 90.

states wrongly that the crocodile has no tongue, but this assertion is partly qualified in his second part. He was acquainted with the differences in gill structure between the cartilaginous and bony fishes, the latter alone having an operculum, except the sturgeon, which he places, not unreasonably, among the cartilaginous fishes. An operculum is, however, present in the sturgeon, although structurally simplified by degeneration. Rondelet noted the pouch-like gills of the lamprey, which he likened to tubes. He rejects the belief that fish hear by their gills — a belief perpetuated in their French name of *ouïes*. Gills, he says, are peculiar to fishes, and mark them off from other animals. He suggests that the stream of water which passes over the gills enables them to function as cooling organs, which explains why they are so near the heart. In the cartilaginous fishes he was attracted by the spiracle, and observed how it was opened and closed, and how its action was correlated with that of the mouth. This little investigation directed his attention to the neighbouring eye, and he describes the peculiar curtain-like iris which occurs in the skates. The current beliefs that fishes absorb water through the gills, or only extract the air that is *in* the water, are discussed by Rondelet. He rejects the opinion of Aristotle that fishes cannot respire in water any more than man or beast. All aquatic animals, he says, have need of respiration, and do respire, whether they have lungs or gills. If you keep fishes in an open vessel they may live for days, months or even years, but if the mouth of the vessel be closed so that there is no access of air to the water, they are suffocated. This would certainly not happen if water alone were necessary for their existence. Therefore they require free air which must penetrate into the water in which they live. He now proceeds to qualify this sound conclusion by assuming that the aquatic mammals and the cephalopods respire air *and* water, the latter being thrown out and the air retained. On the other hand he understands the differences between the respiratory mechanism of the Cetacea and the aquatic Carnivora.

Rondelet refers to the belief that animals which respire by organs other than lungs do not require or possess a diaphragm. He declines to accept this doctrine, and asserts that almost all fishes have a diaphragm.[1] The scaly fishes have an " imperfect "

---

[1] Following the translators of Aristotle and Galen, Rondelet refers to the diaphragm as the Septum transversum.

or membranous and non-muscular diaphragm. This pericardio-peritoneal septum or " diaphragm " is complete in the scaly, but perforated in the cartilaginous, fishes. His reasons for stating that the scaly fishes have a large stomach because they have an " imperfect " diaphragm are unconvincing. Although Rondelet describes the gut of various fishes, he left the spiral valve to be discovered by Severino. We know that he dissected a number of elasmobranchs, and it is evident that these researches could not have been carried very far. He remarks that all fishes do not possess an obvious spleen, and he seriously discusses the fable that in the anchovy the gall is produced in the head, which, he says, explains why their heads are cut off when being prepared for the table, lest the bile should taint the dish. The true fishes, birds and reptiles, he wrongly adds, have no kidneys [1] or bladder, but later he found both these structures in marine tortoises. He concludes that they form an inseparable pair and cannot exist apart. The aquatic mammals and tortoises, however, having a greater abundance of blood and excretions, must on this account possess both kidneys and bladder. Rondelet, like Aristotle, devotes some attention to the pyloric caeca, which he points out vary in number and are found only in the true fishes. He was the first to describe the characteristic pyloric caeca of the turbot. But his most important discovery in the anatomy of fishes is one which he himself briefly dismisses, and the significance of which was hidden from him. In fresh-water fishes, he says, there is found an extra bladder full of air which is double in some species.[2] It is also present in certain marine fishes, and in others there is in place of it a space full of air between the backbone and the peritoneum. The latter structure is the swim-bladder as it occurs in such forms as the codfish. It is remarkable that this prominent and vital structure should have evaded the scalpels of Aristotle and Galen. Rondelet thinks the swim-bladder might be a respiratory organ — a kind of lung.[3]

In the scaly fishes Rondelet says that the heart consists of three

---

[1] This error was corrected by Severino in 1645.

[2] He is referring here to the constriction which divides the swim-bladder into anterior and posterior regions.

[3] In 1555 J. Sylvius described a " lung " in *Salmo fario* which was doubtless the swim-bladder. In 1642 the swim-bladders of *Esox* and *Cyprinus* were figured by Ambrosinus.

parts, which are evidently (1) the sinus venosus + auricle, compared by Rondelet with the right cavity [ventricle] of the heart of other animals ; (2) the thick-walled ventricle, which is identified with the left cavity [ventricle] of the mammalian heart ; and (3) the fibrous bulbus arteriosus, from which arises the ventral artery passing between the gills and giving off branches to them. The " Fishes " which breathe by means of lungs have a heart and vessels similar to those of terrestrial beasts. The heart of [true] fishes is not fatty as in land animals, because they live in a humid and colder medium, and the heart cannot become dry. Rondelet had no conception of the circulation either in the fish or mammal. His views on the subject are Galenic. The liver is the source of the veins and the fountain of the blood, and he even distinguishes right and left sides in the heart of fishes. His description of the shark *Lamna cornubica* is marred by a mysterious oversight. He says that instead of the optic nerves which pass from the brain to the eyes in all other animals there is in this fish a hard cartilaginous substance. Here he has missed the optic nerve and found the cartilaginous optic stalk or peduncle. He is happier in his account of the ovoviviparous sharks, and describes how the young are reared in the uterine cavity of the parent. A foetus of *Mustelus laevis* is figured, showing its connection with the parent by way of an " umbilical " cord. He says that when the egg itself is " used up ", the relations of foetus and parent are " as in the quadrupeds ". This implies the existence of a placenta, although he does not specifically claim that a placenta is developed.

A few anatomical observations on frogs appear in Rondelet's second part. He noted the peculiar anterior attachment of the tongue in the mouth, in which, he says, the frogs differ from all other animals. He describes the vocal sacs, but does not observe that they are confined to the males. In their heart frogs resemble quadrupeds rather than fishes. The hyoid bone is mentioned.

After the fishes Rondelet's interests are centred on the aquatic mammals. He dissected a dolphin, the structure of which is compared with that of the pig and man. He describes the gut and its appendages, the renal system and genitalia, and inclines to the conclusion that the dolphin is an aquatic quadruped rather than a fish. He saw the small external auditory meatus, which was overlooked by Aristotle, and associated it with the ear. " In the bone of the head ", he says, " one sees clearly the parts of the organ of hearing." The lungs are more dense than in terrestrial animals, but otherwise these two types of lung structure agree in every respect. He thinks the difference is possibly due to water passing into the trachea, which might enter the lungs if they duplicated the soft and spongy texture characteristic of the land mammals. The kidneys, like those of the otter, resemble a bunch of grapes. The phoca, calf and tortoise also have compound kidneys, consisting of a variable number of small kidneys, and so also has the newly-born human infant. Then follows one of his truly remarkable statements. The brain, he says, has transverse but no longitudinal divisions, and is hence indivisible into right and left portions. This of course is quite wrong. The antero-posterior concentration of the brain of the dolphin, which results in giving the width a preponderance over the length, may have deceived him into regarding the lateral as the antero-posterior surfaces. Such might well happen at this early period if his description had been based on a brain *removed from the skull*, or on an artist's drawing. Rondelet quotes with approval Aristotle's description of a marking experiment in which the tails of living dolphins were incised by fishermen and the animals then returned to the sea. The recapture and recognition of these individuals long after established the fact that dolphins could survive twenty-five to thirty years.

Rondelet describes the castoreum or aromatic secretion of the preputial glands of the beaver. The ancients believed that these glands were the testes, and speak of one pair only. This is explained by the fact that they are enclosed in a common integument, as figured by Rondelet himself, and therefore appear as a single pair until separated by dissection. There are actually four pairs of glands, of which the posterior three are bound together, and may be, and often have been, mistaken for one. The most anterior pair were called by Hunter the castor or preputial glands, and the three posterior the anal glands, the assumption being that only the first pair opened into the preputial cavity. Pliny quotes Sextius Niger (*c.* first century B.C.) as correcting the error that the castoreal glands are the testes, and as recognizing the true testes of the beaver. Nor, says Sextius, can they represent the bladder, because they are paired, albeit having a single attachment. Rondelet compares them with the uropygial gland of birds, and gives good reasons for the belief that they cannot be the testes — for example, he says, they have no connection with the urethral canal. It is, however, evident that Rondelet himself was confused by the presence of the posterior series of glands, and was not certain what *were* the testes in the beaver. A more complete and accurate account of the testes and castoreal glands, accompanied by an excellent figure, was given for the first time by the Parisian anatomists in 1669, and later by Wepfer in 1671.

III

THE DEVELOPMENT OF CRAFTSMANSHIP

IX
COITER

MORE important than Belon and Rondelet as a comparative anatomist, since he was the first to elevate this study to the rank of an independent branch of biology, is the Frisian Volcher Coiter, whose relevant works were published in 1572 and 1575.[1] In these classic memoirs " he constantly urges the comparison of human anatomy with that of beasts as an occupation worthy of a philosopher " (Singer). He was a product of the school of Fallopius at Padua, and also studied under Eustachius at Rome, Aldrovandus at Bologna and Rondelet at Montpellier. Eustachius, he says, was his " excellent friend ", and Aldrovandus " was to be cherished with perpetual regard ". With such a training it is not surprising that, whilst he must have been well grounded in human anatomy, he preferred to *explore* the comparative field. He was also the first after Aristotle to study the development of the chick, a subject which has stimulated the curiosity of naturalists at all times, his own enquiries on this inevitable topic having been undertaken in 1564 and published in 1572. Coiter is no ineffective scholastic, and his work on the chick is more complete and exact than the descriptions of Aristotle. It is even an advance on those of his successors Aldrovandus and Fabricius, the former of whom, however, suggested this research to him. In addition, he gives a good account of the genitalia of the hen, but missed the ostium of the oviduct and the vestigial oviduct of the right side. It is strange that he makes no attempt to explain how the eggs pass from the ovary into the oviduct — a point made clear by Aldrovandus in 1600. Coiter also failed to recognize that the ovary of the bird was homologous with the " female testis " [ovary] of the mammal — a deduction first established by Steno in 1667, although he and others mistook the Graafian follicles for ova. The blastoderm of the developing bird was discovered by Coiter.

[1] Cf. B. W. Th. Nuyens, *Bijdr. Gesch. Geneeskunde*, **13** (1933).

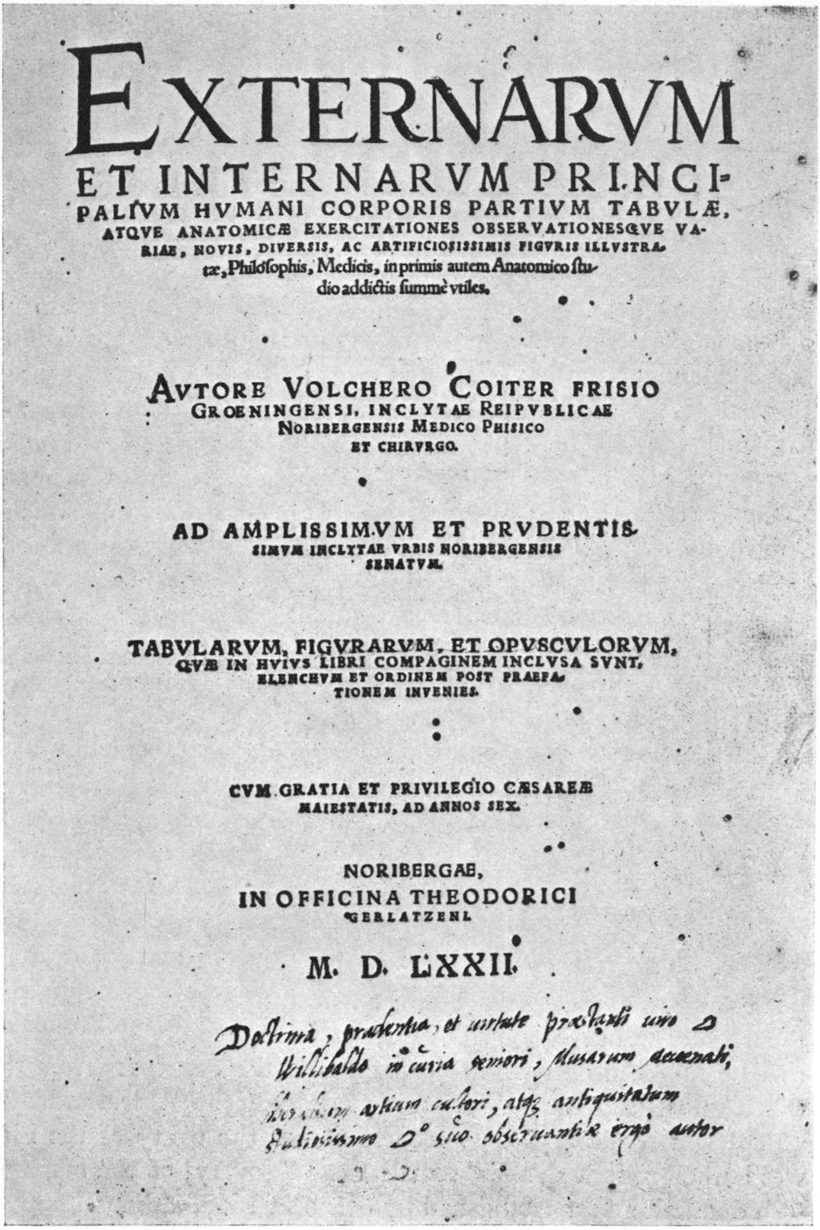

EXTERNARVM

ET INTERNARVM PRINCI-
PALÍVM HVMANI CORPORIS PARTIVM TABVLÆ,
ATQVE ANATOMICÆ EXERCITATIONES OBSERVATIONESQVE VA-
RIAE, NOVIS, DIVERSIS, AC ARTIFICIOSISSIMIS FIGVRIS ILLVSTRA-
tæ, Philoſophis, Medicis, in primis autem Anatomico ſtu-
dio addictis ſummè vtiles.

AVTORE VOLCHERO COITER FRISIO
GROENINGENSI, INCLYTAE REIPVBLICAE
NORIBERGENSIS MEDICO PHISICO
ET CHIRVRGO.

AD AMPLISSIM.VM ET PRVDENTIS-
SIMVM INCLYTAE VRBIS NORIBERGENSIS
SENATVM.

TABVLARVM, FIGVRARVM, ET OPVSCVLORVM,
QVÆ IN HVIVS LIBRI COMPAGINEM INCLVSA SVNT,
ELENCHVM ET ORDINEM POST PRAEFA-
TIONEM INVENIES.

CVM GRATIA ET PRIVILEGIO CÆSAREÆ
MAIESTATIS, AD ANNOS SEX.

NORIBERGAE,
IN OFFICINA THEODORICI
GERLATZENI.

M. D. LXXII.

Doctrina, prudentia, et virtute praeſtanti viro Wilibaldo in curia ſeniori, Muſarum avenati, iis ... artium cultori, atq; antiquitatum ... D. ſuo obſeruantiae ergò autor

FIG. 21.—Title-page of Coiter's first work on comparative anatomy

Before proceeding to examine Coiter's investigations as a comparative anatomist it is appropriate to refer briefly to his work on osteology, which was originally published in 1566. He was the first to study the growth of the skeleton as a whole in the human foetus, but Fallopius and Eustachius had to some extent prepared the way. Coiter points out that the bones are preceded by cartilages which afterwards become " transmuted " into bone — a delusion accepted universally by anatomists until 1736, when it was overthrown by Nesbitt. He notes the parts which first become bony, and shows correctly how the spreading of centres of ossification, fusion and remodelling account for the form of the bones as seen in the adult. The auditory ossicles are wrongly excluded from this general truth, because, he says, they are from the beginning osseous and do not increase in size. He established that the teeth calcify from without inwards, and that the root is formed later than the crown. Teeth, he adds, are not preceded by cartilage like bones. They are not bones and consist of a different substance. It is remarkable that he fails to mention the existence of a separate pre-maxilla in foetal man, indications of which had already been recorded by Vesalius and Fallopius in 1543 and 1561.

Coiter appears to have dissected animals under the direction of his teacher Aldrovandus, who exercised a benevolent but critical influence over his work. All the plates are signed " V. C. D.", and are therefore the work of the pupil. The types investigated cover almost the whole vertebrate series except the fishes, of which only the eel was examined in any detail. Invertebrates are not included. The Amphibia are represented by the newt and frog, reptiles by the crocodile, lizard, tortoise and viper, birds by the starling, woodpecker, wryneck, parrot, cormorant,[1] crane, great diver and fowl, and mammals by the pig, goat, ruminants, horse, squirrel, rabbit, mouse, rat, wolf, fox, dog, cat, badger, marten, hedgehog, mole, bat, tailed and tailless apes, and man. The section on the anatomy of birds is especially admirable. He expresses the differences between the birds known to him in tabular form, and this is almost the first attempt at a general classification of the group. He is also the first to give detailed and accurate figures of the osteology of many mammals, birds

---

[1] In this bird Coiter figures the singular sesamoid bone, found only in cormorants, which projects backwards from the occipital region of the skull.

and " Amphibia ", which compare very favourably with others published long after his time. He follows Belon in comparing the human skeleton with that of other vertebrates, but Coiter selects as his examples a higher and lower ape, the fox, wolf and dog. He emphasizes, however, points of difference rather than homologies, and has therefore not fully grasped the purpose and significance of the comparative method. He describes and figures for the first time the complete skeleton of the frog, but he is puzzled by the elongated tarsus and offers no interpretation of it. He became interested in the lungs and mechanism of respiration in frogs and lizards, and sought to interpret the structure of the mammalian lung in terms of the simpler organ of the lower vertebrate. The skeleton of the newt is compared with that of the lizard, and the latter again with that of the crocodile. In his anatomy of the skeleton and viscera of the tortoise he noted the spongy nature of the lungs and described the testis, but unaccountably missed the kidneys. He gives an account of the poison glands of vipers almost a century before Redi, who is usually credited with having been the first to do this.

The anatomy of birds receives attention in both the 1572 and 1575 works of Coiter. Apart from some attempts to unravel the histology of the " stomach ", the approach is purely anatomical. The skeleton, and especially the muscles, are well described, and he discusses the physiology of the muscles of flight. He observed the ring of sclerotic plates found in the eye of most birds, and showed that the lightness of the avian skeleton is due to the diploë construction. He was the first to mention the air sacs of birds, which were probably known to Aristotle, and were re-discovered by Harvey in 1651, and to find that the lungs adhere to the ribs and vertebrae. The cerebral hemispheres of the brain, he says, are quite smooth and the cerebellum very large. There is a tympanic membrane and cavity but only one auditory ossicle [columella], which he wrongly compares with the malleus of the mammal.

Coiter's most interesting contribution to avian anatomy, however, is his independent discovery of the very curious tongue and hyobranchial apparatus of the woodpecker, which had already attracted the attention of Leonardo.[1] He gives an ample and sound

---

[1] Coiter, however, was the first to discover a similar mechanism in the wryneck.

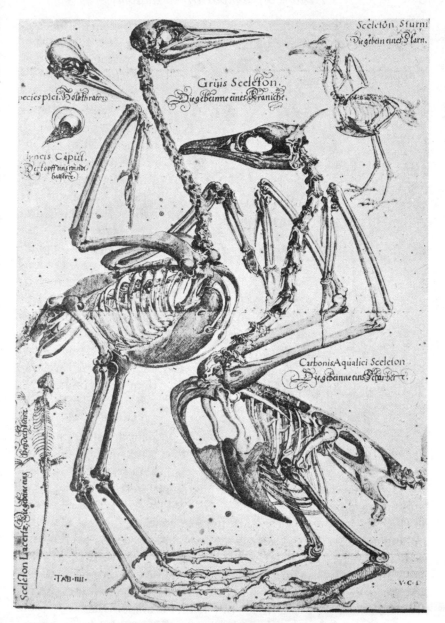

FIG. 22.—Coiter, 1572. Skeletons of lizard, starling, crane and cormorant, and skulls with hyoids of woodpecker and wryneck

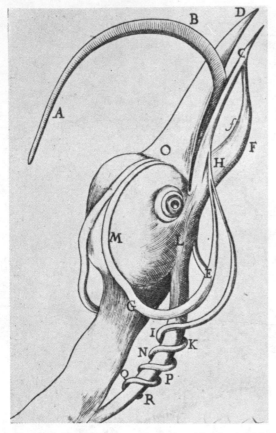

FIG. 23.—Borelli, 1681. Tongue and hyobranchial apparatus of woodpecker

description of the osteology and myology of this region, and shows how the long tongue is operated by the protractor and retractor muscles, but fails to observe the extraordinary and inexplicable fact that *both* posterior cornua of the hyoid are inserted into the *right* nostril. In attempting to explain how this unique contrivance is used in feeding he gave free play to his imagination, with the result that his conclusions are more picturesque than convincing. It was not until more than a century had passed that the structure was re-examined, this time almost simultaneously by Jacobaeus and the Parisian anatomist Perrault (1680), the latter regarding the movement of the tongue from a purely mechanistic point of view, but his explanation, though plausible and ingenious, is based on an inadequate knowledge of the anatomical parts. His posterior cornua also are represented as symmetrical, but this has no bearing on the validity of his speculations. A year later, in 1681, Borelli independently reviews and figures the mechanism of the tongue, but again his anatomy is incomplete and faulty, and he mistakes the large salivary gland for one of the muscles.[1] In 1709 Mery criticizes and amends the work of

[1] At the time of writing, the paper by La Hire (1695), whose analysis of this complex is also mechanistic, had not been seen.

Perrault and Borelli. He, and later Waller (1716), saw that the posterior cornua were inserted into the right nostril, and he correctly identified the salivary glands. His account of the whole apparatus and its mode of action is the fullest and most accurate up to his time. It is remarkable that Perrault, Borelli and Mery were apparently all unaware that they had been anticipated by Coiter. Finally Waller revises and supplements previous descriptions, all of which are discussed, although he is familiar with Coiter's work only through the abstract in Blasius. Some of his criticisms of Mery were stressed in ignorance of the specific differences which this singular organ is known to exhibit.

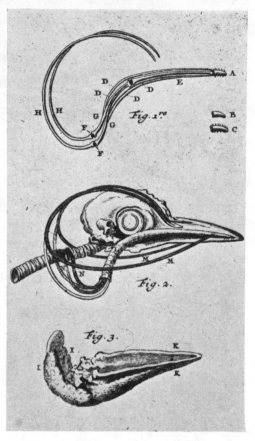

FIG. 24.—Mery, 1709. Tongue, hyobranchial apparatus and salivary glands of woodpecker. 2 illustrates *both* posterior cornua passing into the *right* nostril. 3 shows the very large sublingual glands, *I*, and their openings, *K*

Coiter's contributions to mammalian anatomy relate chiefly to the skeleton of a number of types. Other features to which he devotes attention are the thoracic and abdominal viscera of the goat, the characteristic dentition and orbicular muscle of the hedgehog, the mammae of the rat, the os penis of the marten, and the foot of the bat. He was the first after Aristotle and Leonardo to describe the four-chambered stomach of ruminants, a poor figure of which by his master Aldrovandus was published in 1613, and a still poorer one in 1642.[1]

[1] Both posthumous. Aldrovandus died in 1605.

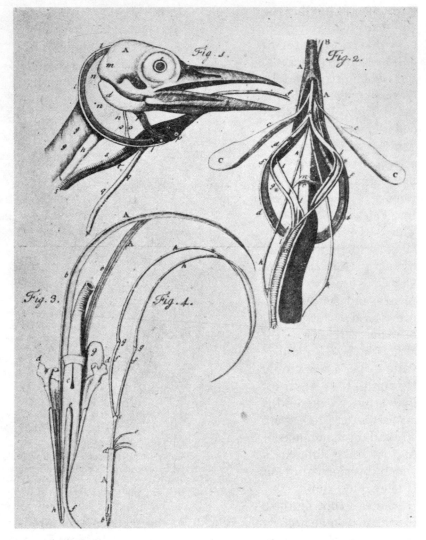

FIG. 25.—Waller, 1716. Skeleton and muscles of tongue of woodpecker. In 1 the
sublingual glands are shown at *d*, in 2 at *c*, and in 3 at *g*. 4, tongue and hyobranchial skeleton

His work on the nervous system resulted in the discovery
of the dorsal and ventral roots of the spinal nerves, and he
saw also the grey and white matter and the central canal of
the spinal cord. The last-named feature, however, had been
previously discovered by Estienne in 1545, and it is interesting

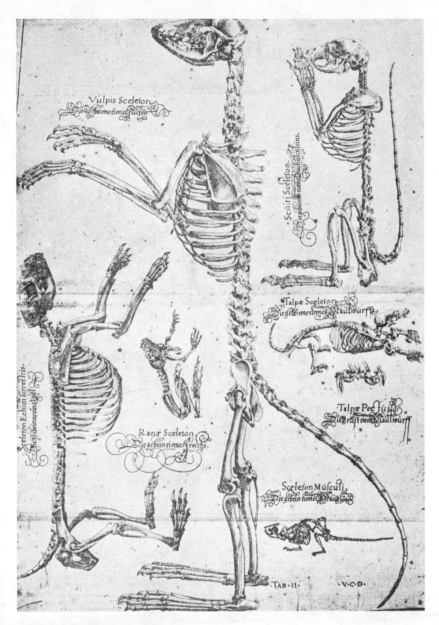

Vulpis Sceleton

Scüiri Sceleton

Talpæ Sceleton

Talpæ Pes

Sceleton Echini terrestris

Ranæ Sceleton

Sceleton Músculi

TAB. II.

V·C·D·

FIG. 26.—Coiter, 1572.   Skeletons of frog, fox, mouse, squirrel, mole and hedgehog

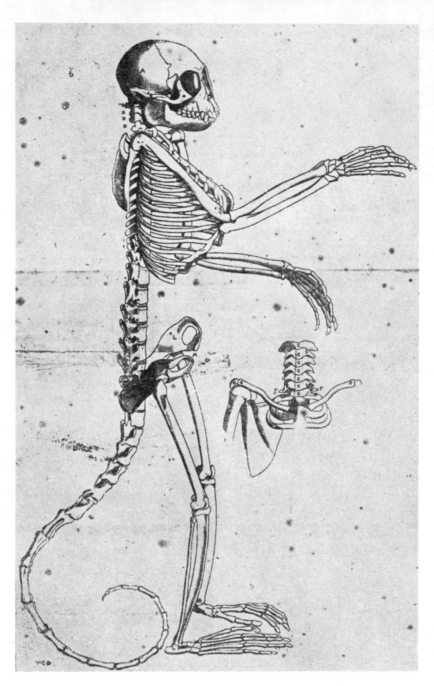

FIG. 27.—Coiter, 1572. Skeleton of a Capuchin monkey, *Cebus* sp.
Note inaccurate orientation of pelvis

to record that as late as 1844 Todd was actually disputing its existence.[1] Coiter denies that the rete mirabile of the brain, which he found in the ox and wrongly regarded as nervous in character, occurs in man. Finally, attention must be directed to his important observations on the beat of the heart, which he studied in the eel, frog, lizard, viper, chick and cat. He removed the heart, and studied its pulsations outside the body. He noted the sequence of contraction of the different chambers, and the changing form of the heart as a whole during its motions. Much later Harvey was making use of similar methods, but he seems to have been familiar only with Coiter's observations on the development of the chick.

X

RUINI

In 1598 Carlo Ruini, a senator of Bologna, completed his great work on the anatomy and diseases of the horse, and is thus the author of the first comprehensive monograph on the anatomy of an *animal*.[2] Practically nothing is known of the life of this remarkable man — except that he was possibly murdered. He was born *c.* 1530 and died on February 2 or 3, 1598 — about a month before his work was published. One plate is dated 1590, which would indicate that the book had been in preparation for some years. Bayon has recently revived the suggestion that Ruini may be credited with the text, but that the figures are those drawn by Leonardo to illustrate his own projected treatise on the anatomy of the horse. No evidence can be produced for the latter statement, which is inconsistent with the well-grounded belief that Leonardo never wrote such a treatise, nor, assuming that Ruini's woodcuts are reasonable reproductions of the original drawings, would any historian of art recognize in them the craftsmanship of Leonardo. It is true that the last figure of the superficial muscles of the horse in Book V is well posed and discovers some artistic feeling, but it has not the subtlety of the art of Leonardo. Moreover, it is impossible to avoid the conclusion that Ruini's work is the direct and

---

[1] In 1845 Todd and Bowman copy a figure by Stilling which shows the central canal, but its reality is still questioned.

[2] On the title-page of the first edition, and there only, Anatomia is spelt Anotomia, which is unusual but permissible. Also on the title-page, and again in the dedication, Cardinal Aldobrandini's name is misspelt Aldrobandini.

DELL'ANOTOMIA,
ET DELL'INFIRMITA
DEL CAVALLO
DI CARLO RVINI
SENATORE BOLOGNESE.

ALL'ILLVSTRISS.ᴹᴼ ET REVERENDISS.ᴹᴼ
MONSIGNORE CARDINALE
PIETRO ALDROBANDINI.
Nepote del Santissimo CLEMENTE OTTAVO Sommo
Pontefice, & dello stato, & essercito Ecclesiastico
soprintendente Generale.

IN BOLOGNA,
Presso gli Heredi di Gio. Rossi. MDXCVIII.
Con licenza de' Superiori.

FIG. 28.—Ruini. Title-page of first edition of his *Anatomy of the Horse*

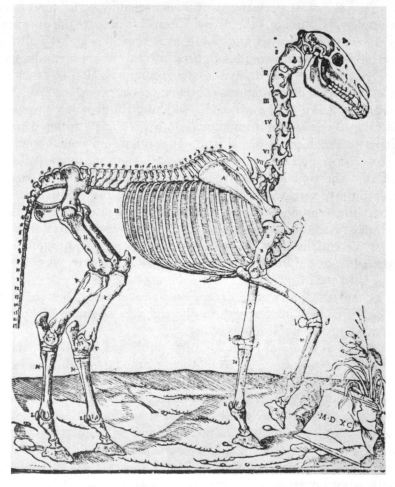

FIG. 29.—Ruini, 1598.   Skeleton of the horse

logical outcome of the *Vesalian* tradition, since it resembles, if it does not equal, the masterpiece of the founder of anatomy in almost every detail.   It is instructive to trace the parallel between these two works.   In both cases we observe a steady resolve to exhaust the anatomy of *one* type, and to avoid digressions by the way.   Ruini's treatise, as we should expect from the cumbrous nature of his subject, is the more topographical, but so far as possible he works through the animal system by system in the same patient and exhaustive manner.   We know the anxiety of Vesalius to secure the most perfect illustrations available at the time, how he employed a pupil of "the divine Titian" to prepare the drawings for the wood engravings, and indulged a capricious and not always amiable fancy of throwing his figures into expressive attitudes and supplying them with a panoramic background. In all this, provided we exclude the pirated figures engraved on copper which have not the artistic merit of the original woodcuts, Ruini is his close but not altogether successful imitator. Both anatomists suffered from persistent and flagrant plagiarism. It is often said that this was a custom but not a crime in the seventeenth century, in spite of the fact that the practice was frequently condemned, and in many cases bitterly resented. Thus shortly before his death in 1691 Robert Boyle proposed to the Council of the Royal Society " that a proper person might be found out to discover plagiarys, and to assert inventions [discoveries] to their proper authors " — a proposal assented to by the Society but apparently not acted upon.[1] In 1694 Cowper was complaining of the scarcity of original works and the prevalence of copying and stealing, but in 1698 he had become a plagiarist himself, and was stigmatized by his victim as a robber and highwayman.

Snape's anatomy of the horse, first published in 1683, is based on Ruini. None the less its author claimed the honours of a pioneer, for, he says, none had gone before or showed him the way ! Ruini's name is not even mentioned, although Snape's plates are close copies of Ruini's figures, notwithstanding his assertion that he has " by a curious draught or delineation represented to you such observations as are made in true dissections ". One of his plates representing the entire skeleton has, he claims, been " drawn exactly by one that I keep standing in a Press ", but it is difficult to believe that this skeleton in the cupboard could have been as unlike a horse as Ruini's figure which Snape has copied. In another of Snape's plates the only original feature is the addition of a superfluous dragon-fly to the background, nor can we excuse the subtle dissimulation which warns us " not to trust too much to these *copies*, as I may call them, without practising upon the original body itself ". It is worth noting that Snape himself was plagiarized, and so *ad infinitum*.

A French plagiarist of Ruini was Saunier (1734), who had the

[1] Weld, i, 329.

effrontery to label his plates " Déssiné dapprés Natture ", and claimed in the preface that they represented the life-work of himself and his son, and were prepared at the cost of incessant study and great expense. These transactions, and the early literature of biology is full of them, recall the indignant rhetoric of Robert Knox : " As to the hack compilers their course is simple : they first deny the Doctrine to be true ; when this becomes untenable they deny that it is new ; and they finish by engrossing the whole in their next compilations, omitting carefully the name of the author ". He might have added a fourth chapter to this tale of obliquity, in which the discovery is attributed to *another* worker. For some time now we have borne with numerous and determined attempts to deprive Harvey of the discovery of the circulation of the blood. On one of these attempts Daremberg makes the following satirical comment : " I have been singularly disappointed ", he says, " to see such an imposing array of citations brought into the service of an indefensible cause, and to learn that of all the ancient and modern writers it is Harvey who has played the *smallest* part in the discovery of the circulation ! " [1]

In addition to the unwelcome attentions of the plagiarist Ruini's work has not escaped the more insidious activities of prejudiced commentators. According to one of them Ruini did

---

[1] Thomson had previously remarked : " The attempts to deny the reality of this great discovery, and, when that was impossible, to rob Harvey of the honour of it, were innumerable ".

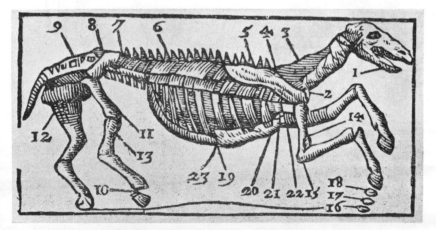

Fig. 30.—Ferrari, *c.* 1560. Skeleton of the horse

not write the Anatomy, another does not believe that he wrote the Diseases, and a third accuses him of stealing the illustrations. These charges, which, if sustained, would dispossess Ruini of *any* share in his own work, may or may not be true, but this much we can say — there is no *evidence* in support of any of them. Criticism of this type provokes the reflection that, if these spurious anticipations of classic discoveries justified the interpretation now put upon them, their fate was singularly and invariably unfortunate, for at the time they were written they convinced no one.　Only when the facts have been firmly established by others are the merits of these simulacra " drawn from their dread abode ".

Harvey entered as a student at the University of Padua in the year 1598,[1] a date which coincides with the publication of Ruini's work at Bologna.　The history of this book shows that it must have been well known, and its influence considerable, although Harvey makes no reference to it in any of his writings.　In the preface to Jourdain's edition of Ruini's anatomical plates of

[1] This date is subject to correction — 1597 and 1600 have also been maintained.

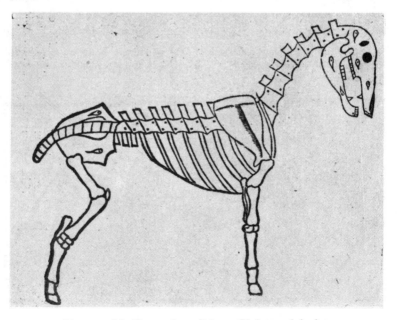

FIG. 31.—Markham, 1631 edition.　Skeleton of the horse

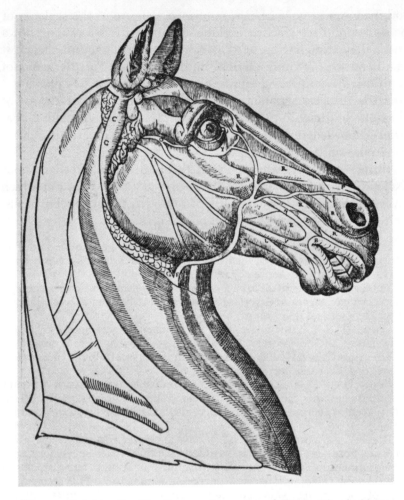

Fig. 32.—Ruini, 1598.   Muscles, vessels, nerves and glands of head of horse

1647 [1] it is stated that these detailed and accurate drawings were made from Nature by the very famous master painter Titian, but no authority is given for this highly improbable conclusion.

Before the time of Ruini the anatomy of the horse was practically unknown, and the grossest caricatures, such as those of Ferrari, *c.* 1560, and Scaccho of 1591, were published and accepted as serious representations of its structure.   Even after Ruini's

[1] Jourdain's work was re-issued twice — in 1655 and 1667, but they are only title page editions.

work had appeared Markham published a figure of the skeleton of the horse in 1610 which it is difficult to regard as anything but a jest.[1] Leonardo having left his work on the horse unfinished, it was therefore, as Sir Frederick Smith remarks, " at the hands of Ruini that the subject of equine anatomy jumped at a single bound from the blackest ignorance to relative perfection, the degree of which it is difficult to exaggerate ". It should be added that Ruini's knowledge of human anatomy was very helpful to him in his studies on the horse.

Ruini's treatise,[2] which passed through some fifteen editions (exclusive of pirated versions) between 1598 and 1769, but which nevertheless is little known, is divided into the following five books, each with its own series of wood engravings :

BOOK I.—" Animal Parts." *Head* : mouth, tongue, teeth at different ages, muscles, vessels, glands, skull, brain with its membranes, nerves and ventricles, sense organs.

BOOK II.—" Spiritual Parts." *Neck and Thorax* : skeleton, muscles, hyoid, larynx, nerves and vessels of neck, heart and its valves, trachea, lungs and pleurae, diaphragm.

BOOK III.—" Nutritive Parts." *Abdomen* : gut and its glands, peritoneum and mesentery, spleen, renal organs, great vessels of abdomen and portal vein, abdominal muscles, vertebral column and spinal cord, sacrum, pelvis and tail.

BOOK IV.—" Generative Parts." *Genitalia* : reproductive organs of both sexes and their vessels, development of horse, gravid uterus and placenta, foetus — its envelopes and circulatory system, foramen ovale.

BOOK V.—" Outlying Parts." *Limbs* : bones, muscles, vessels and nerves of fore and hind limbs. Added to Book V are seven figures of a diagrammatic character summarizing the more important features of the skeleton, veins and arteries, nerves and muscles. The reconstructions of the arteries, veins and nerves recall vividly the least inspired and convincing efforts of Vesalius. This is especially true of Ruini's schematic figure of the portal vein.

Ruini is one of the many anatomists who has had thrust upon him the honour of having demonstrated the circulation of the blood before Harvey,[3] and in 1869 a tablet was erected in the School of Veterinary Science at Bologna recording this fiction.

[1] In the 1631 edition an attempt is made to improve this figure, but it is still a fantastic libel, albeit viewed by its author as the " most perfect Anatomy of the bones of a horse ".

[2] The first edition is one of the greatest rarities of early zootomical literature.

[3] Cf. E. B. Krumbhaar, *Ann. Med. Hist.* N.S. 1 (1929).

Fig. 33.—Ruini, 1598.   Superficial muscles of horse

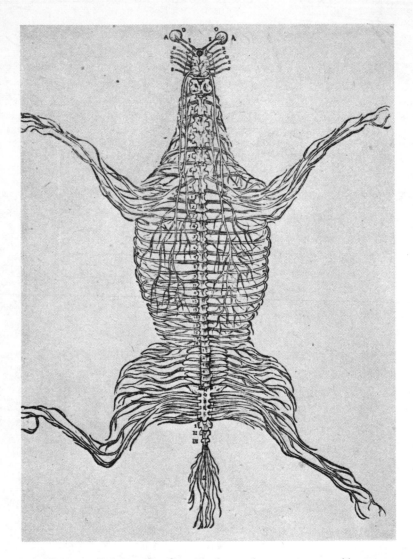

FIG. 34.—Ruini, 1598. General scheme of nervous system of horse

There is, however, no doubt that Ruini, as regards the circulation, was still dominated by Galenic ideas, and therefore has no claim to be associated with Harvey in this discovery. He has in fact very little to say on the circulation, and his language is not only brief but by no means free from ambiguity. Nowhere does he show any appreciation of the crucial fact that the object of the valves of the heart and vessels is to ensure an ever *onward* flow of the blood. As regards the heart he suggests that the function of the auricles and their valves is to protect the caval and pulmonary veins from the pressure exerted by the ventricles. He believes that blood passes from the right ventricle to the lungs and returns to the left ventricle, whence it is dispatched via the aorta to all parts of the body, but he proceeds to discount this by stating that the pulmonary vein also transmits blood from the left ventricle *to* the lungs. Therefore in the pulmonary vein there is a to-and-fro movement of the blood. So far he is further from the truth than his predecessor Servetus, who correctly outlined the pulmonary circulation in 1553 in clear and unmistakeable language.[1] How the blood is returned from the body, and the part played by the veins in the circulation, is not explained by Ruini. But the claim that he anticipated Harvey is refuted by his views on the portal system. The portal vein, he says, collects food material from the gut and conveys it to the liver, which is the organ chiefly concerned in the manufacture of the blood. From the liver it is distributed to certain parts of the body by the veins. This is pure Galenism, and it is not the only passage in which he expresses his belief in a *peripheral* flow of venous blood. Thus a vein, which is evidently one of the factors of the azygos — the superior intercostal vein, is said to *nourish* the intercostal spaces, and the coronary *vein*, wrongly figured as opening into the postcaval, *nourishes* the heart.

The method followed by Ruini in the dissection of his large subject was to begin with the parts most deeply seated and gradually work towards the surface. The absence of a comparative bias explains his difficulty in identifying and naming the muscles, which he distinguishes by numbers, and is also responsible for his failure to recognize the absence of the clavicle, with the result that

---

[1] It is commonly stated that Wotton, in 1694, was the first to direct attention to this passage. It was, however, read by " Mr. Hill " at a meeting of the Royal Society held on April 7, 1686. According to Meyerhof (*Isis*, **23**, 1935) the pulmonary circle was known to Ibn an-Nafîs in the thirteenth century, *c.* 1270.

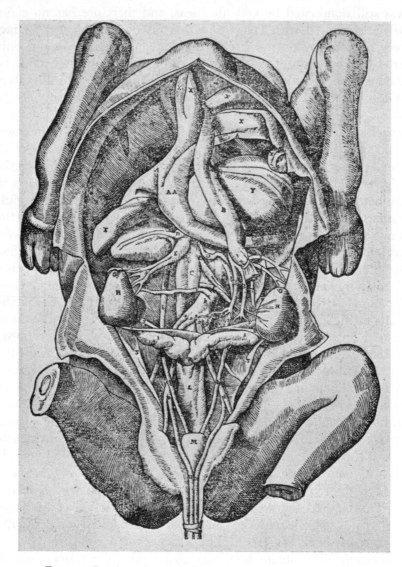

Fig. 35.—Ruini, 1598.   Abdominal viscera of an intra-uterine filly

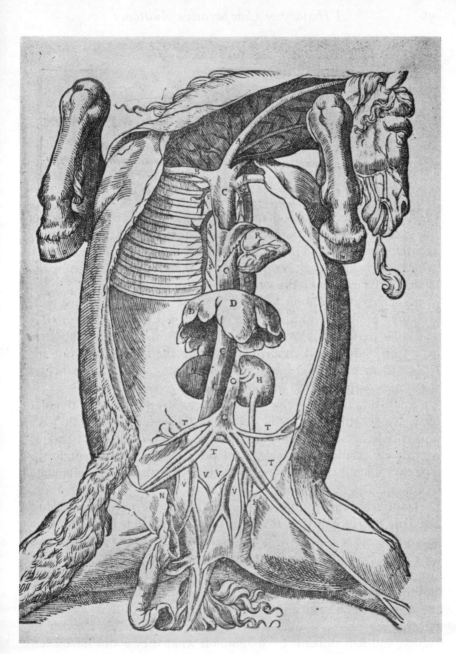

FIG. 36.—Ruini, 1598. Heart and great vessels of abdomen and thorax of horse

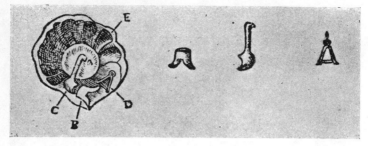

FIG. 37.—Ruini, 1598.   Auditory ossicles of horse

the first rib is regarded as that bone.   His description and figures of
the gut, liver, spleen and urogenital system are detailed and very
accurate, but he could not have examined the lining of the stomach
or he must have seen the two compartments, sharply defined by the
changing character of the mucous membrane, representing a left
or oesophageal dilatation and a right true stomach.   These facts
remained unnoticed until they were described by Severino in 1645
and more particularly by Home in 1807.   Ruini noted the absence
of a gall-bladder and described the bile duct correctly, but his
failure to mention the pancreas is surprising, since the pancreas
is a large compact gland in the horse.

He distinguishes developmentally between pre-molar and molar
teeth by showing that only the anterior molariform teeth have
milk predecessors.[1]   The laryngeal region is well described as
regards its cartilages, muscles and the function of the epiglottis.
The thyroid and the components of the hyoid also receive adequate
treatment.

He traced the course of the recurrent laryngeal nerve, but
erred in concluding that it was a branch of a spinal nerve.   His
account of the sense organs is admirable for the time.   The eye and
its muscles and the ear with its ossicles are closely examined,
although the auditory nerve is not correctly identified.   His first
auditory ossicle is the incus, the second the malleus and the third
the stirrup.   The brain and its membranes are thoroughly explored.
His figures of the cranial nerves, however, have no relation to
reality, and it is difficult to imagine on what they could have been

[1] This discovery was not established until 1767, when it was communicated by
Tenon to the French Academy of Science, but not published until 1797.   The authority
of Aristotle was against its acceptance.

based. Ruini was familiar with the essential facts of the peculiar structure of the horse's foot, to which he had evidently given close attention, but he figures only the bones and muscles.

In January 1599, shortly after the publication of Ruini's work, there appeared in Paris a treatise on the osteology of the horse by Heroard. It is highly improbable that Heroard had heard of Ruini, whose name is not quoted. His work is detailed and accurate, and, apart from its small scale, worthily approaches the standard set by his Italian contemporary. It is also important as indicating that the descriptive anatomist was at last conscious of, and attempting to emulate, the high purpose in research of Vesalius. Heroard gives an excellent description and figures of the three auditory ossicles, which he says are con-

FIG. 38.—Ruini, 1598. Brain and cranial nerves of horse

sidered to be the principal agents in hearing, but he repeats Ruini's mistake of identifying the first rib as the clavicle.

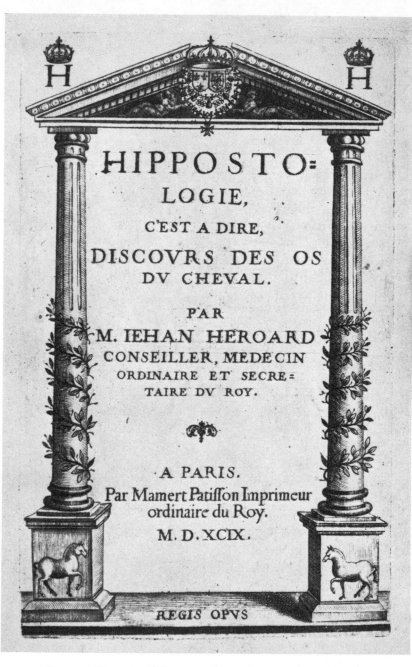

FIG. 39.—Heroard.   Title-page of his work on osteology of horse

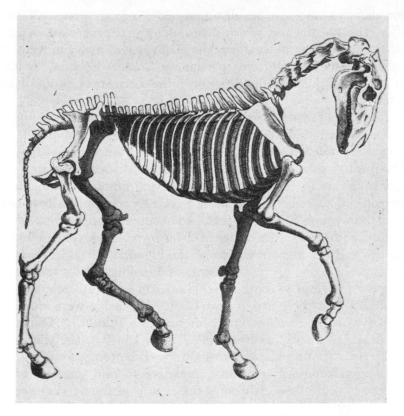

FIG. 40.—Heroard, 1599. Skeleton of horse

XI

FABRICIUS

A contemporary of the Dutch master Coiter, but one who considerably outlived him, was Jerome Fabricius, an anatomist grateful to Englishmen as the teacher of William Harvey. By demonstrating the valves of the veins to his English pupil, Fabricius played an unwitting but considerable part in the discovery of the circulation. Records of the Anatomy School at Padua go back to 1387, and perhaps even to 1252 ; but its most brilliant period was in the sixteenth and seventeenth centuries. At that time Vesalius, Columbus, Fallopius, Fabricius, Casserius, Spigelius and Veslingius successively occupied this famous chair of anatomy, and students assembled from all parts of Europe to profit by the lectures and demonstrations of so distinguished a professoriate.   Thereafter came the decline, which made such rapid progress that early in the seventeenth century the University was omitting its public dissections " from parsimony ", and Gibbon, writing in 1764, was justified in describing the University as a " dying taper ".

Time has failed to endorse the abilities of Fabricius as worthy of the exalted traditions of his chair.   He was a convinced and devout Aristotelian and Galenist throughout the whole of his life. Their works are quoted at length and on all occasions with the reverence due to Holy Writ — himself the humble commentator thereon.   This reluctance to bring a severely critical faculty to bear upon the writings of the ancients, and to subordinate his respect for classical authority to the evidence of his own senses, robbed him of the discovery of the circulation of the blood, and was a weakness inexcusable in one of the successors of Vesalius.   His work, therefore, is an attempt to combine scholasticism with research, and the possibility that these two approaches to knowledge were mutually exclusive never seems to have crossed his mind.   " Thus far, Fabricius, pleasantly and elegantly," says Lluelyn, the translator of Harvey, " but his arguments are not well bottomed."   None the less original enquiry was more attractive to him than teaching, and if he cannot be acclaimed as the creator of the *science* of comparative anatomy, he must be assigned an honourable position among the first explorers of *animal* structure.

In 1594, when Fabricius had been professor of anatomy for thirty years, he built at his own expense what was then regarded as an " ample and splendid " theatre to accommodate his large classes — a gloomy, curious, but highly interesting erection which has happily survived the neglect of posterity.   In this famous room, says Albinus, Fabricius was the first to dissect the human body before a great concourse of spectators.   Franklin has recently published measured drawings of this anatomical relic.   It is small, slightly oval in ground plan, and the six galleries rise as nearly vertically as possible.   The audience, which could not have exceeded three hundred, had to stand.   The advantage of this type of construction was that the furthest spectator was only twenty-five feet from the body undergoing dissection.   In the theatre as originally constructed there were no windows, and it was lighted only by candles and lamps. Until 1872 the building was still being used for the teaching of anatomy, but it is now preserved, almost in its original state, as a national monument.

The publications of Fabricius on comparative anatomy cover the last nineteen years of his long life, but they are not the product of his old age. Haller remarks that this explains why these senile works, printed long after the research itself was completed, are not truer to Nature. Moreover they were only a part of a detailed but uncompleted series of comparative studies. The tract on the development of the chick, for example, was published in 1621, two years after his death, but we know that the work was in progress long before then. Nicolaus Fabricius, Lord of Peiresc, a pupil of the anatomist, visited him in 1601, and was permitted to study his embryological methods. An egg taken from a sitting hen was opened every day " that he might thereby make observation of the formation of the Chick, all along from the very beginning to the end ".

It must be admitted that Fabricius scarcely deserves the commendation which his researches on the chick have received. He asserts that the liver, heart, veins, arteries, lungs and all abdominal organs are produced together, but that the bones and framework are formed before the soft parts. The latter statement is not based on observation but on analogy, the building-up of the foetus being compared with the construction of houses and ships, in which the framework must come first. Hence he supports neither epigenesis nor preformation, but a modified Galenic scheme. The anatomy of the female genitalia of the fowl receives close attention, and he describes for the first time the organ now known as the bursa Fabricii. He was not aware that it occurred in both sexes, or he would not have concluded that it was a receptaculum seminis, for which he was justly criticized by Harvey. Nor is it ever a " *duplex vesicula*."

Fabricius is more successful on the development of the mammalian foetus (1600), and even the critical Haller is constrained to describe it as " splendidum opus ". The sheep is the type most carefully investigated, but other vertebrate species examined include the cow, pig, horse, guinea-pig, mouse, dog, cat, snake and dog-fish. This treatise has been described as the first work on comparative embryology — a designation which is not undeserved.

The compound cotyledonary placenta of ruminants was known to Aristotle, Galen, Leonardo and Vesalius, but Fabricius was the first to print a detailed description and figures of it.[1]   He confirms previous statements that each cotyledon consists of interlocking but separable maternal and foetal portions, and adds details of their arterial and venous supply.   He re-discovered, and detached from the chorionic sac, the very long bifid allantois of the sheep, and inflated it with air.   He noted that it was devoid of cotyledons, but nevertheless correctly described it as the allantois.   At first sight this appears to be a contradiction in terms.   It must be remembered,

[1] A similar figure of the allantois and chorionic sac of the sheep was prepared by Eustachius in 1552, but first published in 1714.   The closely twisted condition of the umbilical cord figured by Eustachius is not a normal occurrence in the sheep—an error repeated by Fabricius in one figure.

FIG. 41.—Fabricius [? 1604].   Cotyledonary placenta of sheep, showing long bifid allantois, *E*, and untwisted condition of umbilical cord.

however, that in his time the relations of the mammalian allantois to the placenta were unknown, and hence the apparent complete independence of chorion and allantois would not strike him as remarkable. The peculiar allantois of the sheep and cow baffled Harvey, who concluded that it was either the chorion, or an abnormal structure, or even an invention on the part of Fabricius.

The ductus arteriosus is well figured in the sheep and man by Fabricius, and it is stated to be large and functional in the foetus but converted into a solid cord at birth. These facts, and also the existence and occlusion of the foramen ovale in the heart of the mammalian foetus, were not discovered by Fabricius, but were known to Galen.[1]

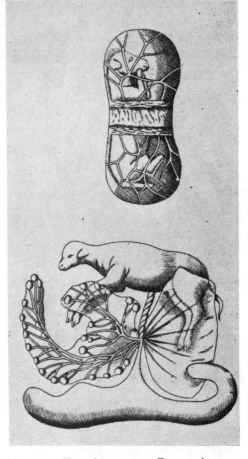

Fig. 42.—Eustachius, 1552. Zonary placenta of dog and cotyledonary placenta of sheep

The ductus arteriosus was re-discovered in modern times by Fallopius in 1561, Botallus saw the foramen ovale in 1564, whilst in the latter year both structures were described by Vesalius and Arantius. Fabricius adds a negligible account of the visceral anatomy of an advanced sheep embryo, and he notes that the placenta of the pig is not cotyledonary. His figure of the female genitalia of the pig is reasonably good and recognizable for its genus, and is specially interesting as showing an isolated ovisac on

[1] According to Franklin the term foramen ovale was introduced by Harvey in 1628, and Fabricius was the first to *figure accurately* the foramen ovale and ductus arteriosus.

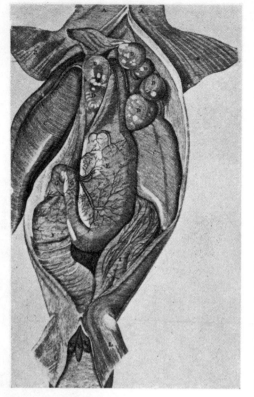

FIG. 43.—Fabricius, 1600.   Abdominal viscera of *Mustelus vulgaris*.   Note indications of the spiral valve

one side. Descriptions of the zonary placenta of the dog,[1] which he calls a girdle placenta, and of the discoidal placenta of the mouse and guinea-pig are included, the two latter animals being recognized as belonging to the same natural group.

The figures of the development of the serpent (1600) show very little. This animal seems to have been a grass-snake of the genus *Tropidonotus*. The only interesting feature established is the occurrence of the paired penial sacs at the sides of the cloacal opening, which, before they are introverted, are very conspicuous in the late unhatched embryo. Fabricius, having evidently the mammalian scrotum in view, wrongly calls these structures the testes.[2] Such are the consequences of exceeding the limits within which the comparative method legitimately functions.

Fabricius now extends his embryological enquiries to the Selachian *Mustelus* (1600).   He

[1] Eustachius figured this placenta in 1552, but there was no publication until 1714.

[2] This error was corrected by Vesling in 1647.

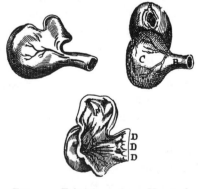

FIG. 44.—Fabricius, 1600.   Heart of dog-fish

describes the gut, spleen, gall-bladder, bile duct and female genitalia, but does not mention the caecum or the oviducal gland.[1] Indications of the valve in the spiral intestine are shown in one figure. This fish, he says, is oviparous and viviparous. He wrongly regards the ovary as paired, and overlooked the ostium of the oviduct, finding instead a more posterior opening which is certainly not there. The uterus is figured opened up so as to expose its contained embryos. As these embryos show the yolk-sac to be quite free, and without a placenta, his species must have been *Mustelus vulgaris* and not *M. laevis*. He figures also the heart of (presumably) a dog-fish which possessed a single " vesica " [auricle] and ventricle, and a conus (called the " large vein ") with three valves. This may well be the first published figure of the structure of the heart of a fish.[2] Other early dissections of the fish heart (*Salmo*) were the work of F. Schuyl, and were figured in his Latin translation of Descartes of 1662.

The most famous of the publications of Fabricius, however, is his folio tract on the valves of the veins, published in 1603. He first saw the valves in 1574, but he is not their discoverer, although he was the first to give adequate descriptions and figures of them, previous records of these structures being vague and imperfect. They were first figured by Alberti in 1585. The re-discovery of the valves is at once the high achievement and the tragedy of the labours of Fabricius. Chance had placed in his hands the key to the problem of the circulation of the blood, but the greatest opportunity of his life was allowed to pass. The history of science teaches us that the most formidable obstacle which the discoverer had to overcome is that innate timidity or conservatism of the human mind which hesitates to disturb settled convictions, and, what is still more fatal, holds back when the final and decisive step alone remains to be taken. This was the barrier which Fabricius was unable to thrust aside. He realized that the valves would *impede* a peripheral movement of the blood in the veins, but he did not perceive that they blocked the passage of *any* blood in that direction. The function of the valves of the veins, he says, is to counteract the effect of gravity, and to prevent blood flooding the feet and hands and accumulating there. The valves are the

---

[1] Nevertheless there is evidence that Fabricius saw this gland.

[2] Severino is severe on Fabricius for describing this heart as double, but Fabricius makes no such statement, and Severino has misread the passage in question.

" doorkeepers of the many parts to intercept the movement of nutriment downwards until the parts above have acquired their fitting share of it ". He believes, therefore, in a to-and-fro motion of the venous blood, the valves to some extent controlling the delaying action of gravity on the blood stream. His memoir relates almost entirely to man, but he also found valves in the veins of the ox.

Coming now to the contributions of Fabricius to comparative anatomy, we find that he compares man with lower animals with the object of assessing the points common to all of them, and also the minor differences which discriminate those of only specific rank. Osteology, myology and the sense organs are particularly investigated, on material which had in part been procured from foreign lands. In the skin (1618) he distinguishes between dermis and epidermis, and devotes his attention specially to the products of the latter, such as the hairs, bristles and spines of terrestrial animals, feathers of birds, scales of reptiles and fish, wings of insects, exoskeleton of insects and crustacea, and the shells of Mollusca. He concluded that the " cornea " of the eye of the serpent was a modified part of the skin, since it was shed with the rest of the epidermis. Judging from the context he probably saw the very curious orbicular muscle in the integument of the hedge-hog, but it is not specifically described. In this, however, he had been anticipated by Coiter in 1572.

In 1600 Fabricius gives a good and full account of the comparative anatomy of the laryngeal region of the sheep, ox and birds (goose, turkey and fowl), but the pig, horse and ape also receive attention. He describes correctly the epiglottis, glottis, ventricles of the larynx, recurrent laryngeal nerves, muscles of the neck, the thyroid, cricoid and arytenoid cartilages, and the comparative anatomy of the hyoid. He is often tempted to carry comparison too far, as when he homologizes the laryngeal cartilages of birds and mammals, admitting, however, that the epiglottis is absent in birds. In later publications of 1601 and 1603 he ventures into the difficult field of phonetics, and discusses at length the physiology of voice production, the functions of the various parts of the larynx, the uses of voice to the animal, and the reasons why the larynx is situated in the neck and at the summit of the trachea. The speech of man and animals is compared, and some experimental evidence

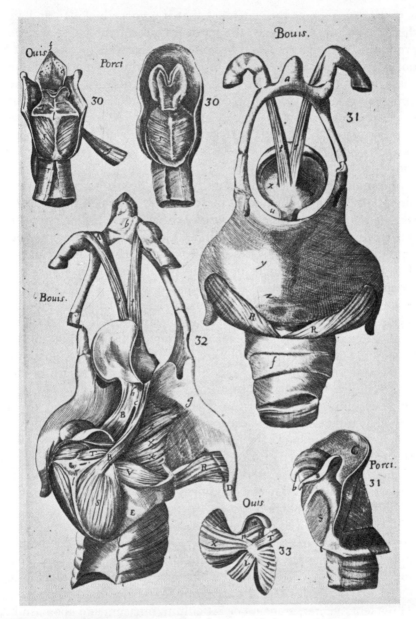

FIG. 45.—Fabricius, 1600.   Larynges of pig, ox and sheep

is produced, such as the statement that if a goose be killed and air then blown into its trachea the natural voice of the goose will be produced by the larynx,[1] and he concludes that it is the conformation of the median fissure of the larynx which determines the character of the sound. By larynx, however, Fabricius means the upper or true larynx. He was not acquainted with the organ of voice in birds, which is situated in the lower larynx or syrinx, as was discovered by Perrault in 1680, and re-discovered by Duverney in 1686 (published 1733), by Girardi in 1784 and by Cuvier in 1795.[2]

Fabricius published his lengthy tract on the structure and physiology of the alimentary canal in 1618, in which he strongly commends the practice of *comparative* anatomy. Haller refers to this treatise as a " senile work full of Galen's ideas, arguments and repetitions ". In a long series of comparisons Fabricius points out that the caecum is single in quadrupeds, double in birds and absent in fishes — in the last-mentioned group overlooking the rectal gland of Selachia. The stomach is reviewed as regards the subdivisions of its cavity, its macroscopic structure, size and situation, in man, quadrupeds, horned animals (ruminants), birds, fishes and some invertebrates. Similarly the macroscopic anatomy of the wall of the intestine, and the relation of the length of the gut to the size of the body, are studied in a variety of types. Misled by an inadmissible comparison with ruminants, he assigns three stomachs to birds, with the exception of the crane, stork and heron which have only one, the three corresponding with the crop, proventriculus and gizzard of modern terminology. The relation of the pleurae and mesentery to the heart, gut and blood vessels is understood, as indeed it was by Galen, and he investigated also the omentum and its function.

Following the lead of Aristotle, Leonardo and Coiter there is a long discussion on the comparative anatomy of the compound stomach, and the physiology of rumination, in the sheep, ox, goat and deer. It was Gaza in 1476 who converted (badly according to Fabricius) Aristotle's terms for the four chambers into the Latinized equivalents venter, araneum *or* reticulum, omasum and abomasum.

---

[1] Similar experiments were described by the Amsterdam anatomists in 1667, by Perrault in 1680 and by Muralt in 1683, their subjects being the duck, goose and kite.

[2] In 1674, and more definitely in 1681, Blasius associates voice production in birds with the *lower* larynx.

Fabricius considers the four chambers and their linings from six points of view, after the manner of Aristotle : (1) animals which have horns and no upper incisors have many stomachs ; (2) animals which have no upper incisors have many stomachs ; (3) animals which have no upper incisors ruminate ; (4) animals which do not masticate their food have many stomachs ; (5) animals which live on prickly and woody foods have many stomachs; (6) animals which ruminate feed on prickly, rough and hard foods. Fabricius does not accept all these propositions, basing his disagreement on a wider comparative knowledge, nor does he recognize, as Grew did in 1681, that the compound stomach of ruminants has been built up by the addition of three *new* chambers or outgrowths to the normal mammalian stomach represented by the abomasum. His discourse on the physiology of the stomach and intestines is verbose, repetitious and rashly speculative. More research and fewer words might have produced some enlightenment, even at a time when any major contribution to these difficult problems was impossible.

The structure of the human eye, the comparative anatomy of the eye, the uses of its various parts and associated glands, the physiology of vision and the methods to be employed in dissecting the eye are some of the topics examined by Fabricius in his work on the sense organs published in 1600. He describes six eye muscles in man, apes and fishes, but in quadrupeds which move with the head in the prone position there are seven, as had previously been noted by Galen and Vesalius. This extra seventh or choanoid muscle, he says, is not present in fishes, such as the pike, and may consist of two, three or four parts, which encircle the optic nerve and eyeball within the normal series of eye muscles. In the sheep, gazelle and ox the seventh muscle is well developed, especially in the gazelle, where it forms a continuous unbroken cylinder. In the sheep it is in four sections.[1] Fabricius upholds the view that each sense organ has its own specific function, and can discharge no other. On the ground that a light-perceiving organ must itself be transparent, he locates the seat of vision in the lens, which, he thinks, alone receives and discerns visual impressions. The optic nerve is insensitive except to a stimulus which can be communicated to it only by the lens.

---

[1] This muscle is usually not fasciculated in ruminants.

FIG. 46.—Fabricius, 1600. Eye muscles of sheep. *G, H, L, M,* seventh or choanoid muscle in the fasciculate condition

The type selected by Fabricius for his investigation of the ear is man — apparently at the infant stage, as the figures show a separate bony tympanic ring. He says that this ring is conspicuous in children and can be easily detached, but that later it fuses with the neighbouring bone. The last statement is incorrect for the normal human infant, where the fusion occurs before birth. In this memoir Fabricius has produced a useful general and comparative discussion of the auditory organ as a whole, but it contains little that is new. He deals with the auditory ossicles and the tensor tympani muscle, which latter he claims to have discovered in 1599, although Eustachius had described and figured it in man and the dog in 1563, and the account of it published by Casserius in 1600 (? 1601) was contemporaneous [1] with his own work and the product of his own department. Fabricius could not have detected the vestigial condition of the human pinna or he would not have attempted to understand, as also did Casserius, why it was im-

[1] It has been stated that Casserius has priority of publication over Fabricius, where both carry the date of 1600.

mobile. Neither did it occur to him that the pinna might be a new formation characteristic of mammals. Whilst, therefore, it would be legitimate to enquire into the causes of its disappearance in some mammals, there was no occasion to attempt to explain its absence in birds, reptiles and fishes, where it had never existed. These enquiries, however, are worth recording if only because they imply that the comparative studies of Fabricius had suggested to his mind vague possibilities of the mutability of species. It is astonishing that he should have had no definite knowledge of the auditory labyrinth, since the three osseous semicircular canals (" cuniculi ") had already been clearly described by Fallopius in 1561, of whose work he could scarcely have been ignorant. The true or membranous semicircular canals were seen in 1600 by Casserius in the pike, and later in 1645, by C. Folli in the mammal.

In 1614 Fabricius was writing on the morphology and functions of muscles, their vessels and nerves. This work is comparative, and embraces the structure of muscle in various insects and crustacea, in which animals he dwells on the fact that the skeleton is external to the muscles whereas in the vertebrates it is internal to them. Against this sound observation we have to balance the statements that arthropod muscles have no tendons or nerves, and that in these " bloodless " animals there is no heart or vascular system. As regards animal locomotion he discusses the osteology and myology of walking in mammals and flying animals, and also flight, swimming and crawling in birds, insects, millipedes and other types. He compares the skeleton of the limbs of the horse and man, and corrects the age-long misinterpretation of the position of the elbow and knee in the horse, as Leonardo had done before him. In this treatise the comparative anatomy of birds has a prominent place, and he considers at length, from the point of view of gravity, the balance of the body on the legs when the bird is standing. Even the humblest creatures are impressed to take a modest part in these deliberations. When worms creep, he says, the circular muscles are responsible for the forward movement, whilst the body is shortened by the longitudinal muscles. In the same year (1614) Fabricius issued his discourse on the comparative anatomy and physiology of joints, but his memoir on respiration, although written in 1599, was not published until sixteen years later. In this latter memoir the respiratory function and its mechanics in birds and mammals are considered.

If, however, the work of Fabricius was lacking in philosophical interest, it exercised a valuable influence on the future of anatomical science. The *Fabrica* of Vesalius had concentrated attention on *human* anatomy, and the dissection of monkeys, dogs and pigs as substitutes for the human body was discouraged to such an extent that the investigation of animal structure was in danger of neglect. Fabricius attempted to restore the balance, and succeeded not only in virtue of his own researches, but also as an influential and popular teacher he was instrumental in diverting into comparative channels the labours of others far abler than himself. Whatever we may think of his own works, as the instructor of Casserius and William Harvey it is impossible that he will ever be forgotten.

XII

CASSERIUS

Early in the seventeenth century (1601, 1609) Casserius, first the domestic servant, then the pupil, and finally the understudy of Fabricius at Padua, published his classic works on the organs of sense and voice.[1] He definitely repudiates the practice, which, owing to the example of Fabricius, was no longer rigidly observed, that human anatomy should constitute the sole charge on the time of the professor, and his own works owe their value to the fact that they are largely comparative. Although he is a great craftsman rather than a thinker, he nevertheless endeavours to explain the fabric of man by reference to that of the lower animals. His works may be limited in scope, but they represent the most ambitious and detailed investigations on comparative anatomy carried out at the time, and for long afterwards. They attain a high standard of descriptive accuracy and artistic merit, and had the powers of his mind been commensurate with his skill as an anatomist he would have overshadowed Vesalius and Harvey, and stood out as the originator of the comparative method.

Of the two works on comparative anatomy published by

---

[1] There is no printed title to the first and more important of these works, and the engraved title is undated. The colophon of Part I gives 1601, and of Part II 1600, but in the text of Part II he refers to observations made in 1601. The latter, therefore, is probably the correct date of publication of *both* parts.

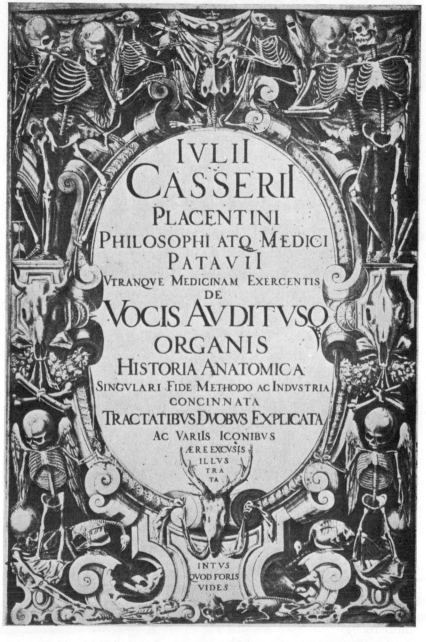

IVLII
CASSERII
PLACENTINI
PHILOSOPHI ATQ MEDICI
PATAVII
VTRANQVE MEDICINAM EXERCENTIS
DE
VOCIS AVDITVSQ
ORGANIS
HISTORIA ANATOMICA
SINGVLARI FIDE METHODO AC INDVSTRIA
CONCINNATA
TRACTATIBVS DVOBVS EXPLICATA
AC VARIIS ICONIBVS
ÆRE EXCVSIS
ILLVS
TRA
TA

INTVS
QVOD FORIS
VIDES

FIG. 47.—Casserius, 1601. Engraved title-page of his work on the larynx
and auditory organ

Casserius, the second, printed in 1609, may be briefly dismissed. It deals with the muscles of the hand and foot, " not too accurately " according to Haller, the muscles of the hyoid, tongue and nose, the turbinal bones, brain, and muscles of the eye. Casserius independently demonstrated that the choanoid muscle did not occur in the eye of man, as was implied by Galen and Vesalius. His first monograph of 1600–1601 aroused the admiration of Haller, who says that it combines most effectively human and animal anatomy and is illustrated with very beautiful plates. Casserius examined the glottis, larynx, chest and ear with great care, but on the nose and eye he is less thorough. He pledges himself to employ only the simple unpretentious language appropriate to philosophic truth — an admirable resolve which, however, is no guarantee that the thinking will be clear and the exposition intelligible. A long list of errata testifies to the absence of the author as the work was passing through the hands of the printer, and also to the difficulty of detecting *all* the errors which the craft and subtlety of that artist had introduced.

Casserius divides his work into three parts : (1) *structure* : (2) *action* — how the parts function ; (3) *uses* — what that function is. His method is first to describe the human condition in foetus and adult, and then to follow the organs under examination through a long series of animals, and his practice in dissection is to start with the superficial structures and work inwards. According to Casserius all the various senses are modifications of the sense of touch, in which, therefore, the origin of human consciousness is to be sought. It has been stated that during his investigations on the ear Casserius discovered incidentally, before Steno, the duct of the parotid gland. Whether this is so may be doubted. He has certainly not published any recognizable description of it, but in one of his figures of the cranial muscles of man the point at which the duct perforates the buccinator muscle to enter the mouth is indicated. Steno found the duct in 1660, and published a description of it in 1661. Needham, however, had previously *seen* it in 1655, but does not refer to it in his writings until 1667.

In the monograph on the larynx the following types were investigated : man, ape, cat, dog, pig, goat, ox, sheep, horse, rat, rabbit, turkey, goose, cormorant, heron, frog and the insects

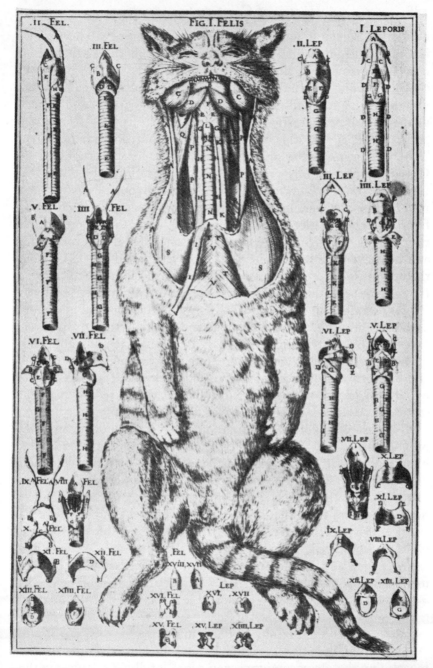

FIG. 48.—Casserius, 1601.   Larynx, hyoid and associated muscles of cat and rabbit

cicada, cricket and locust. The recurrent laryngeal nerves are carefully examined and correctly derived from his sixth pair of cranial nerves [the vagus], but they do *not* give off the nerves to the diaphragm, which latter, therefore, he confuses with cranial nerves. The recurrent nerves are accurately described and figured as looping round the subclavian artery on the right side and the aorta on the left, but this had been known since the time of Galen. That the larynx is the principal organ of voice is recognized by Casserius, and he examines fully and carefully the laryngeal cartilages, and the muscles of the larynx, hyoid, epiglottis and neck. The aberrant hyoid of the pig, however, is not correctly figured. The anterior cornua are omitted, except for short ventral portions in one figure, and he shows only the thyrohyals, basi- and hypohyals, all in one piece. The thyrohyals do *not* arise from the epiglottis — a minor error, having regard to the conditions as they are found in the pig. He illustrates in this animal the epiglottis (badly described in the legend), the thyroid, cricoid and arytenoid cartilages, and the hyo-epiglottideus and crico-thyroideus muscles. His dissections of the neck and larynx of the cat and rabbit are particularly elaborate, and demonstrate the cartilages and intrinsic and extrinsic musculature of this region. The skeletal parts of the hyoid and larynx of the goat and sheep are figured *in situ* and separately, and the intrinsic muscles are also illustrated.

The dissections of the dog are thorough, and include the muscles of the throat and neck, the hyoid, larynx and the mandibular lymph glands, the laryngeal cartilages being dissected out and described separately. The ox is treated in the same way, but not so fully. In the thyroid cartilage of the latter animal, and also the horse, he found the large foramen thyroideum, through which the superior laryngeal nerve passes to the interior of the larynx. Casserius, however, wrongly states that the nerve in question is the recurrent nerve. The foramen of the thyroid occasionally present in man transmits an abnormal branch of the superior laryngeal *artery* and is therefore not a comparable structure, but the internal laryngeal branch of the superior laryngeal nerve does, as a very rare variation, pierce the thyroid cartilage in man.

Casserius discusses at length the metaleptic action of the muscles on the glossocomium, or in more intelligible language the permutation of muscular motion in the laryngeal machine. In this he is following Galen, who, misled by the anomalous course of the recurrent nerves, the embryological explanation of which was not then available or suspected, compared the nerves with the cords stretched round pulleys in a well-known surgical appliance, and devised a mechanistic theory of voice production which illustrates the sandy foundations on which such doctrines are usually based. In this problem the position of Casserius is further compromised by his error of leading the *recurrent* nerve through the thyroid foramen, thereby introducing a second metaleptic point which would in fact neutralize the first.

In the frog, Casserius confuses the genio-hyoideus and sterno-hyoideus muscles, and his dissections of the larynx reveal only the arytenoid cartilages and a portion of the highly elaborate crico-trachealis cartilage of that animal. These figures should also have shown the location of the posterior cornua of the hyoid relative to the larynx, which is so obvious that his omission of the hyoid must have been deliberate. On the other hand his account of the muscles of the hyoid and larynx of the turkey is excellent, and compares favourably with modern descriptions. He says nothing of the posterior larynx or syrinx, but notes that the epiglottis is absent in birds and frogs.

A long philosophical discussion on the definition, nature and causes of voice and sound is unaccompanied by experimental evidence. He says that the glottis and its muscles, by which, following Galen, he means the larynx, is the efficient cause of voice, and neither the expansion and contraction of the lungs, nor any other organs exercise a regulating influence over voice production.

A happy excursus among the Invertebrata provides us with the first description and figures of the sound-producing organs on the abdomen of that noisiest of insects, the cicada. Long afterwards, in 1740, Réaumur described the mode of action of this complex organ, in ignorance apparently of the work of Casserius. The vague reference in Aristotle scarcely deprives Casserius of the honour of the discovery. He recognizes that the apparatus includes a sound-producing structure, the timbal, a resonator, the folded membrane, and a sound-receiving organ, the tympanum. He saw also, and figured, the muscles of the timbals, and noted that in some species the apparatus was exposed, and in others

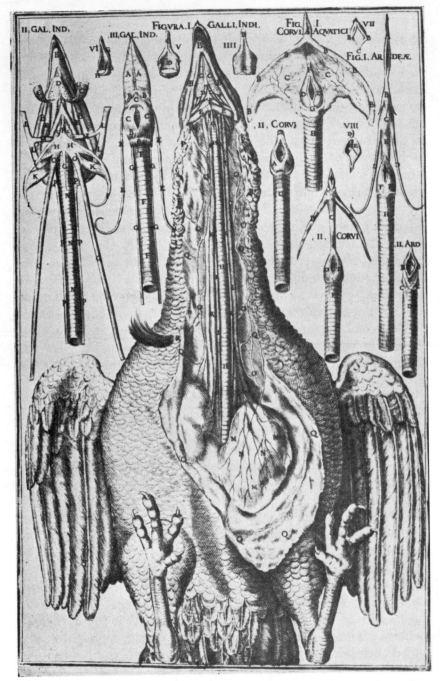

FIG. 49.—Casserius, 1601. Larynx, hyoid and associated muscles of turkey and other birds

concealed under an operculum.[1] He was also aware that in certain locusts and crickets stridulation is produced by friction of modified areas of the surfaces of the wings, but he does not associate sound production with sex.

Casserius is somewhat uneven in his treatment of the mammalian auditory organ. Although he has some highly important observations to record on the structure of the internal ear, he has devoted most attention to the external and middle ears. In the section on the comparative anatomy of the pinna he describes its cartilages, muscles, vessels and nerves, and the pinna of man is compared with that of the pig, ox, goat, sheep, cat and dog. He claims that the muscles of the external ear have not previously been fully described in man, where they vary in number, and that they are much better developed in quadrupeds. A very good description of these muscles in the ox and cat is given, and he even attempts to institute a comparison with the cranial muscles of birds. His account of the intrinsic muscles of the pinna is less satisfactory, and he succeeded in finding only two in various brutes and one in man. The tympanic ring is carefully described, and its state in infants and adults compared. Casserius was particularly interested in the structures associated with the tympanic region, such as the chorda tympani, already discovered by Fallopius in 1561 but not recognized as a nerve, the auditory ossicles and their two muscles. For his material in this work he drew upon man in various states of development, the monkey, pig, horse, sheep, ox, calf, cat, rabbit, goose and turkey. In the last-named bird the columella auris is homologized with the malleus of the mammal — a conclusion inevitable at the time, and not subject to revision until the comparative embryologist demonstrated the morphology of the auditory ossicles in modern times. Casserius claims to have discovered the stapedius muscle in the horse and dog in 1601, but it had already been described by Varolius in a work issued posthumously in 1591. Casserius states also that the auditory ossicles have the same size and hardness in infants newly born as in adult man — a circumstance which greatly surprised him, as well it might, although it had previously been maintained by Coiter.

---

[1] Some confusion must be admitted — due either to careless lettering of the figures or misprints in the legends.

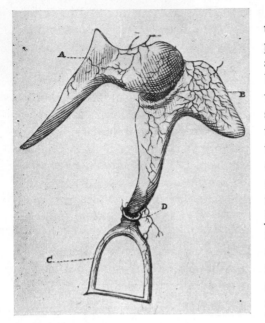

FIG. 50.—Ruysch, 1697. Injection of vascular periosteum of the auditory ossicles of man

We owe to Casserius the first detailed and comparative account of the auditory ossicles. He believed that the ossicles were uncovered, and considered it essential to their transmitting function that they should be so. The ossicular periosteum, however, does exist, and was first revealed by the injections of Frederik Ruysch in 1697. Casserius points out that the ossicles are not mentioned by Galen, and suggests that he could not have seen them. This, he adds, is because Galen dissected monkeys, where the ossicles do not occur ! [1] Vesalius missed the stapes because, according to Casserius, he worked with congested material, and that ossicle, owing to its position, would be hidden under a blood coagulum. Casserius figured a very interesting series of auditory ossicles in man, calf, horse, dog, rabbit, cat, rat,[2] pig, sheep and goose, each triad being shown *in catena*, and separately. In some of these figures the tensor tympani and stapedius muscles are represented.

Casserius did not discover the osseous labyrinth of the mammal, but he gives us the first adequate description and figure of it.[3] The membranous labyrinth of the fish, however, is described by him for the first time.[4] In the mammal he points out that there are only

[1] This astonishing error was corrected by Riolanus *fil.* in 1614.
[2] According to T. Bartholin (1654) the " Mus " of Casserius was the lemming.
[3] The mammalian cochlea was known to Empedocles and Fallopius, but Eustachius was the first to figure it in 1552 (published 1714).
[4] It has been asserted that Swammerdam (*c.* 1670) was acquainted with the internal ear of fishes, but all he *says* is that " in fishes the entrance of the ear " (*ingang des Oors*) " is a very pretty structure ". This the Latin translator, followed by the English, renders : " fish have a wonderful labyrinth of the ear ".

FIG. 51.—Casserius, 1601. Series of mammalian auditory ossicles, with the tensor tympani and stapedius muscles of horse and dog

FIG. 52. — Casserius, 1601. Bony labyrinth of human infant, showing cochlea, oval and round fenestrae and the three semicircular canals

three semicircular canals, which (and here he has Fabricius in mind) are not so numerous or on the other hand so complex that they cannot be counted ! These three canals occupy different planes of space. He saw also the connection of the utriculo-sacculus with the cochlea, and the scala vestibuli and scala tympani of the latter organ. He reproduces a highly interesting small but accurate figure of the bony labyrinth of the human infant, which shows the cochlea, oval and round fenestrae (previously described by Fallopius in 1561), and the spatial relations of the three semicircular canals, but no ampullae.

Fishes, says Casserius, have no lungs but possess a concealed ear which no one has yet investigated. Some fishes, he adds, have no ears. His plate of the pike is a remarkable historical document, despite the fact that it is a record of error as well as of discovery. The pituitary fossa in the base of the cranium is said to transmit the cerebral excreta conveniently to the palate, although the fossa is excluded from the mouth by the parasphenoid bone. There is a passable figure of the dorsal surface of the brain, the parts of which could scarcely have been accurately identified in the sixteenth century.[1] The ear, however, invited, and received, a more satisfactory treatment. It should be remembered that Casserius was breaking entirely new ground. Not only was the ear of fishes unknown, but the very possession of an auditory sense had been denied to these animals. Also the only knowledge of the hearing organ available at the time related to the mammalian ear. This consequently provided the type to which the ears of other groups might be expected to conform. When, therefore, Casserius discovered the internal ear of the pike he was induced to look for the external and middle ears. " Seek and ye shall find ! " In the nasal organ Casserius imagined that he had found the missing parts. The olfactory membrane scored with radiating lines was a presentable tympanum, and the olfactory tracts, fusing to form a single

---

[1] The three figures of the brain are the first representations of the brain of a fish.

vessel,[1] were canals representing the middle ear which conveyed sound to the main auditory organ, situated far behind at the side of the brain. Hence Casserius recognizes in the nose of the pike a true olfactory organ, combined, however, with the external and middle chambers of the ear.

Casserius gives two admirable figures of the ear of the pike — one showing the organ *in situ* and the other dissected out and somewhat mutilated. In these figures we can identify the sacculus and lagena, utriculus, utriculo-saccular canal, vestigial endolymphatic duct, the three semicircular canals with their ampullae, and the otoliths of the utriculus, sacculus and lagena. The relations of the horizontal semicircular canal are not correctly shown, but in the pike this canal, unless closely examined, might appear to pass from the anterior to the posterior semicircular canal as figured by Casserius.

The monograph on the ear is concluded with a discussion, intuitive rather than experimental, on the nature and propagation of sound, and the functions of the various parts of the auditory mechanism. For example, his confused explanation of the perforation of the stapes is based on the probable effect on the conduction of sound which a structure of such a shape might be expected to produce. The stapedial artery does not persist in the types examined by Casserius, and hence its relation to the form of the stapes was unknown to him. The lack of respect for methodology which discriminates too often the modern scientific memoir is painfully characteristic of the works of the old masters. They could or would not see that any failure to define in rigorous terms the conditions which must be satisfied before a secure decision can be reached, was to abandon demonstration for guesswork, nor did they recognize that even the science of words cannot afford to dispense with a valid basis of observation.[2] It is true that guesses are occasionally right, and sound conclusions drawn from bad or deficient observation. Nevertheless such unsupported results do not convince, but succumb to the assaults of the first destructive critic. To take an example. Duverney in 1683 maintained that as fishes have no cochlea the seat of hearing must reside in the semi-

---

[1] There is, of course, no fusion, but only close apposition.

[2] " Let all reasoning be silent when experience gainsays its conclusions " (Fabricius, quoted and approved by Harvey).

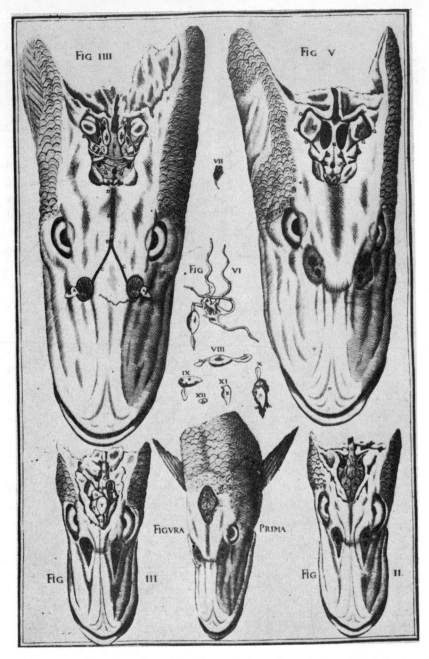

FIG. 53.—Casserius, 1601. Pike, *Esox*. Brain, and membranous labyrinth of
ear *in situ* and dissected out

circular canals. Thus he not only takes the phenomenon to be explained for granted, but overlooks any part which might be played by the sacculus. To this Camper replied in 1767 that as the Cetacea have a well-developed cochlea and "probably" *no* semicircular canals, the auditory receptor must be in the cochlea. Therefore he makes the same assumption as Duverney, but bases his argument on a serious misstatement of fact, which latter was corrected by Monro in 1785, John Hunter in 1787 and Cuvier in 1796.

HARVEY. THE ENCYCLOPAEDISTS

XIII
HARVEY

I⊤ is appropriate and inevitable that the early anatomists should have been human anatomists. The human subject, as being of supreme interest to man himself, was dissected first, in spite of the difficulties and perils attached to so unpopular a study, and a nomenclature rapidly grew up around that study. The same terms were then applied to the lower animals when it was their turn to be investigated. It is not sufficiently realized that the basic terminology of anatomical science is iatric in origin — the work of the human anatomist and medical man, who in fact is the founder of the biological sciences. He it was who gave to the parts of the body of an insect the same names he had previously applied to the analogous regions of man. So far the student of animal structure had combined in his own person the functions of physician, anatomist and physiologist, but so vast a body of research and commentary had accumulated as the result of the publication of the *Fabrica* of Vesalius, that it was becoming increasingly difficult for one man to cope with it.

The professor of the Galenic period, content with the exposition of the classic authors, averse from touching a body with his own hands, and whose scholastic discourses were illustrated by the dissections of ignorant hirelings on dogs and pigs, was already almost a legendary figure. Within a brief period of some fifty years from the printing of the *Fabrica* the time had arrived for biological science to segregate into autonomous units, and for the biologist himself to specialize in one of those units. The man to take the first step in the new direction was an Aristotelian — William Harvey. As he advanced in years and experience it is manifest that no one saw more clearly than he did the direction in which anatomy was to develop, but the tyranny of the Greek tradition was too firmly established to be completely cast aside.

In his last work its influence was again as potent as ever, and so he remained a Peripatetic to the end of his days. In his discourse on the circulation and in the two letters to Riolan Harvey makes ninety-seven references to twenty-one authors, and it is significant that, of these, fifty-six are drawn from Aristotle and Galen. On the other hand he proclaims emphatically that he learns anatomy not from books but from the fabric of Nature, and as regards accepting obediently the works of the Ancients he says : " the facts manifest to the senses wait upon no opinions, and the works of nature bow to no antiquity. For indeed there is nothing either more ancient or of higher authority than nature."

It is not easy to assess the genius of Harvey, and it is only possible to form a just estimate of it by studying *both* the work on the Circulation (1628) and the treatise on Generation (1651).[1] The author of the former is one of the immortals, but the " Generation " is the work of a sadly puzzled human being. And yet when he published it, twenty-three years after the appearance of his *magnum opus*, his knowledge was greater and his powers more mature. The explanation is quite simple. On the circulation, he is, as it were, putting the finishing touches to the edifice, but on generation he is laying the foundations. In the first case others had gone before and shown him the way, but in the second he is encountering all the trials and perplexities of the pioneer. No great discovery is ever the work of one man — not even when that man is William Harvey. It is only when we take stock of the limitations of his time that the merits of his discoveries appear in their true perspective. What were those limitations ? The compound microscope had been invented, but was not in general use — at all events Harvey did not use it.[2] Blood corpuscles were unknown. Blood capillaries had not been suspected, and would probably not have seen if their existence had been deduced. No sections. No methods of injection. No dissections other than those made with the unassisted eye or with the help of the simplest of magnifying glasses. Material for dissection difficult to obtain. The study of human anatomy regarded by an unenlightened public with loathing and contempt, and in some quarters its practice

[1] These are dates of publication. Harvey, however, was working on generation before 1628.

[2] For a long time after this period the compound microscope was far inferior in performance to the simple microscopes of Leeuwenhoek.

fraught with danger to the anatomist. Even Robert Boyle approached the dissecting room "with anxious and reluctant steps". " One would think ", he says in 1663, " that the conversing with dead and stinking Carkases (that are not onely hideous objects in themselves, but made more ghastly by the puting us in mind that our selves must be such) should be not onely a very melancholy, but a very hated imployment. And yet . . . I confess its Instructiveness has not onely so reconciled me to it . . . that I have often spent hours much less delightfully, not onely in Courts, but even in Libraries, then in tracing in those forsaken Mansions, the inimitable Workmanship of the Omniscient Architect."

For centuries the blood had been supposed to oscillate in the veins, and to pass from one ventricle of the heart direct into the other. Such beliefs had become an inveterate tradition of the medical schools — the heritage of thirteen centuries of passive credulity. It needed no small effort of courage and genius to eradicate doctrines woven into the texture of the human mind, but to succeed in so doing could not fail to inaugurate a new method of enquiry. Harvey's contribution to the thesaurus of biological learning is in fact not so much the discovery of the circulation of the blood as the vindication of the comparative and experimental methods. The demonstration of the circulation, important as it is *per se*, becomes nothing more than an illustration of the value of the method by which alone such problems can be solved. It was therefore the institution of the *method* which marked an epoch in the history of the biological sciences.

What now is Harvey's connection with comparative anatomy ? Galen and Vesalius employed monkeys, dogs and pigs as *substitutes* for the human body. They did not consider that dissections of these animals had any special interest or importance of their own — they were only stages on the way to man. Still less did they endeavour to advance the status of human anatomy by linking it with the structure of the lower animals. Morphology was not yet conceived, for the first intention of the sixteenth-century biologists, such as Sylvius, Estienne, Vesalius, Columbus, Eustachius and Fallopius, was to perfect the current system of *human* anatomy. Here, however, Belon, Coiter, Fabricius and Casserius are arresting exceptions, and Fabricius is an exception all the more important since he was the teacher of Harvey ; for although no anatomist is handled so critically (and justly) by Harvey himself, it is clear, and in fact Harvey was wont to admit, that his old teacher of anatomy at Padua exerted a moulding influence over his studies.

Unfortunately the greater part of Harvey's researches on comparative anatomy — and they must have been considerable — were lost or destroyed in the savage tumults of the Civil War. He used to tell John Aubrey that of all the trials he had sustained no grief had been so crucifying to him as the loss of these manuscripts, and, as it is, his published work on the circulation includes notes on sponges, zoophytes, earthworms, three types of Crustacea, five insects, six Mollusca, four fresh-water fishes, two Amphibia, three reptiles, eight birds, and ten mammals representing four of the orders. At least fifty species must have been examined, but Harvey's definitions are too general to enable an exact figure to be given.

It will be seen that this list covers a considerable part of the animal kingdom. Even his detailed study on generation, based as it is largely on the chick, includes numerous acute and original observations on other animals, and is thus one of the earliest essays in *comparative* embryology. It is also not purely embryological. For example, we find in this work an account, not, it is true, the first, of the air sacs of birds. Many years afterwards this discovery was the occasion of a priority dispute between Peter Camper and John Hunter,[1] both of whom were unaware that the air sacs had already been described — imperfectly by Coiter (1572), but more exactly by Harvey (1651), the Amsterdam anatomists (1667) and Redi (1682). Their right to the discovery is indisputable, but some vague passages in Aristotle seem to suggest that he also had seen the air sacs. Harvey describes the abdominal air sacs in the ostrich and other birds, and their connection with the trachea, and he compares them with the air bladder of fishes. He inflated them with a pair of bellows from the trachea, and named them the *membranous abdominal cells.*

Some of the sixteenth-century anatomists got as far as recognizing that the structure of animals should be investigated, and that it might even have some bearing on the fabric of man, but the

---

[1] This dispute had reference more particularly to the penetration of the air sacs *into the bones*, first observed by Frederick II.

idea of a system of comparative anatomical *science* was never present in their minds, and it was reserved for the prophetic genius of Harvey to detect the value of the new method, and to test its possibilities. Harvey in fact not only *emphasizes* the importance of the anatomy of the lower animals, but he goes considerably beyond that. He *makes use* of them to overcome the difficulties presented by the structure of man. To him an insect is worth studying for its own sake, and still more so for what is to be learned of the fundamental truths of biological science. He seizes every opportunity of illustrating his views, and stimulating his imagination, by appeals to the " viler creatures " as he calls them, and the beating of the heart of a Crustacean is not only interesting as such, but to him it throws a powerful light on the beating of the heart in man. Indeed, he set out to examine the action of the heart not only in man but also in all animals that have hearts. Thus, he says, you will reach the truth by inspecting many and various animals, and collating the numerous observations so made. " If, as Aristotle tells us, the immortal gods were present in the kitchen of Heraclitus, let us not despise the study of the humbler animals. For the great and almighty Father may dwell even in the smallest and seemingly despicable creatures." Consequently he is disposed to criticize those anatomists who "confine their researches to the human body alone, and then when it is dead ".

Harvey is naturally interested in the comparative anatomy of the heart. He says that all animals have hearts — not merely those which have red blood, but also molluscs, Crustacea and insects. In the insects the heart has but one chamber and is situated in the dorsal region of the " tail ". He watched its movements with a magnifying glass. A pulsating heart can also be seen in the louse, and in a " small transparent shrimp " found in the Thames and the sea [? Mysid] the motions of the heart may be viewed with the greatest distinctness. He is, however, wrong in denying that oysters [1] and mussels have hearts, and in asserting that in other Mollusca the heart consists of a single vesicle, the auricle, the ventricle being absent. The beat of this heart, he says, was slow and irregular. He found that the auricle of the heart in some fresh-water fishes was fleshy and looked very

---

[1] Willis described the heart of the oyster in 1672.

like a lung.[1] Fishes, frogs, lizards and serpents have a two-chambered heart, made up of a ventricle and a bladder-like auricle. The auricle contracts first and the ventricle later. Harvey is obviously mistaken in assigning one auricle only to these animals, except the fishes. Also he does not appear to have noticed the sinus venosus and the conus and bulbus arteriosus. The ventricle of an eel, he says, continued to contract after removal from the body, and even when cut into several pieces the movement was continued in each piece. In comparing various hearts he concludes that the existence of right and left ventricles is dependent on the presence of lungs, and where there are no lungs there is no right ventricle, which, he assumes, is hence wanting in fishes. All animals which have a ventricle possess also an auricle, and where there are two ventricles there are two auricles. This statement explains his error that the lower vertebrates having only a single ventricle are restricted to a single auricle. Among other scattered comparative observations Harvey failed to discover the ventricles of the brain in birds, and assumed that insects respire because they emit bubbles of air " from their tails " when they are submerged. From this brief discussion the conclusion is justified that the comparative method had been thought out and put into practice by Harvey as early as 1628, and it cannot be questioned that in him we recognize and acclaim the prophet of comparative anatomy.

We should expect that the effects of Harvey's work would be of a twofold nature. First, the attention of anatomists would to some extent be diverted from man to the lower animals, now shown to possess an interest and importance worthy of closer study ; and second, the profound impression which so dramatic a demonstration as the circulation of the blood had produced would tend to focus contemporary research on that method of *experiment* by which it had been achieved. Anatomy began to move forward again, just as it had after the publication of the *Fabrica* of Vesalius, and so formidable a body of knowledge was built up that the seventeenth century stands out as the most constructive and significant period in the history of modern biology.

[1] He evidently has in mind the simple or amphibian type of lung. The comparison is not inept.

XIV
SEVERINO

At about this time the literature of anatomy had assumed such proportions that it began to attract the attention of compilers and commentators. Many works of an encyclopaedic character were published, but it is sufficient to consider only the most important of them. The earliest comprehensive treatise on comparative anatomy is the *Zootomia Democritaea* of Severino, printed in 1645. The author was an almost exact contemporary of William Harvey, whose contribution to the circulation is welcomed as a " golden book ", without, however, its conclusions being accepted. Severino was an Antiperipatetic, which explains why Democritus rather than Aristotle appears on his title-page. He also regards Democritus as the first zootomist. There is a persistent tradition, based on nothing more convincing than probabilities, of a meeting between Democritus and Hippocrates. The legend does not gain in credibility by the fact that modifications of it introduce the persons of other actors such as Aristotle and Heraclitus. It is interesting as an "abuse of the privilege of fiction", but also because it is the subject of the engraved title of the rare, highly curious and original work now under consideration, of which only one edition appeared.[1] The laughing philosopher, having retreated from the world to pursue his studies among the amenities of the neighbouring cemetery, becomes the object of concern to the citizens of his native Abdera, who were apprehensive of his reason. They draft a letter, of which the extant copy is a medieval forgery attributed, among others, to Epictetus, invoking the assistance of the Father of Medicine. The engraving represents in the background the lively distress of the " stupid Abderitians ",[2] whilst in the foreground, working in the shadow of his plane tree without the city wall, we observe Democritus, surrounded by the *membra disjecta* of his anatomical recreations, awaiting the composed and stately figure of the physician. He is asked why he occupies himself with the dissection of the viler creatures, and he replies in philosophic idiom that his object is to discover the cause of folly, the seat of

---

[1] In 1675 T. Bartholin, on being asked why he was preparing a new edition of the *Zootomia*, replied that he had never, even in his dreams, contemplated doing so.

[2] This was their reputation, which Severino accepts without question.

FIG. 54.—Portrait of Severino from the *Zootomia Democritaea*

FIG. 55.—Engraved title-page of the *Zootomia* of Severino, 1645, illustrating the meeting of Democritus and Hippocrates. Other representations of this meeting occur in the works of Cheselden (1740) and Albinus (1749)

which he suspects is in the bile.   The conference [1] speedily
establishes the wisdom and sanity of the patient, and Hippocrates
retires to allay the anxiety of the citizens.   Or, in the energetic
language of Severino — " Says Hippocrates, ' By Jove, O Democritus, thou speakest truly and wisely ' ".

As a surgeon Severino favoured the stern and ruthless school of
iron and fire — a school of the blackest medieval cast, imbued with
all the terrors of torture and mutilation.   He accuses his brother
surgeons of timidity and cowardice in failing to resort to drastic
operations in certain diseases, and even the hapless patients are
unreasonably abused for their reluctance to face these appalling
ordeals.   We search in vain in his animal anatomy for the strength
and boldness which his reputation as a surgeon would lead us to
expect.   He is crude, diffuse and superficial, and his work consists
largely of brief catalogues of the coarser anatomical facts, which
are not brought into relation with each other or with corresponding
structures in other animals, and therefore do not call for special
comment.   His work might have been written in the preceding
century, before the possibilities of anatomy had been revealed by
Vesalius, Coiter and Casserius.   Not that he is unfamiliar with the
works of his predecessors, all of whom are quoted with the surprising and important exceptions of Ruini and Belon.[2]   Nor are
his criticisms of others tempered by the calculated restraint of one
who anticipates that he himself must in due course be called upon
to defend his own mistakes.   Severino is essentially a sceptical and
at times a contemptuous critic, and even the ancients do not always
escape his invective.   And yet he falls occasionally into flagrant
and inexcusable error.   For example, he refers to the common
opinion of men that the tentacles of snails have eyes at their
extremities.   This is a belief which admits of easy verification, but
Severino prefers to dismiss it in favour of the opinion of Pliny that
the tentacles are tactile organs which take the place of eyes.   " Vir
acris ingenii ", says Haller, but adds in extenuation " potissimum
animalium incisor."

Severino is disposed to employ curious and far-fetched similes
which enrich the text without enlightening the reader.   The same

---

[1] Cf. J. B. F. van Gils, *Bijdr. Gesch. Geneeskunde*, **6** (1926) ; **16** (1936).
[2] Severino discusses Belon's work in his later treatise on the viper published in
1651.

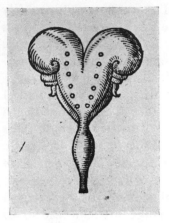

FIG. 56. — Severino, 1645. Female genitalia of calf. Note the number and arrangement of the uterine cotyledons, known to Aristotle

striving after unusual effects finds expression in his figures, some of which are so original as to be unlike the objects they represent. Rather do they suggest quaint decorative symbolic devices, as if the author had in view the aesthetic and mystical rather than the anatomical implications of his work. If they had appeared in a treatise on ornament their zoological source would not have been suspected. Thus in the figure of the uterus of the calf the paired circles, which presumably were suggested by the uterine cotyledons, are translated to the *exterior* of the uterus and arranged to produce an artistic effect, to the prejudice of their true number, distribution and purpose. Of the same character are the figures of the brains of the hawk and a colubrine snake, and the heart of a crab. Again it may be questioned whether there was any necessity to illustrate the liver of the hare, unless the caricature of it that is printed is its own justification. Some of the figures are simply inscrutable, and we can only guess what they are supposed to illustrate, and wonder what was in his mind when he devised them. All these difficulties are greatly enhanced by his failure in most cases to provide any descriptive explanation of the figures in the text.

Human and comparative anatomy are distinguished by Severino as "Andranatomy" and "Zootomy". In a long general section, which embraces also the anatomy of plants, he recognizes the unity of the vertebrate animals, including man, and regards divergences from the type as due to disturbances of function. The general similarity he attributes to Divine design. He discusses whether man or animals should

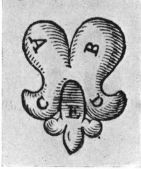

FIG. 57.—Severino, 1645. Brain of hawk

be investigated first, and notes that Sylvius would concentrate on man and Rufus of Ephesus on the brutes. He proceeds to argue the point, as if human and animal anatomy presented separate and unrelated problems, which were comparable only from such points of view as relative interest and complexity, and it does not seem to have occurred to him that they might be mutually illuminating. Nevertheless he accepts the general and obvious similarity of man, apes and brutes in respect of their limbs and other parts, but is not impressed by the more subtle resemblances between man and birds, although he quotes with approval the statement in the *Timaeus* of Plato that a unifying principle pervades all vertebrates, of which the common denominator is man himself. Hence the discovery that man is *one* of the final products of a culminating series, but *not* the basic symbol of the whole, completely eluded him. It is surprising that he does not quote Belon's essay on the comparative anatomy of the avian skeleton, and we can only conclude that it was not known to him.

In comparing the structure of the ape and man Severino considers that their affinity is so patent that the ape should be exploited for medical purposes, and therefore stress is laid only on the points of difference. As an anatomist, however, he prefers the pig as a substitute for man, and hence subscribes to a tradition so ancient that for centuries the pig had enjoyed the doubtful honour of being the anatomical domestic animal. He strongly supports de Zerbi in recommending repeated dissections of a number of different species, because a structure that is not well developed in one may be found more perfect in another. It is clearly apparent, he says, that zootomy is necessary to round off the study of andranatomy, and until zootomy has been fully explored our knowledge of andranatomy cannot be complete. He returns to this point more than once, especially in a long defence of the importance of a knowledge of animal structure to the human anatomist. None the less, he adds, with native perversity, man is the " archetype " of the living world — the microcosm in which all other animals are epitomized. Severino raises one problem which, could he have solved it, would have justified us in ranking him with Belon as a comparative anatomist. The superficial resemblance of the viper and the eel entraps him into arranging their distinguishing features in parallel columns. The statements of fact in these columns are not always sound, and although he himself expresses no opinion on the result of the comparison, leaving the reader to draw whatever conclusions he can from this irregular alliance, it may be inferred that he does *not* regard these two forms as related anatomically.

The following types were dissected by Severino, but it was not the practice of his time to describe adequately the species examined, and hence some of the identifications must be ranked as doubtful :

MOLLUSCA : slug, snail, *Purpura, Loligo, Sepia, Octopus.* CRUSTACEA : crab, *Leander squilla* and the Isopod *Aega.* INSECTA : *Gryllus,* two beetles, two Lepidoptera, looper caterpillar. ARACHNIDA : scorpion, spider. PISCES : *Raja, Torpedo, Galeus, Pelamys, Balistes, Lophius, Box salpa,* gold-fish, *Hippocampus, Uranoscopus, Muraena, Anguilla,* hake, *Trigla,* herring, tunny, *Sparus melanurus.* AMPHIBIA : toad. REPTILIA : lizard, marine and terrestrial tortoises, two snakes, viper, *Coluber.* AVES : owl, duck, goose, hawk, quail, raven, crow, common coot, turkey, meadow-pipit, three types of pigeon, magpie, ? buzzard. MAMMALIA : mole, hedgehog, horse, ass, goat, pig, chamois, ox, sheep, dog, cat, fox, weasel, rabbit, hare, porcupine, rat, Indian rat, marmot, mouse, bat, long-tailed monkey.

This list discloses a preponderating interest in vertebrates, represented by some sixty-three species, mostly fishes, birds and mammals, and includes only seventeen species from the higher invertebrate groups. The field covered, therefore, is a large one, particularly for that time, but the results are weak in philosophic interest, and discover a mind averse from cultivating the imaginative aspects of science, or even of understanding them when revealed by others.

We may now consider in some detail typical examples of Severino's zootomical researches, selected to cover most of the groups investigated by him. He re-discovered the heart in the higher Crustacea, and gives one of his small but attractive figures of it, which, however, only remotely suggests the shape of that organ. He watched it beating, and established the absence of red blood.[1] In the cricket, *Gryllus,* Severino notes the composite nature of the cornea of the eye and remarks that it resembles the seed of *Faeniculum* — meaning evidently a cluster of the umbel.

---

[1] Harvey (1628) makes a passing reference to the Crustacean heart, which he says consists of an auricle without a ventricle. Sachs (1665) describes the pulsating heart of the crab, and Willis (1672) gives a good figure of the heart and arteries of the lobster, and a less satisfactory one of the heart of the oyster, showing the auricle and ventricle. In both these animals Willis saw the heart in action.

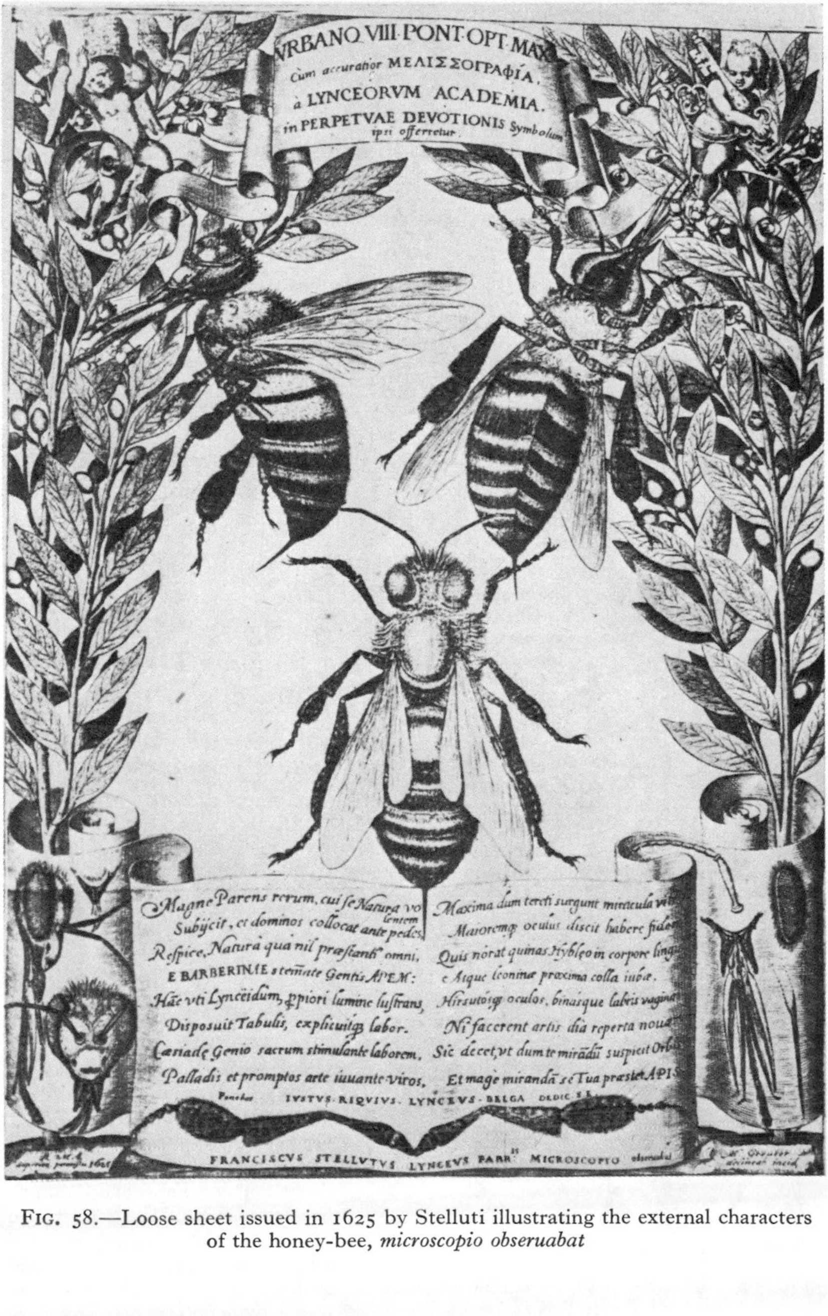

FIG. 58.—Loose sheet issued in 1625 by Stelluti illustrating the external characters of the honey-bee, *microscopio obseruabat*

The simile, as usual, is more picturesque than appropriate. He was not the first to describe the faceted eye of Arthropods, being preceded in 1644 by Hodierna, who counted thirty thousand "little squares" in the eye of a fly, and in 1625 and 1630 by Stelluti,[1] who was the first to publish a description and figure of the compound eye — in the honey-bee. His observation was made even earlier, in 1618. Severino's remark that the "aspera arteria" [trachea] of Insecta is extremely slender may mean that he had seen a part of the tracheal system before Malpighi. He considers that insects are just as wonderful in their structure as the higher animals, and therefore well worthy of study. Such work, however, requires a microscope, and hence the small animals should be examined last.

Severino's most ambitious and difficult dissections continue and extend the work of Aristotle on the Cephalopoda, the types being *Loligo*, *Sepia* and *Octopus*. His figures, which are unaccompanied by any descriptions, are the first representations of the anatomy of these animals, and they are not unreasonably defective as regards the facts, but his interpretation is vitiated by expecting to find in a Cephalopod the essential features of mammalian organization. Thus in the

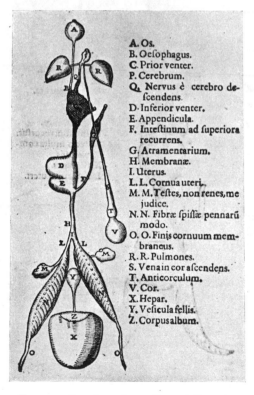

A. Os.
B. Oesophagus.
C. Prior venter.
P. Cerebrum.
Q. Nervus è cerebro descendens.
D. Inferior venter.
E. Appendicula.
F. Intestinum ad superiora recurrens.
G. Atramentarium.
H. Membranæ.
I. Uterus.
L. L. Cornua uteri.
M. M. Testes, non renes, me judice.
N. N. Fibræ spissæ pennarū modo.
O. O. Finis cornuum membraneus.
R. R. Pulmones.
S. Vena in cor ascendens.
T. Anticorculum.
V. Cor.
X. Hepar.
Y. Vesicula fellis.
Z. Corpus album.

FIG. 59.—Severino, 1645.   General dissection of a male *Sepia*

[1] Only a very few copies of the 1625 folio sheet appear to have survived, but it must have been distributed at the time, as it is discussed by J. Faber and Columna in 1651, Derham in 1726, P. van Musschenbroek in 1734, and Parsons in 1752, the last-named author publishing a full description of a copy *in his own possession*.

figure of the female genitalia of *Sepia* the uterus and its
cornua are actually parts of the gut — probably the stomach
and caecum, the membranes of the uterus are the posterior
mantle veins, and the testis [1] and gall-bladder are the nida-
mental glands.   His figures of the male genitalia are more
successful, and the diagram of the unravelled male genital duct
shows what we now call the genital orifice, Needham's pocket,
seminal vesicle, vas deferens and accessory gland.   In this figure
he correctly names the ventricle of the heart and the ctenidia,
and as regards the latter structures it is strange that he should
identify what are manifestly the same features in other species as
the cornua uteri.   Such a failure to exploit the comparative
principle, in a case where it was so obviously applicable, is damaging
to his claims to be included among the comparative anatomists. [2]
The figure of the gut of *Octopus* is recognizable as such, but again
the interpretation is faulty.   His cerebrum may be a part of the
central nervous system or the anterior salivary glands, the lungs [3]
are the posterior salivary glands, the " vena " is the oesophagus,
the liver is the crop, the " saccus multa complectens " is the
stomach, the kidneys are the branchial hearts and the cornua uteri
are the ctenidia.   His most detailed figure (59) is that of the gut,
probably of *Sepia*, but here also he failed to grasp the meaning of
what he had seen.   His cerebrum may have been a part or the whole
of the viscero-pedal system, the lungs are the posterior salivary
glands, the first stomach is probably a part of the liver, the ink-sac
is confused with the rectum, the second stomach and appendage are
the stomach and caecum, the cornua uteri are the ctenidia, the
testes (previously regarded as the kidneys) are the branchial hearts,
the gall-bladder is the vas deferens and the liver is the testis.
Against this formidable list of hapless guesses his only success is
that he correctly identifies the ventricle of the heart, [4] and if the
cephalic artery is described as " vena in cor ascendens " it is at all
events recognized as a blood vessel.   Such are the difficulties to be

[1] Severino's female gonad.

[2] Swammerdam, in his chapter on the structure of *Sepia*, probably had Severino
in mind when he condemned the " errors and follies " of his predecessors in this field.
Unfortunately he then proceeds to record a few errors on his own account.

[3] He flatters himself on having " discovered " lungs in Cephalopods, and claims
that the lungs of " Fishes " were first found in these animals.

[4] In his list of Errata, which itself is open to correction, Severino says : " *Sepia
corde caret apud Erasmum ; cui contradicit idiographia nostra* ".

surmounted when a highly specialized type is being investigated almost for the first time.

The valve in the spiral intestine of Elasmobranchs was discovered by Severino, and he claims that it had not been seen by Rondelet or Fabricius. The latter, however, figures the windings of the valve so far as they are visible on the external surface of the gut in *Mustelus vulgaris* in 1600, but apparently did not find the valve itself (cf. Fig. 43, p. 104). Severino figures but does not adequately describe the " admirable structure " of the spiral intestine in *Torpedo*, the dog-fish and the shark *Galeus*, and he perceived that it exercised a delaying action on the passage of the food. In a later section of the work reference is made to the " spira coracoida affabra " with which the lining of the posterior gut of *Torpedo* is provided. After Severino the spiral intestine was examined, described and figured in various Elasmobranchs by Steno in 1664 and 1673, in *Alopias* and *Carcharias* by the Parisian anatomists in 1667 and 1680, and in *Acipenser* and the skate by Swammerdam in 1673. The observation of the last-mentioned is dated 1671, and in the gut of the sturgeon he found a " beautiful " ligamentous spiral valve, afterwards named " skrew-gut " by Grew in 1681. Swammerdam, who seems to have been ignorant of the work of his predecessors on this structure, gives a

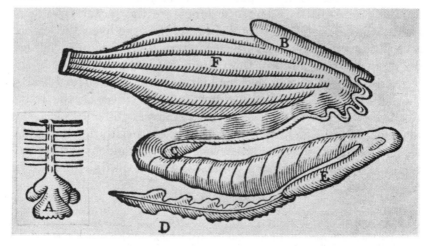

Fig. 60.—Severino, 1645. Gut of *Galeus* showing stomach, spleen, pancreas, spiral intestine and rectal gland. Also separate figure of heart, ventral aorta and afferent branchial vessels

good description and figures of the valve, and shows how the food is made to travel in a spiral down this part of the gut. Severino also discovered and figured the caecal or rectal gland of the same fishes in *Galeus*, but committed the elementary blunder of regarding it as the uterus. In the same dissection he exposed the heart and afferent vessels,[1] and saw the spleen and pancreas, but was doubtful as to the interpretation of the latter organs. In a series of anatomical notes on *Torpedo*, written in collaboration with Volckamer, a small cavity in the cardiac

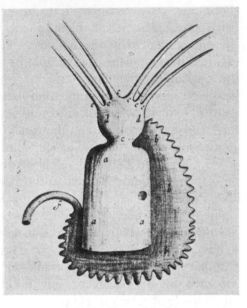

FIG. 61.—Ent, 1668. Heart of *Lophius piscatorius*. *a*, ventricle and on right position of auriculo-ventricular aperture; *b*, auricle; *c*, *d*, bulbus arteriosus; *e*, afferent branchial vessels; *f*, precaval vein. Sinus venosus not shown

region was inflated with air, which latter escaped into a much larger space identified as the lung. There being no lung or its homologue in *Torpedo*, the spaces in question must have been blood sinuses.

The swim-bladder of fishes is described as the "utriculus natatorius" by Severino. He says it is situated in the abdomen and is full of air, and he believes that its function is to assist the fish in swimming, just as men use bladders when learning to swim. In another passage he refers to it as a "vesicula" which is hidden below the spine and is an aid to buoyancy in swimming — a point of particular importance in the marine species. He returns to the discovery of the swim-bladder in his *Antiperipatias* of 1659, and now claims that it was first seen by himself, whereas it had been found and described long before by Rondelet, Sylvius, Gesner and

[1] A comparison of the following three early figures of the heart of fishes is sufficient to illustrate the leisurely development of the comparative anatomy of this organ : Fabricius (*Acanthias*, 1600) ; Severino (*Galeus*, 1645) ; Ent (*Lophius*, 1668).

Fabricius. He considers it to be either a lung or analogous to the lungs, and noted its blood plexus, which he assumed acted on the air contained in the bladder. He observed that it was subdivided in some fishes and simple in others, but failed to work out its relation to the blood stream, which partly accounts for his involved views on the circulation of the blood in fishes. Another feature which attracted Severino's attention in fishes was the mechanism of erection and depression of the anterior dorsal fin in *Balistes* or trigger-fish. The spines of this fin, the first of which is a very large denticulated lethal weapon, act in concert, and the fin can be suddenly elevated and remain firmly locked and immovable, whatever force is brought to bear on the formidable first spine. If, however, the slightest pressure is exerted on the small second spine, which acts as a locking device, the first spine is released and the fin collapses. Severino was evidently even more familiar with the anatomy of this very curious and effective contrivance than its discoverer Salviani, who gave a good description of its mode of action in 1554. In his dissections of the eel *Muraena* Severino examined the abdominal viscera, heart, afferent vessels and swim-bladder, and if the figure is crude the sounder identification of the parts is a welcome compensation.

Severino tells us that he dissected the viper on April 29, 1616, which would indicate that his zootomical studies extended over a considerable period.[1] He reproduces the figures and description of the anatomy of this animal by Baldus Angelus Abbatius of 1589,[2] and adds some notes of his own. He found that the heart consisted of two auricles [3] and one ventricle, and emphasized that the viscera were modified in conformity with the vermiform organization of the type. He indulges his favourite practice of comparing animal structures with extraneous objects when he likens the testis of the viper to a pine cone. His examination of *Coluber quaterradiatus* is more detailed. He gives an account of the larynx and respiratory organs, noting that the lungs are vesicular, and found three nestling birds in the large stomach. The heart, great vessels and biliary

---

[1] The only other date mentioned in the *Zootomia* is February 1, 1642, when he dissected *Torpedo* with the assistance of Volckamer.

[2] Böhmer and others record an earlier edition of 1587, but no copy of it can be found.

[3] This seems to be the first recognition of the existence of two auricles in a lower vertebrate.

apparatus were investigated, and he describes and figures two gall-bladders, the anterior of which, however, was evidently the distal expanded retiform portion of the hepatic duct. The same figure shows the pancreas, or, as he calls it, possibly correctly, the spleen. His diagram of the male genitalia has been simplified to bring out two points neither of which has any relation to reality. A small rough drawing is labelled " Cerebri forma haec est ". Assuming an intention to illustrate the dorsal surface of the brain, it may represent the elongated olfactory lobes, the hemispheres wrongly united across the middle line, and the optic lobes, the cerebellum and medulla being ignored.

The analysis of the skeleton of the wing of a bird betrays Severino's weakness as a comparative anatomist. Omitting the phalanges of the second and third digits, he distinguishes three large and five small bones, using a terminology partly his own and partly derived from classic human osteology. The bones are : humerus, cubitus [ulna], radius, two ossicles articulating with the cubitus and radius [free carpals], plectrum and tellina [second and third carpometacarpals], and an ossicle pointed like a reed-pen [first digit]. The smaller bones are compared, on grounds which must remain mysterious, with the strings of a musical instrument and its plectrum. This performance displays little genius for philosophic anatomy on the part of Severino, and compares unfavourably with Belon's previous solution of the same problem, with which he should have been familiar. A more useful contribution to avian anatomy for which Severino is responsible is his discovery of the well-developed bony ring of sclerotic plates in the birds of prey.

Coming now to mammalian anatomy, Severino devoted particular attention to the structure of the pig, dog and cat. He describes the viscera, the short spiral caecum of the dog first figured by Blasius in 1673, biliary apparatus, heart and great vessels, azygos vein, phrenic nerves, and male and female genitalia. His account of the relations of the great vessels to the heart is confused, and he does not hesitate to speak freely of the association of veins with ventricles and arteries with auricles. Nor does he understand the functions of the valves of the heart. This is all the more remarkable when it is remembered that he is writing after Harvey, and had studied his work. His perverse inaccuracy is again in evidence when he contradicts the readily demonstrable statement of Casserius that the stapes of the cat is perforated. As an offset to these errors he must be credited with the discovery of Peyer's glands, which he found in the gut of the dog and cat in 1645. Peyer first saw these patches in 1673,[1] and described them in 1677. Grew included an account of them in his lectures of 1676, but his observations were not published until 1681. Severino was also the first to describe the pancreas of the horse, and to note that the stomach of the same animal consisted of two parts distinguished by the character of the lining, one having the pearly white and the other the reddish mucosa. Both these features had been overlooked by Ruini. It has already been pointed out that Grew had correctly compared the fourth chamber of the compound stomach of ruminants with the simple stomach of man, and although he was anticipated in this by Severino, the latter does not recognize so fully the implications of the comparison. As regards the physiology of the compound stomach, which had been previously discussed by

[1] Steno observed the glands in the small intestine in 1673, but did not publish his results.

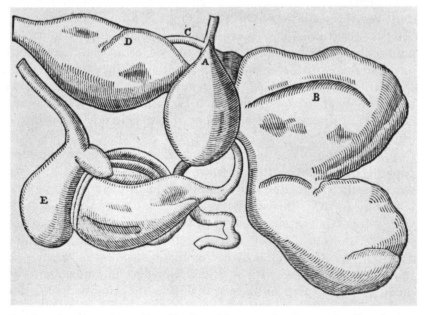

Fig. 62.—Severino, 1645.   Gut of lamb.   *A*, honeycomb ; *B*, paunch ; *C*, psalterium
and ? oesophageal groove ; *D*, abomasum ; *E*, caecum

Aristotle, Coiter (1572), Aldrovandus (1613) and Fabricius (1618), Severino is quite sound. He says that the food is regurgitated from the penula [paunch] and ollula [honeycomb] into the mouth, where it is masticated. It then descends into the conclave [psalterium], and finally reaches the fourth chamber or " *ventriculus* (*propriè dictus*) ".

In a brief account of the general anatomy of the mole Severino is specially interested in the eye. He says it is in its normal position, but is covered by a thin dark skin like the seed of the flea-bane. It had no optic nerve as far as he could see, nor was he certain that it was able to function as an eye. The dark covering was not the only impediment to vision, but other necessary structures were also wanting. These vestiges, he says, have rather the nature of eyes without being true eyes, in the production of which Nature seems to have been disporting herself, and to be showing that she could produce eyes of whatever type she willed. Severino had a good opening here, but it is clear that although he describes the eyes of the mole as " vestigia ", he was not using this term in the modern sense, nor did he recognize that these structures were true but *degenerate* mammalian eyes, still possessing all the essential parts of such organs, including the optic nerve and eye muscles. One final point selected from Severino's mammalian researches. He remarks that the extremity of the penis of the ram resembles the head and neck of a snail with the second tentacle removed. He was evidently the first to see the processus urethrae or filiform appendage of the very singular penis of that animal.

Severino reprints in his work some observations on comparative anatomy published by J. Sylvius in 1555. They relate to *Simia*, the muscles of which are compared with those of man, the genitalia and cotyledonary placenta of the sheep and cow, the genitalia and placenta of the pig, and anatomical notes on the dog, deer, lion and *Salmo fario*. He says the gut of the trout has many " loose duo-denums " [pyloric caeca], and that he found a diaphragm and lung in that fish [pericardio-peritoneal septum and swim-bladder].

The *Zootomia* is completed by a section on methods. He contends that the older methods employed in the examination of bodies are inappropriate and useless. He demands first of all a strong natural inclination for the work. The anatomist must not abominate the human body, nor must he be reluctant to explore with his own hands the secrets of living things. He must also by constant practice acquire the rare ambidexterity of the dissector, who with one hand controls the pen and with the other plies the knife. If he wishes to represent Nature truly he must go straight to the original or " archetype ", and not to books and pictures, and he must develop firm and skilful hands. It is not, he says, with genius alone that the final victory lies. " Labor omnia vincit improbus." In this section of his work Severino discusses the employment of assistants, the instruments required, including one " invented in our lifetime called microscopium ", rules for the dissection of the larger animals and the detailed examination of the smaller ones and smaller organs, and finally rules specially applicable to individual cases, such as the dissection of a Cephalopod and the air-bladder of fishes. It should be mentioned that he claims to have discovered in 1617 the excarnation method of studying the geography of the blood system, in which the vessels are filled with a stiff solidifying medium, often a fusible metal alloy, and the soft parts afterwards removed by maceration or corrosion. The method was developed by Spigelius and Glisson in 1627 and 1654, and particularly by Bidloo in 1685. It became very popular in the eighteenth century.

Severino published three other works which include researches on comparative anatomy. The first, printed in 1651, is his monograph on the viper — a " senile and garrulous work " according to Haller. Among the many issues traversed is the question whether poisonous snakes were allowed to enter Noah's ark, and if so how the activities of such dangerous companions were controlled in that capacious vessel. Severino holds that the viper does not normally secrete venom any more than that man is not normally enraged. It acquires the habit as the result of certain environmental conditions, and any excitation to which it may be subjected, or when it is bilious and angry. Then the venom which is engendered in the liver passes to the vesicles of the mouth and is ready to be discharged. This treatise includes also useful contributions on the anatomy and generation of the viper by Vesling (1644–7) and Hodierna (undated). The structures in the unhatched snake interpreted by Fabricius as the testes are correctly assessed by Vesling. He says that in the male there are in the umbilical region two vascular globules which to others are the testes, " but to me they are clearly the very spongy bifid penis not yet withdrawn within the body ". Vesling was thus induced to look for the true testis of the viper, which he found and described.

The other two works are posthumous, Severino having died in 1654. The *De Piscibus* appeared in 1655. It is concerned with respiration in fishes, and how they can live out of water, but as the aquatic mammals are brigaded with the fishes the issue is from the outset compromised. He allows, however, that *Phoca* is more of a quadruped than a fish. Anatomical notes on all the systems of *Phoca* and *Delphinus*, and on the caudal serrated spine of the sting-ray, *Trygon*, are also added. The *Antiperipatias* is dated 1659.[1] In it are discussed various questions of importance concerning the anatomy and physiology of fishes. Respiration, he says, is certainly the function of the gills, which extract the air and vapour mixed with the water.[2] The biology of intertidal fishes and eels which can live for some time out of water is considered. Water, he thinks, is unsuitable for respiration, but water *vapour* is an efficient medium. The structure and true uses of the branchial apparatus are debated. He recognizes three chambers in the heart of fishes, which correspond with the modern sinus venosus, auricle and ventricle. Fishes lack the *right* ventricle, according to Severino, who, however, believing as he did in the presence of lungs in fishes, was deprived of the basis on which others supported this view. He refers frequently to the occurrence in fishes of a " diaphragm " [pericardio-peritoneal septum], and discusses its functions. The lamprey, he claims, has lungs [respiratory or sub-oesophageal canal] and a cartilaginous " pericardium ", and the large or " perfect " fishes have two respiratory organs — gills and lungs, in order that they may breathe in and out of water. The lungs referred to are the " utriculus " or swim-bladder, which he repeats is full of air. He is disposed to believe that all fishes have lungs of one kind or another. His account of the circulation of the blood in fishes is imaginative and needlessly complicated, and shows that he had never carefully dissected the heart and great vessels of a fish.

---

[1] No complete copy of this work has come to light. The copperplate illustrations are invariably wanting.

[2] This was known to Galen.

XV
BLASIUS

Gerard Blaes or Blasius, who died in 1692 according to Nuyens, the date of his birth being unknown, published three works on comparative anatomy under his own name in 1673,[1] 1674 and 1681. The anonymous *Observationes* of the Amsterdam anatomists printed in 1667 and 1673 were partly contributed by Blasius, and some of them are claimed by him in his later publications.[2] The first two of the works just mentioned record his own researches, but the last is largely a compilation or textbook, based on some ninety authorities and illustrated by sixty useful plates. It includes also his own results more or less revised, abbreviated or extended. In the 1673 volume Blasius describes the anatomy of the following fourteen types : tortoise, duck, pigeon, ox, sheep, pig, dog, cat, civet-cat, fox, rat, rabbit, hare and monkey. In the following year he revises seven of his previous descriptions, and omits seven, but adds new ones on the snake, heron, hedgehog, and the spinal cord and nerves of the horse. In the textbook all the preceding memoirs, except those on the monkey, cat, pig and horse, reappear with certain modifications, and a description of the dissection of a male tiger undertaken in 1677 is also printed, and for the first time. His most detailed investigations were confined to the dog, civet-cat, hedgehog, heron, tortoise and snake.

In the 1681 treatise 119 types are selected for treatment, and as many as seventy-eight of them are represented in the illustrations. The number 119 is made up as follows : MOLLUSCA, 2 ; CRUS-TACEA, 2 ; INSECTA, 9 ; ARACHNIDA, 1 ; PISCES, 29 ; AMPHIBIA, 3 ; REPTILIA, 7 ; AVES, 27 ; MAMMALIA, 39.

Blasius is particularly interested in embryology, and deals with the foetal membranes and development of ruminants, rabbit, cat, horse and the chick. The illustrated monograph on the chick is a very careful and thorough piece of compilation, and may be recommended as a competent exposition of contemporary embryology. It is evident that the textbook was planned on a generous scale, and may be accepted as the first comprehensive manual of

---

[1] A second extended edition with new engraved and printed titles appeared in 1676, but pp. 1-288, which include the comparative anatomy, are identical in both editions.

[2] Swammerdam also claims part authorship of this work (*Bib. Nat.* 2, 890).

comparative anatomy based on the original and literary researches of a working anatomist. It is thus a valuable introduction to the history of zootomy, since most of the relevant literature published up to that time is skilfully handled by one of the best known of its exponents. On the other hand the student must not expect to find in this work a review of comparative anatomy as it is understood to-day, or even as it was understood by the more philosophical workers of the period. In all the writings of Blasius we search in vain for any discussion or integration of the data which have been so laboriously assembled. The manual especially is a catalogue of anatomical observations, and never reaches a higher level than the inanimate records of the pure descriptive anatomist, consisting as it does of a series of independent descriptions of animal types based on the most reliable sources of information then available.

Some light is thrown on the character of Blasius by his un-provoked attack on the guileless Steno. He dismisses his victim, who was only twenty-three years of age at the time, as a "wretched boy", and accuses him of deceit, ingratitude, bad manners, injustice, blundering, foolishness, perfidy, incivility, unveracity, treachery, calumny, scoffing, malice, arrogance, perversity, audacity, shamelessness, impudence, fatuity and depravity. The occasion of this exhaustive and relentless indictment was the somewhat trifling dispute as to which of them had been responsible for the discovery of the duct of the parotid gland. The

Fig. 63.—Blasius, 1673. General dissection of tortoise, showing heart and respiratory tract, gut with caecum displayed to the left, and male genitalia

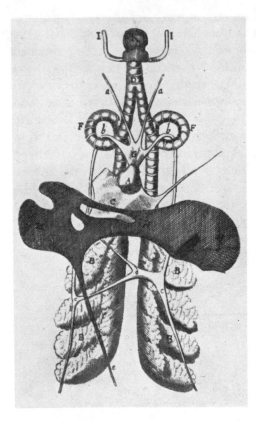

Fig. 64. — Blasius, 1673. Heart, respiratory organs and liver of tortoise. The figure shows the right and left aortae threaded through loops of the bronchi

judgment of posterity, however, leaves the issue in no doubt, and the invective of Blasius is condemned as the measure of his own mendacity in laying claim to this modest discovery.

Some reference may now be made to those sections of Blasius' manual which are based on his own researches. The heart of the tortoise, he says, consists of but two chambers — an auricle and a ventricle. His figures of this organ vary from edition to edition, and there may be some lack of congruity even in the same plate. He evidently saw the origin from the ventricle of the common pulmonary artery, left aortic arch and innominate artery. His vein which passes "from the auricle into the liver" must have been the postcaval. The elaborate figure of the heart and respiratory organs with the head in the withdrawn position represents the aortic arches as being threaded through loops in the bronchi — a condition which would make the protrusion of the head a difficult and dangerous proceeding. His illustrations of the male genitalia do not agree, but assuming that the last edition expresses his final views, the figure is reasonably accurate as regards the facts, but the parts are not correctly identified. His vesicula seminalis is the epididymis, and the parastata [epididymis] is the kidney. The ureter and vas deferens are distinguished but not lettered. In the gut the caecum is shown although its existence is denied in the text, and his detailed account of the "diaphragm" only convinces us that he saw the peritoneum. In the serpent, Blasius describes the asymmetry of the organs associated with an elongated body form, but finds only a single auricle in the heart, the great vessels of which were imperfectly examined and most of them overlooked. His account of the biliary apparatus is more satisfactory, both as regards observation and interpretation. He noted that the heart continued to beat for some time after decapitation, and even after it had been removed from the body.

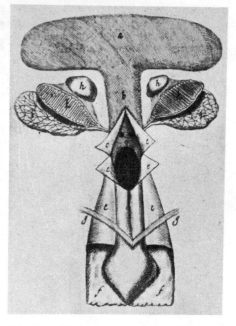

FIG. 65.—Blasius, 1674. Urogenital organs of tortoise. *h*, testis; *k*, epididymis; *i*, kidney

Blasius was the first to associate the production of voice in birds with the lower larynx. In his account of the dissection of the heron of 1674, but more clearly in the manual of 1681, he remarks that "just at the place where the trachea divides into two branches there exists in each branch a single cleft, which is responsible for the production of sound". This reads as if he had seen a syrinx tracheo-bronchialis, and his crude and simple figures seem to bear this out. But compare Perrault's more detailed results of 1680 (p. 430).

The unique and astonishing orbicular muscle of the hedgehog is described, and for the first time figured, by Blasius in 1674, the figure, however, only indifferently suggesting the muscle it is designed to represent. The skin and muscle, he says, adhere so closely to each other that they are almost inseparable, but the spines are not directly involved. The muscle itself had been previously dissected by Coiter in 1572 and by Steno in 1672.[1] One of Blasius' most successful efforts is his dissection of the male

[1] Not published by Steno, but reported by Borrichius in 1673.

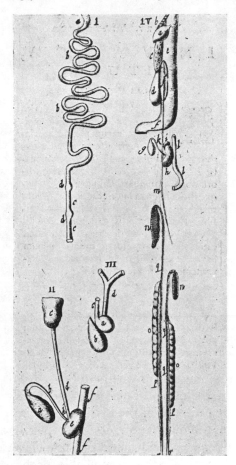

Fig. 66.—Blasius, 1674. Anatomy of snake. I, gut; II, biliary apparatus and pancreas; III, heart; IV, heart, portions of gut, gall-bladder, pancreas, spleen and male urogenital organs

genitalia and os penis of the rat. In his time it was not possible correctly to interpret the parts of such a preparation, but nevertheless he does not go badly astray. His " three muscles " represent the crura penis and corpora cavernosa, and the " glandulae " are the prostate and Cowper's glands, which latter he found before Cowper, who described, but did not discover, them in 1699 in man. Blasius is mistaken in stating that the vasa deferentia open into the bladder, but he established the presence of glands on these ducts, and he saw the large seminal vesicles. His figure shows the position of the testes reversed, unless he has in error derived the vasa deferentia from the caput instead of the cauda epididymis. Blasius was apparently the first to describe the mammalian ovarian bursa or sac, his material being the civet-cat. This sac is a small enclosed portion of the coelom which lodges the ovary, and in some mammals, such as the rat, it communicates *only* with the Fallopian tube. In his later works he gives two modified versions of this figure, which are less accurate than the original.

Blasius' observations on human anatomy are followed by eighty-five pages devoted to the anatomy of the dog. This is the first comprehensive and original treatise on a vertebrate since the publication of Ruini's volume on the horse in 1598. His reason for writing it was that owing to the difficulty of obtaining human

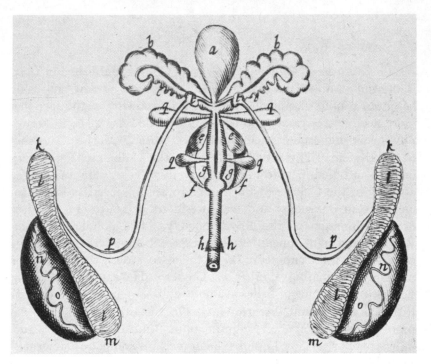

FIG. 67.—Blasius, 1674.   Male genitalia of rat

bodies for dissection substitutes had to be found, of which the dog was one.   It was hence imperative to acquire an exact knowledge of the points in which the dog resembled and differed from man.   His approach to the dog, therefore, is not as a comparative but as a human anatomist.   Not that he is indifferent to the anatomy of the dog as such, but if, as he says, the human anatomist has to accept the dog in place of man, the structure of this animal must be better known, since the current anatomical texts and tables are exclusively human.   He then proceeds to work through the dog system by system, making comparisons with man where necessary to his purpose.   In this small but noteworthy monograph Blasius is one of the early describers of the ligamentum teres uteri.   He gives a good figure of it in the dog, which shows the tubular process of the ligament passing into the inguinal canal or canal of Nuck.

XVI

COLLINS

The *Systeme of Anatomy*, by Samuel Collins, published in 1685, is the only work of its justly neglected author. Some confusion has been produced, and indeed still exists, owing to the fact that there were three contemporary Samuel Collins, all well-known physicians, and each of them has been credited with the authorship of the System. The actual author was the Collins who was born in 1618 and died in 1710 in his ninety-third year. He was elected a fellow of the College of Physicians in 1668, served as its censor on fourteen occasions, and was president in 1695, but failed to obtain admission to the Royal Society. His ponderous folios, which at first sight appear very impressive, become less so the more closely they are examined. They failed to arouse the admiration of so discriminating a critic as Cuvier. He says that Collins' figures of the brains of fishes are poorly drawn, and explained with little knowledge and discernment. On the other hand Haller is more complimentary. " Magnum opus ", he exclaims, " vastum opus, parcius est in hominis anatome, in comparata uberius." He adds that the work is based on the researches of the author himself and his friends, *particularly Edward Tyson*,[1] and that on the anatomy of animals none deserve more praise than Collins. But Hutchinson will have none of this. He condemns the book outright as " of less value than the head that is placed before it " — referring to the beautiful line engraving of Collins by Faithorne which constitutes the frontispiece.[2] The criticism is harsh but not wholly undeserved.

Collins tells us that the work is the result of his studies after retirement, and that he has " with great Care and Faithfulness laid open various kinds of Creatures to inspect their *Viscera*, which I have ordered to be curiously drawn with a Pencil from the Life ". W. Faithorne, sen., was engaged to work on the plates, and three of them bear his name as artist and engraver. Some ten or more of the others betray his workmanship, but the remainder suggest an

---

[1] Haller's knowledge is often uncanny, and it is surprising that he should have known this.

[2] Of this portrait it may be said, as of another, that it is " endowed with every merit, excepting that of likeness to the original ".

inferior craftsman. There is evidence that the liaison between author and engraver was not complete. In many cases the legends only partly tally with the figures, which themselves are often imperfectly lettered. Again, figures may be described which are not in the plates, and the dissection figured may not be the one which appears in the explanation. In the figure of the gut of the skate the parts have been so disarranged that the ends are transposed, and the spiral intestine is described as the oesophagus. The literary style of the work suffers from disabilities no less considerable. Collins believes that the most trifling investment of fact is consistent with unlimited returns of speculation. His capacity for producing on demand vague subjective generalities provokes the suspicion that these clouds of verbosity serve only to disguise his own ignorance. Here is an example in which Collins is attempting to demonstrate the function of the nerve plexuses of the abdominal region. He says : " I humbly conceive, these eminent Plexes are designed to some other use, to convey nervous Liquor into the *Viscera*, lodged in the lower Apartment, and serveth as a Ferment, to prepare the Chyle in order to Concoction in the Stomach, and to its farther elaboration, and refinement in the Guts, and Glands of the Mesentery, and to meliorate the Blood, by rendring it more exalted, in assisting the Secretion of the bilious Recrements, from its more noble Particles in the Glands of the Liver, and to help the separation of the serous Faeculencies from the Blood, made in the Glands of the Kidneys ".

It is difficult to understand, from what we know of the abilities of Collins, how he could have produced a work of this range and pretension, and if we subscribe to the general impression that Tyson's contribution to it was outstanding, as seems probable, we still have to explain why he submitted in silence to an act of plagiarism published twenty-three years before his own death. Collins mentions having examined a lion which had been dissected by Tyson and Slare, and there are also three references to Tyson in the text. Plate 12 is copied without acknowledgment from Tyson's monograph on the porpoise published in 1680, but Plates 22, 39 and 48, illustrating the gut and male genitalia of the heron, the anatomy of a crab and the base of the human brain, are admitted to be the work of Tyson. Collins' indebtedness to Tyson, however, is nowhere adequately recognized, nor are his borrowings confined to that admirable anatomist. Observations by Malpighi, Swammerdam and others are likewise appropriated, and to such an extent that it is unwise to attribute any discovery to Collins without searching for it in the works of his predecessors.[1] This to some extent explains his scanty discussion of anatomical methods. He appears to have sparingly practised the art of injection, and states that he had examined the structure of the oviduct " by the helpe of a Microscope ", which instrument is mentioned in a very few places only. It is true that he generally scrutinizes the finer structure of the parts under examination, but only in so far as it was ascertainable by histological dissection *without* the use of the microscope.

Can we describe Collins as a comparative anatomist ? " I am ", he says, " a great Servant to Comparative Anatomy, which hath cost me a great deal of Time and Money, in the procurement of various Animals, which I have opened with great Pains and Care, and inspected their inward Recesses." Nevertheless his intentions, good though they be, are constantly finding expression in comparisons which are destitute of scientific value or application. Thus he frequently likens the animal body to a stately house of three stories, comprising pleasant apartments or allodgments adorned with household stuff, rich and noble furniture and utensils, fine carvings, seelings and floors. The mouth is the dining-room with the lips as folding doors. " This lower Story is encircled with great abundance of good Furniture." When he descends from the heights of metaphor to the sober realities of anatomy he sees in the vesicles of the ovary of *Homo* the ovarian eggs of birds, and he claims that they may truly be called eggs. They contain a liquor which if coagulated with heat resembles in appearance and taste the white of a bird's egg. This is not too far from the truth, although the idea is not his own. In his handling of anatomical verity, however, the result is less impressive, and he gives us only lists of morphological characters in a number of animals, without

---

[1] It would be a laborious and merciless task to trace to their sources all the discoveries which Collins is supposed to have made, but it is significant that the result of many specific attempts to do so was to reveal the authority which had been laid under contribution in every case. Further, his manifest inability to vindicate a questionable practice by the profitable use to which it is put, and the lack of knowledge and anatomical vision which discriminate those parts of the work which are clearly his own, make it difficult to believe that there is a single original observation of importance in the whole of this grandiloquent but meretricious compilation.

instituting any comparisons between corresponding features in the types selected, with the result that no conclusions emerge from this mass of data. Even when obvious and favourable openings present themselves he consistently fails to take advantage of them, which implies that his knowledge and understanding were inadequate to the task. Thus in his lengthy exordium on the skull of quadrupeds and birds he concentrates on trivial and meaningless details, and fails to draw attention to such outstanding features as the manner of suspension of the jaws in these groups, and the existence of double or single occipital condyles. Again, he rejects the possibility of comparing the blood vessels of fish and terrestrial animals on the wholly inadequate ground that fish have no lungs. Indeed, he regards such a comparison as only less speculative than an attempt to align the blood vessels of a vertebrate and an insect. Admitted that the task of homologizing the circulation of an aquatic and a terrestrial vertebrate would have been difficult in Collins' time, it could at least have been attempted. A beginning had in fact already been made by Harvey and the Parisian ana- tomists. But the perversity of his point of view mingles with everything that he does, and Collins is the hapless victim of his own ineptitude. His object in instituting comparisons was not better to *understand* the fabric of the human body, but to " render it more *illustrious* ". " Man, the master-piece of the Creation ", " the exquisite standard giving rules and measures to other animals ", " the complement of the Creation, and Epitome of other Creatures " — this is the theme uppermost in his mind to the exclusion of less sublime but more relevant enquiries. But however this may be, in one respect Collins has achieved origin- ality. He is the inventor of the anatomical prayer, and he concludes each section of his treatise with a paean to the Deity in which anatomical allusions play a leading and topical part. We hear the Creator thanked for a variety of mercies, but it was the privilege of Samuel Collins to thank him for the neatness and skill with which the viscera were stowed away in the abdominal cavity. None the less, Collins, the arch-plagiarist, complimenting the Divine Artificer on his work, is not an edifying spectacle.

Although the plan of the *Systeme* is professedly comparative, man occupies the central position, and is the only type whose structure it is his purpose to exalt. Hence man is always considered first, then beasts, birds, fishes and insects in the order mentioned. The lower animals, he says, are satellites whose sole interest is their relation to man — the noblest subject of anatomy. This is going back on the comparative method even as it was understood and practised at the time, and when it was but a first step taken towards the possibility of a second. The chief sections comprised in Collins' work are : skin, including the feathers of birds, scales of fish and the " hair " of insects ; muscles, muscular actions and movement in quadrupeds, birds, creeping animals, fish and insects ; mouth, stomach, gut, pancreas, spleen, liver and biliary apparatus, kidneys, ureters and bladder ; genitalia ; thorax, including diaphragm, thymus, pericardium, heart and great vessels, lungs, trachea and gills ; skull ; brain, cranial nerves and rete mirabile. There is no comparative anatomy of the sense organs, muscles and osteology, except the skull. The types which are more or less described by Collins include more than 115 species as follows :

CRUSTACEA : lobster. ARACHNIDA : mite. INSECTA : locust, ant, " library-moth ",[1] silkworm, palmer-worm (Lepidopterous cater-pillar, perhaps *Lasiocampa rubi*), blow-fly, gnat, flea. PISCES : lamprey, dog-fish, monk-fish, the shark " Galaeus Laevis " (? *Mustelus laevis*),[2] skate, sting-ray, sword-fish, torpedo, sturgeon, barbel, bream, brill (*Rhombus*), burbot (*Lota*), carp, cod, dab, eel, fishing frog, flounder, gar-fish, grayling, green-cod, gudgeon, gurnard, halibut, herring, horse-mackerel, John Dory, mullet, perch, pike, plaice, pope (*Acerina*), red-gurnard, roach, salmon, smelt, sole, tench, trout, turbot, whiting. AMPHIBIA : newt, frog, toad. REPTILIA : lizard, chameleon, turtle, land-tortoise, viper, " snake ". AVES : brant-goose, bustard, crane, cuckoo, curlew, daw, duck, eagle, fowl, godwit, goose, hawk, heron, kestrel, kingfisher, nightingale, ostrich, owl, partridge, pheasant, pigeon, snipe, stork, swan, teal, turkey, woodcock. MAMMALIA : por-poise, " whale ", camel, deer, goat, horse, ox, pig, sheep, alpine-mouse, beaver, guinea-pig, hare, mouse, porcupine, rabbit, squirrel, bear, cat, civet-cat, dog, hyaena, lion, otter, polecat, tiger, weasel, hedgehog, ape.

Before proceeding to consider some representative contribu-tions to comparative anatomy for which Collins is responsible, reference may be made to his views on certain problems of com-parative physiology. On the physiology of the heart he has no clear ideas to offer, in spite of the fact that he has the example of

[1] The apterous insect known as the silver-fish, *Lepisma*. Collins has copied his account of this species from Hooke, who calls it a moth.

[2] The figure agrees rather with the ovo-viviparous *Galeus vulgaris*, but Collins states that the foetus is " fastned to the *Uterus* ", which would suggest *M. laevis*.

Harvey before him, and his familiarity with the lower vertebrates should have suggested promising lines of research. His investigation of the skin is equally barren, and discloses no fact of importance discovered by himself, although there is much vague and tentative speculation. Nevertheless he remarks : " Thus I have given you a short view of a Humane Skin, that we may fitly make a comparison of it, with that of other Animals ".

The mite of scabies has been known since the tenth century, and in modern times it was described by Benedictus in 1508, Moufet in 1634 and quaintly figured by Hauptmann in 1657. Shakspeare mentions it in *Romeo and Juliet*, first printed in 1597. Collins, however, has no suspicion of the parasitic nature of the disease, and his discussion of it might have been written centuries earlier. He speaks only of contagious scabby ferments, subtle salt particles, stagnant and concreted nervous and serous liquors, and other meaningless abstractions. Collins attributes the contractile properties of muscle tissue to the tendon, the " carnous fibres " assisting only to a minor extent. He does not agree with Steno that muscle and tendon include but one type of fleshy fibre, in the former part the fibres being " loosely united ", and in the latter, " closely conjoined ".

Unhappily he differs also from Steno when that anatomist is right, as when he argues that the contraction of muscle affects its form but not its bulk. Collins, however, is on firmer ground in refusing to believe that muscle contraction is due to " inflation " by the volatile parts of the nervous liquor or animal spirits, and his views on the relation of muscle and nerve approximate more to modern conceptions. In dealing with muscular motion such as flight in birds and insects, swimming in fish and crawling in reptiles, Collins is apparently unaware that a careful examination of the anatomy of the parts in question is a vital preliminary to the discovery of their function. Instead, he discourses aimlessly of the " nimble contractions " of abductors, adductors, tensors and flexors, with which he has obviously no real acquaintance, and which he " humbly conceives " to conform to a common plan in all these groups. Hence he does not even suspect that the mechanism of flight in birds and insects and of swimming in fish may not be comparable anatomically or physiologically. The result is that he is all the time blindly groping and blundering among strange mechanisms and phenomena, and cannot be expected to explain what he has neglected to understand. It is not suggested that this applies to the whole of Collins' treatise, but assuming that he was in places making use of Tyson's unpublished material, the lack of *personal* familiarity with the subject matter of his work, which is so conspicuous in certain parts of it, is readily explicable.

The two figures of the anatomy of the crab are among the earliest representations of the structure of that animal. One of them shows the testes and vasa deferentia and the anterior modified abdominal appendages of the male, which he calls the " finns ". The gut is also included, but he has missed the mid- and hind-gut caeca. In the second figure a part is labelled " Other Intestina Caeca" which can only have been the heart. He wrongly interprets the terminal section of the vas deferens as a penis, which therefore according to him is double in the crab.

Fishes, birds and mammals are the groups to which Collins devoted most of his attention. In the fishes he describes the maxillary and mandibular breathing valves or internal lips of the halibut. These structures prevent the loss of respiratory water from the mouth on the fall of the gill cover, but Collins imagines that they intercept the passage of water *into* the mouth.[1] He distinguishes correctly between the muscular diaphragm of the cetaceous " fish " and the membranous septum transversum of the true fishes, which latter, therefore, he says, is incapable of motion, and is nothing more than an inert partition between two cavities. In the freshwater fishes Collins found and described the swim-bladder, and he shows how the fish is " boied up " by this structure. The swim-bladder of fish was, however, known to Rondelet long before Collins' time, and it was also investigated by Ray and Redi in 1675 and 1684. In the garfish *Belone* Collins noted that the gut passed straight from oesophagus to anus [2] — a condition found in only a very few species of Teleosts, but Borrichius had already recorded it in 1673. Collins saw and figured the rectal or caecal gland of *Rhina squatina*, which is short, club-shaped, very wide and has a large lumen in this species. He made the strange mistake of regarding it as a kidney, but having found it also in two other Selachians he identified it correctly as

---

[1] An impossible interpretation assuming that he had *seen* the valves.

[2] This is Tyson's observation.

the caecum. The gland was first described by Severino in 1645. An important oversight is his failure to observe the spiral valve of these fishes.[1] In " *Galeus laevis* " he remarks that the [spiral] intestine might be called the colon " did it not want the connivent valves ". This can only mean that he *had* examined the lining of this section of the gut, and how in that case he could have failed to observe the spiral valve can be explained only on the assumption that he is recording observations not his own. He even figures the viscera of a young dog-fish which shows externally the spiral course of the valve, but makes no comment on it. In another place he refers to " Cells like those of a Honey Comb (which have been discovered in the single Gut of Sturgeon) which are instituted by Nature (as I conceive) to give many stops to the overhasty passage of Excrements ". This passage cannot be interpreted as a reference to the spiral valve of the sturgeon, since the lining of almost the whole gut of that fish exhibits the honeycomb structure, as first noticed by the Amsterdam anatomists in 1673.

Collins gives a comparative account of the number of the pancreatic ducts and their relation to the bile duct in mammals, birds and fishes. He states that in the carp and barbel the pancreatic duct opens into the stomach, whereas in those fishes the stomach is absent. He was acquainted with the general scheme of the circulation of the blood in a fish, and the function of the gills as respiratory organs. By extracting the air mixed with water the gills were to be compared with the lungs of the terrestrial vertebrates. He tested the course of the blood through the gills by injections of a " black liquor ". But all this was known before his time. Most of Collins' attention, however, as far as the vascular system is concerned, was confined to the heart. In the skate, he says, this organ has one dorsal auricle, but in the figure the auricle is shown passing *ventral* to the conus, and is, with characteristic inconsequence, labelled the " right ventricle ". The same figure also represents the cardiac aorta as being continued beyond the anterior innominate artery as far as the mouth. This may be an error of the draughtsman. It is not referred to in the text, and the conspicuous thyroid gland, which he does mention, occupies the position assigned to the forward extension of the aorta. The distribution of the anterior and posterior innominate arteries is cor-

[1] Also known before his time.

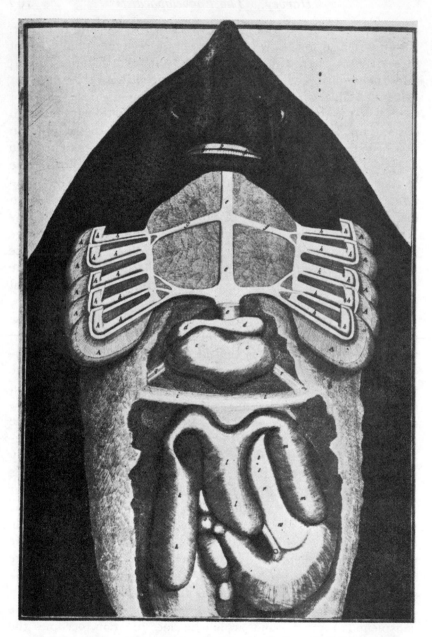

Fig. 68.—Collins, 1685. Skate. Gut, heart and afferent branchial system

rectly figured, but the right half of the sinus venosus is named the "cava" and the left half the "descending aorta", although *both* are stated to be connected with the "appendix of the heart",[1] which is obviously the auricle. The peri-cardio-peritoneal septum is shown. In ?*Galeus vulgaris* there is a better representation of the heart, and only one ventricle and one auricle are described. In the heart of the salmon an auricle, a ventricle and the bulbus arteriosus are shown, the latter having two semilunar valves at its origin from the ventricle. The figure is mislettered, and the bulbus is not mentioned in the text. He attempts to see in this heart a modification of the four-chambered heart of the higher vertebrate.

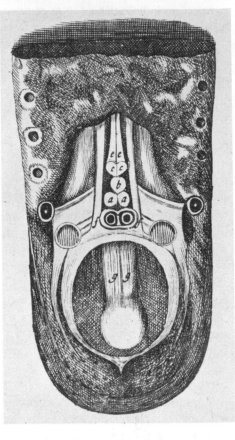

FIG. 69.—Collins, 1685. Brain of lamprey in dorsal view

The heart of *Lophius* is said to have a similar constitution to that of the salmon.

The brain is examined in a large number of cartilaginous and bony fishes, but in most cases Collins does not pursue his enquiries beyond the dorsal surface of the brain. In the lamprey the brain is figured for the first time. It is a small, simple, but quite recog-nizable drawing, and shows the more obvious features which would be exposed in a dissection of the upper surface. Collins' ideas on the development of the brain reveal him as a bio-alchemist rather

---

[1] The older anatomists confine the use of the term "heart" to the ventricular region of that organ.

than an embryologist. He does not even perceive that the brain of a fish can be in a fluid state only as the result of decomposition. " As to the First production of the Brain of Fish and other Animals ", he says, " I conceive it is derived from saline Particles, accompanied with a fluid viscid Matter, easily concreted, so that the seminal juyce, out of which the Brain of Fish is generated, is primarily a fluid transparent Liquor, which afterward growing more solid, loseth its transparency, when it is coagulated by saline parts, in which the plastick virtue is chiefly founded, by whose mediation the loose Texture of the first rudiment of the Brain is consolidated into a firm Cortical and Medullary substance . . . set off in great variety of Geometrical Figures and Magnitudes, which take their first conception and birth from the greater and less proportion of seminal Liquor, impregnated with several Salts, which are the cause of the manifold figuration of the distinct Processes ; because different Animal Salts, as well as Mineral, do shoot their diverse parts of Seminal Liquor, into great variety of curious shapes and sizes." [1]

Collins' description of the brain of a carp may be taken as an example of his work in this field. In such an organ he sees nothing but " processes adorned and beautified with orbicular and tri-angular shapes ". He is a cataloguer pure and simple, and a cataloguer of objects he has neither the competence nor the will to understand. But he has at all events *seen* something. He found the optic nerves, olfactory bulbs and tracts, and corpora striata, but missed the epiphysis. The semi-transparent character of a portion of the tectum opticum allows the underlying structures to be observed, and the appearances so presented have been mis-interpreted by Collins as four small separate lobes situated in the transverse plane, and the valvula cerebelli are shown as forming independent lobes on the dorsum of the brain. Behind this region the cerebellum, the fused facial lobes, and the larger vagal lobes characteristic of the brain of this species, are correctly figured. But all this conveys nothing to Collins, nor does he attempt to throw any light on it by comparing his numerous dissections of the brains of fishes. In the brain of *Mustelus* the olfactory tract and bulb are interpreted as the optic thalami, but in the skate what are clearly the same structures are rightly identified. The optic lobes

[1] Vol. 2, p. 1108.

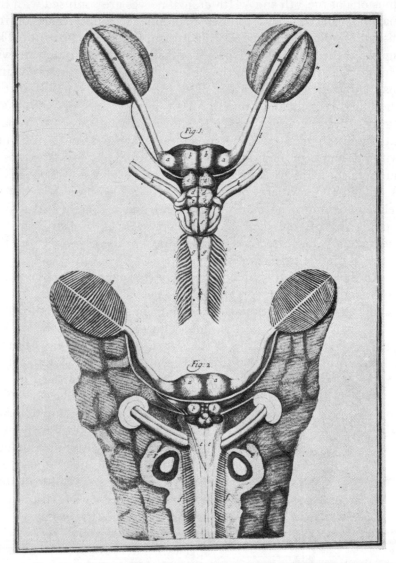

FIG. 70.—Collins, 1685. Brain of skate in dorsal and ventral views

are wrongly separated into nates and testes, and the ophthalmic divisions of the Vth and VIIth cranial nerves are confused with the maxillary nerve — his " upper mandible " being a part of the rostrum remote from the cartilage it is supposed to represent. The figures of the brains of two species of skate are not accurately drawn. He has failed to recognize the optic thalami and optic lobes, and his " optic nerve ", the easiest of all to determine, can only be the combined roots of the Vth and VIIth nerves. In one of the skate figures these nerves are actually shown passing into the eyeball, although the true optic nerve appears in the same figure, but without name or description. The pituitary, saccus vasculosus and inferior lobes are figured, but dismissed briefly as " processes ". A second and smaller " olfactory nerve " is also indicated, but its relations cannot be harmonized with any existing structure. Collins derives the optic and olfactory nerves of fish from the medulla, but he correctly draws attention to the asymmetry of the brain in the Pleuronectidae.

In the gurnard the three free rays of the pectoral fin are exceedingly sensitive tactile organs, and their sensory nerves, which belong to the general cutaneous component, and the corresponding dorsal horns of the spinal cord are therefore greatly hypertrophied. The result is a series of five paired lobes on the dorsal surface of the spinal cord at this level, which give it a very striking and unusual appearance. These lobes are figured for the first time by Collins, although naturally he was unable to grasp their significance. Unfortunately he proceeds to endow the brain of the carp with the same organs, although they do not exist in that species. There are prominent facial and vagal lobes in the carp, as figured by Collins himself, but they are associated with a different neural factor, which is not represented in the spinal region. Collins describes

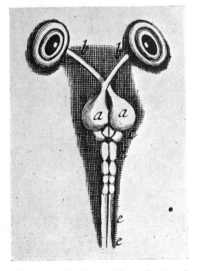

Fig. 71.—Collins, 1685. Brain of gurnard, *Trigla*, showing the tactile centres or lobes at cephalic end of spinal cord

and figures the " uterus " of the ovo-viviparous shark ? *Galeus vulgaris* as a large *diverticulum* of the posterior extremity of the oviduct. The eggs therefore have to descend the whole length of the oviduct, and then pass forwards into the " uterus ". There is no basis for this statement, since the uterus is nothing but a specialized region of the oviduct, and there is hence only a single cavity. He speaks also of a vascularized connection between the embryo and the wall of the uterus, but there are no indications that he visualized this connection as a placenta, nor does he understand how the foetus is nourished before birth. Had he been familiar with the work of Aristotle, Belon, Rondelet and Steno on this point his own account could have been made more informative.

The two classic examples of the intranarial epiglottis, both of them evolved to facilitate simultaneous respiration and feeding under certain specialized conditions of life, occur in some marsupial embryos and the Cetacea. In the latter group it was discovered by Belon, who also suspected its uses, but Ray and Tyson are silent on this curious adaptation. Collins prints a long description of the larynx of the porpoise without noting its intranarial character, but in the crocodile he remarks that the epiglottis is " of a Semicircular Figure, filling up the Interstice of the Fauces", which suggests that the modified type of epiglottis which is found in this animal was perhaps known to him.

Collins distinguishes a " diaphragm " in the tortoise, but believes that the lungs perforate it and project into the abdominal coelom. This diaphragm, he says, stretches from the pericardium to the urinary bladder — an unusual condition, explained by the fact that the pericardio-peritoneal septum of Chelonia, owing to the extension posteriorly of the pericardial cavity, is oblique, and not vertical as it is in fishes. Hence the lungs occupy a space overlapping dorsally the abdominal coelom, which accounts for Collins' assumption that the septum is perforated by the lungs. It is, however, unreasonable to expect that Collins or any of his contemporaries could have had any knowledge of the relation of the septum transversum of the lower vertebrate to the diaphragm of the mammal. On the other hand we cannot excuse the indifferent observation which invites us to believe that the hearts of the land tortoise and the snake have only a single auricle. His predecessors knew better than that.

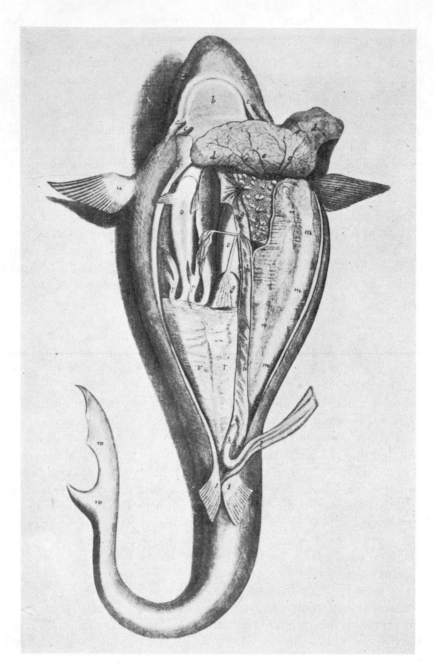

Fig. 72.—Collins, 1685.   Genitalia of the viviparous shark, *Galeus*

Considerable space in the *Systeme* is devoted to the structure of birds, and the mouth, gut, respiratory system, heart and dorsal surface of the brain are examined in some detail. The velum palati, tongue, and the modifications of the mucous surface of the palate in various types of birds — characters which have since been employed by systematists when attempting the classification of this refractory class, receive much attention. There is an elaborate account of the glandular proventriculus and of the muscular structure of the gizzard of the goose, in which a demonstration of the physiology of the milling action of the gizzard is attempted. He mentions the occurrence of small stones in the gizzard and was familiar with their purpose. This is one of Collins' best chapters. The descriptions are ample and the figures good, and in some respects he was on the way towards a comprehension of the mode of action of the avian gastric mill. He also points out the differences between the stomachs of carnivorous and graminivorous birds, the former having small thin membranous stomachs and the latter large thick muscular ones. In some birds he confuses the true stomach with the gizzard, but in others, such as the heron, he distinguishes " three stomachs ", which are obviously the crop, proventriculus and gizzard. He claims that there is a valve between the first and second of those " stomachs ".

Collins, drawing upon Aldrovandus and Sir Thos. Browne, explains the looping of the trachea in the swan by assuming that such a device would act as a reservoir or "cistern of breath" in a species which feeds with its head under water. Had he pursued his comparative studies further he would have recognized that this explanation was untenable. In searching for a diaphragm in birds he found the air sacs, and noted their connection with the lungs. He did not see all the septa defining the various air sacs, but he correctly surmises that in respiration air passes through the substance of the lungs into the air sacs, and is then driven back again, and he even suggests that this process would not affect the bulk of the bird. The air sacs, he says, supply the place of the diaphragm in effecting respiratory movements. Collins describes the right auriculo-ventricular valve of the heart in birds as muscular, and understood its function, which he illustrated by injecting water into the ventricle.

Among his observations on the avian brain he includes an

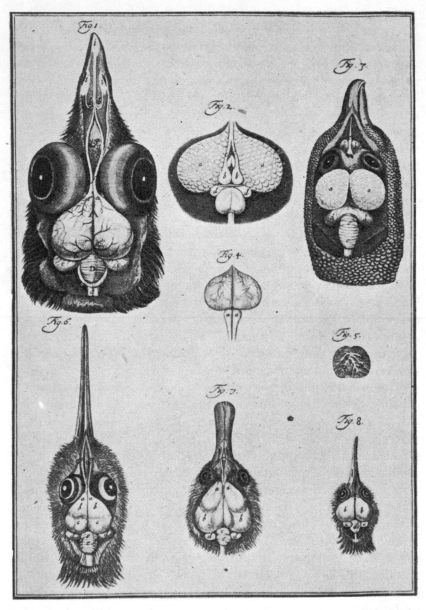

FIG. 73.—Collins, 1685.   Brains of, *1, 2,* bustard ; *3, 4, 5,* turkey ; *6,* woodcock ;
7, teal ; and *8,* snipe

account of the structural relations of the optic lobes, and the connection of their cavities with the ventricular system of the brain. Nevertheless his views on the homology of these lobes are so mistaken that they are compared with a part of the cerebellum, and even with the corpus callosum of the mammal, which latter structure, he says, with an approximation to verity more accidental than intended, is otherwise deficient in birds. In another and happier passage he throws out the suggestion that the optic lobes " somewhat resemble the natiform processes of a human brain ".

Coming now to Collins' work on mammals, he gives a satisfactory description of the mouth and palatal ridges of the calf, sheep, lion, cat, boar and horse, but his account of the curious stomach of the beaver is taken from Blasius, who in his turn was abstracting

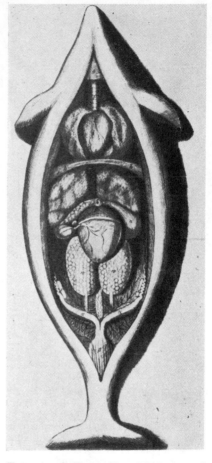

FIG. 74.—Collins, 1685. Thoracic and abdominal viscera of female porpoise

Wepfer of 1671. This stomach is divided by a furrow into two parts, and near the cardiac opening there is a large peculiar cardiac gland to which attention was directed in modern times by Rymer Jones and Owen. The structure and mode of action of the compound stomach of ruminants is discussed by Collins, but his version contains nothing new, and he is wrong in concluding that when the food is swallowed after rumination it passes a second time into the paunch. The views of Glisson on the ruminant stomach are considered, and the compound stomach of the porpoise, described in 1654 by T. Bartholin, and in 1671, 1672 and 1680 by

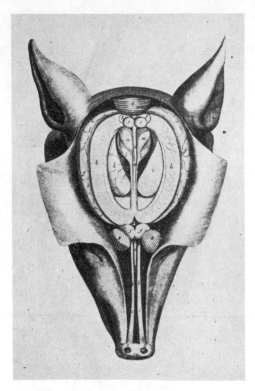

Fig. 75.—Collins, 1685. Dissection of brain of pig. *b*, *c*, branch of V nerve (but course inaccurate) ; *e* ? Harderian gland ; *f*, olfactory bulb ; *g*, *h*, hemispheres ; *i*, *k*, corpora striata ; *l*, fimbria hippocampi ; *m*, hippocampus ; *n*, septum pellucidum ; *o*, *p*, corpora quadrigemina ; *q*, vermis.

Ray, Major and Tyson, is also examined by Collins. The gall-bladder of the lion and cat is stated to have valvular folds in its neck, which give to the lining at this region a sacculated appearance.[1] In man similar structures may be observed forming a kind of spiral valve. As in the case of fishes and birds the mammalian brain is studied only from the dorsal aspect, but he observes that the rete mirabile of the brain is small or absent in the horse and man, although well developed in the calf, fox, sheep and deer. He says it may be studied by injecting a black liquor into one of the carotid arteries, whereupon the coloured mass passes from the side injected through the rete mirabile into the other carotid.[2] Finally Collins describes and figures in " brutes " such as the sheep the seventh or choanoid eye muscle, which was discovered by Galen, and in modern times re-studied by Vesalius and Fabricius.

[1] An instructive example of Collins' impressionist treatment of his authorities. In spite of his claim that these statements are based on his own dissections of the lion and cat, the facts as stated by him are to be found in the Parisian memoirs dated 1667.
[2] This experiment has been copied from Willis of 1664.

XVII
VALENTINI

The *Ampitheatrum Zootomicum* of Valentini was first printed in 1720, and re-issued as a title-page edition in 1742.[1] The engraved title is a plagiarized modification of the title of the English version of the Parisian *Memoires*, which itself is dated 1687. Valentini's work is a compilation, and pretends to be nothing more. Except for a very few notes of minor interest by the author and his two sons, no original observations are to be found in this treatise. Neither is there any comparative anatomy, nor any general discussion of the data collected, which are presented to the reader as a fortuitous concourse of undigested fragments on which he is expected to exert his own constructive genius.

The plates are the work of an indifferent craftsman, and therefore fail to reflect the merits of the originals — in some cases, indeed, such as the sketch of the hyoid of the woodpecker, being little more than caricatures of the parent figures. Nevertheless the volume has its uses and even importance, particularly as a comprehensive guide to the relevant literature of the period, which, however, it must on no account be permitted to supplant.

Valentini has abstracted or reprinted the publications of fifty-eight authors, and describes 169 animal dissections, but the majority of these are the work of three anatomists only — Severino, Perrault and Muralt. The species included are : COELENTERATA, 1 ; ANNELIDA, 1 ; MOLLUSCA, 6 ; CRUSTACEA, 3 ; INSECTA, 21 ; ARACHNIDA, 5 ; PISCES, 20 ; AMPHIBIA, 4 ; REPTILIA, 10 ; AVES, 26 ; MAMMALIA, 50 ; total, 147.

It should be noted that, if we omit the Insecta, the invertebrates are more or less ignored, and that, of the vertebrates, the fishes, birds and mammals are considered to be the most important, the mammals attracting almost as much attention as the rest of the vertebrates put together. These preferences on the part of research workers are the natural consequence of the bias of the times, which placed a heavy premium on all studies having a bearing on man. The Appendix is devoted by Valentini to human anatomy, but it comprises also a useful section on anatomical technique, in which

---

[1] This " edition " is very rare, but the first edition is generally obtainable.

the latest methods for the examination of the lungs, lymphatic and blood vascular systems by injections of fusible metals, mercury and wax are described. The bleaching and articulation of osteological preparations are likewise discussed.

V

THE NEW COMPARATIVE ANATOMY

XVIII
MALPIGHI

WHEN any striking departure in research is afoot it is inevitable that observation must be stimulated before it is possible to formulate any comprehensive views crystallizing the new development and directing the ramifications of the energy which has been set free. Thus we find the biologists subsequent to Harvey, by practising the now-established methods of experiment and comparison, accumulating a vast reserve of facts which were one day to be pieced together into a rational system of comparative anatomy. For at least a century after Harvey the anatomical explorer was extending his horizon as regards matters of observation in many directions, without, however, any instructed attempt being made to collate the results so obtained. We approach a time prolific in detailed anatomical investigation but poor in philosophy. It is this transition period, interposed between the working lives of William Harvey and John Hunter, which may now be considered.

The monographic school of animal anatomy, which had been founded by Ruini, is early in evidence, and its most distinguished representatives in the seventeenth and early eighteenth centuries were Malpighi, Tyson, Swammerdam, Perrault, Martin Lister and G. J. Duverney. Malpighi was born in the year which witnessed the publication of Harvey's work on the circulation (1628) — a century before the birth of John Hunter (1728). His most important anatomical works are :

(1) 1661. Two letters to Borelli on the structure of the lungs.
(2) 1665. Four letters on the structure of the brain, tongue, omentum and adipose tissues, and the skin.
(3) 1666. Discourse on the structure of the liver, cerebral cortex, kidney and spleen.
(4) 1669. Monograph on the structure and life-history of *Bombyx mori*.
(5) 1673, 1675. Dissertation with appendix on the development of the chick.

(6) 1689.   Letter to the Royal Society on the structure of the lymphatic
            glands.
(7) 1697.   Posthumous works.[1]

The *Bombyx* memoir is the first detailed treatise on the anatomy
of an invertebrate.   The manuscript is happily still preserved in
the library of the Royal Society.   The first part of the text is in
Malpighi's neat autograph, and the drawings, presumably also by
him, being executed in sanguine, are hence sketchy in outline.
Both manuscripts of the chick memoirs, the property of the Royal
Society, are dated 1672.   The attractive frontispiece to the *Opera
Omnia* reveals Malpighi in an imaginative and artistic vein.   It was
first engraved for the *Anatome Plantarum* of 1675, and reprinted
in other editions of 1679 and 1687.   The engraved versions of
Malpighi's drawings are not worthy of the originals.

In spite of the possession of a placid and unprovocative
personality, Malpighi was called upon to endure the machinations
of implacable and stupid foes.   On one occasion, in 1689, he was
the victim of a brutal and ribald attack.   His country house was
invaded by masked men, who burnt his papers, battered his
microscopes and left him half dead with fright.   Once before, his
work had suffered severely from fire, and this final crushing blow
reduced him to a condition of profound depression.   Old for his
years, and weakened by sickness, " his head white and his back
weary, he was prepared to go down into the dust of death ".   But
this despairing mood was not to last, and his posthumous biological
works include many important observations made during these
final years.   And even so, numerous others remained unpublished,
the manuscripts of which were only recovered in 1830.   Another
and more serious occasion of dissatisfaction was the imperfection
of his magnifying glasses.   In 1675 and again in 1679 he complains
that these defective tools were seriously hampering his work.   He
was one of the first anatomists to use a compound microscope, and
shares with Leeuwenhoek, who more wisely favoured the single
bi-convex lens, the honour of having established the science of
histology.   And yet the very simple compound microscope at
Malpighi's disposal was but a crude and empirical instrument,
little better than the toys which can be bought for a few shillings

[1] Collected editions of all these works, the *Opera Omnia*, were published at
London and Leyden.

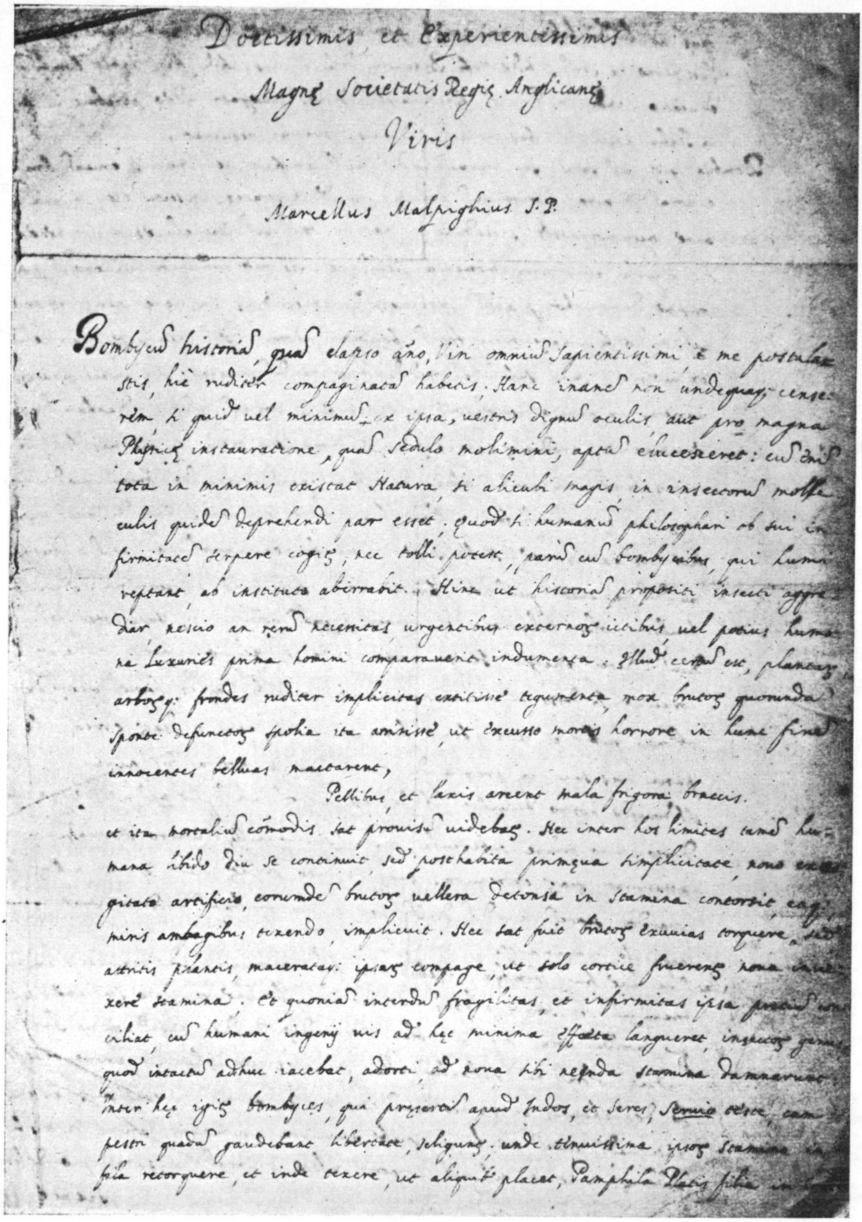

in our own times. The lenses were not accurately shaped, the tube was clumsily constructed, very long, and not inclinable, and the object could be illuminated only by reflected light. Malpighi does not describe his microscopes, but assuming that he employed the most efficient instruments available at the time, such as those manufactured by Eustachio Divini, it is still difficult to understand how he succeeded in seeing what he did. His description of the development of the heart in the chick, stage by stage, is perhaps the most remarkable observational achievement of any biologist in the seventeenth century, since it is not the story of a single chance or happy discovery, but the detailed reconstruction of a series of epigenetic stages. The interpretation is not, and could not be, a modern one, but the facts are well and faithfully recorded and figured. Like the generality of his contemporaries Malpighi takes his methods for granted, and possibly considered it superfluous to discuss or describe them. He was, however, the first, in 1661, to experiment with mercury injections, anticipating Bellini (1662), Swammerdam (February 8, 1672) and Duncan (1678).[1] He injected the pulmonary artery with mercury, whereupon all its ramifications up to the smallest branches acquired a beautiful silver colour.

In his works on the structure of the lungs and glands Malpighi's procedure is based on comparison, and his results are manifestly the consequence of the exploitation of that method. In 1661 [2] he had demonstrated the true nature of the lungs, and had shown that the almost solid lung of the mammal was nevertheless congruous with the bag-like or vesicular lungs of the frog and tortoise. He held that the air in the alveoli of the lung was *not* in direct contact with the blood. It was during this classic research that he discovered the blood capillaries — thus completing Harvey's work, and he afterwards found them also in the mesentery and bladder of the frog. Thus was laid the foundation of our knowledge of the physiology of respiration. When Malpighi's work on the viscera was published in 1666, attention had already been concentrated on the anatomy of the glands, and the more obvious features of these organs were known from the work of a number of observers. The time had in fact arrived for the discovery of their relations and

---

[1] This author attributes the method to Swammerdam.
[2] No copy of this tract has so far been found in England.

FIG. 77.—Malpighi. From the crayon drawing used as frontispiece to his *Anatome Plantarum* of 1675, and also later in other editions of his works. Original in possession of Royal Society

function. Malpighi opened his attack on this problem by studying the comparative anatomy of the glands. He detected the lobules of the mammalian liver, and pointed out how the tubular liver of the invertebrate could be linked up with the parenchymatous liver of the higher animal. He explored the liver in molluscs, fishes, reptiles, mammals and man, and finally reached the sound conclusion that the liver tissue acts on the blood, extracts bile from it, and passes this product down the bile duct, which must therefore be regarded as the excretory duct of the liver. The bile is hence not secreted by the gall-bladder. Malpighi was thus the first to penetrate the mystery of the glands, and he does so not as a chemist but as a comparative anatomist and physiologist. He suspects that they arise by the splitting of an originally single tube, and in this way a number of caecal growths are produced which are closely related to the blood stream, and elaborate from it the secretion appropriate to each individual gland. Ruysch, however, persisted in the belief that the secretion was produced in the blood, the function of the gland being that of a purely mechanical separator. Even Haller in his *Elementa Physiologiae* [1] sums up impressively in favour of this belief, and against Malpighi.

It has already been pointed out that Malpighi, at the outset of his career, was emphatic on the value of the comparative approach to the study of animal structure. This is what he says in his earliest work published in 1661 : " The arcana of nature are so hidden, and their discovery so difficult, that it seems as if our senses were unable to grasp anything completely. And although we persist in this ungrateful task, and maintain our pursuit of natural knowledge, we find ourselves confronted as it were by a great book crowded with puzzles and mysteries. Work as we may to discover the innermost secrets of the viscera of animals, and despite all our efforts, we can assure ourselves of truth only after an infinity of very fatiguing observations, and after dissecting many insects and higher animals, so that we advance step by step in our knowledge of anatomy. For nature requires us to devote our pioneer efforts to simpler types before undertaking more complex works, and indeed we can recognize in the lower animals the faint outlines of the higher." To solve these problems, he explains, he had almost exterminated the race of frogs, and had brought about a greater

[1] T. **2**, Lib. VII (1760).

slaughter than that recorded in the bloody battle of the frogs and
mice, of which Homer speaks.[1] His dissections, he adds, in which
his illustrious colleague Fracassati participated, resulted in the
discovery of marvellous things, which the words of Homer alone
can fitly acclaim : " O, what a truly great work was there before
my eyes ". It will be seen from the above quotation that Malpighi's
conception of the animal kingdom was that of an *échelle des êtres*,
which does not indeed rule him out as a comparative anatomist,
although it bars the way to any advance towards the recognition of
an historical or evolutionary relationship of species.

From the point of view of the present work Malpighi's most
important publication is his monograph on the structure and life-
history of *Bombyx mori*, which seems to have been completed within
a year. " My dissertation on *Bombyx* ", he says, " was an occupation to the last degree laborious and fatiguing, because of the
novelty, minuteness, fra-
gility and entanglement
of the parts. Hence the
prosecution of the task
made it necessary to
develop entirely new
methods. And since I
pursued this exacting
work for many months
without respite, I was
afflicted in the following
autumn with fevers and
inflammation of the eyes.
Nevertheless in accom-
plishing these researches,
which brought to my
notice so many strange
marvels of nature, I ex-
perienced a pleasure which
no pen can describe."
Réaumur (1734) says of

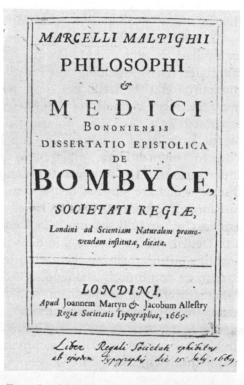

Fig. 78.—Malpighi.   Title-page of first edition
of the work on *Bombyx*

---

[1] Wrongly attributed to Homer.
The second quotation has not been
traced to Homer.

the *Bombyx* memoir : " It is nothing but a tissue of discoveries —
a treatise in which one may obtain a greater knowledge of the
admirable inner structure of insects than in all the works which
have preceded it ". Tyson (1680) is even more emphatic. He
remarks : " Far better a little with accurateness, than a heap of
rubbish carelessly thrown together. *Malpighi* in his Silk-worm
hath done more than *Jonston* [1] in his whole book of Insects."
After the publication of this work in 1669 Malpighi continued to
labour on *Bombyx* in order to clear up some doubtful points and
to confirm others. These final results were brought together in
a letter dated 1670, which, however, was not published until the
posthumous works appeared in 1697, although an abstract of it
was printed at the time it was written. Malpighi anatomized all
phases of the species, but, apart from his very remarkable and
accurate observations on the genitalia of the moth, the larva
claimed the greater part of his attention, and it was on this stage
that his most novel and important discoveries were made.

A good description and figure are given of the head of the
caterpillar, in which the following parts are distinguished : epi-
cranium ; frons, extending downwards at the side of the clypeus as
the adfrontal ; six ocelli, but only five are figured — " which we
may consider, I think, as being eyes " ; clypeus and labrum,
separated by a *flexible* membrane ; antennae ; stipes and lacinia
of the maxilla ; "mentum " of the labium with its spinneret ;
" teeth " or mandibles, which have been rotated outwards, and
their muscles. He contrasts the action of the mandibles of the
insect, which are situated in the same transverse plane, with the
fore and aft jaws of the higher animals. The thoracic legs and the
prolegs are carefully examined and compared. In the latter he is
attracted by the structure, number, arrangement and function in
crawling of the crochets which occur on the volar surface. In 1734
Réaumur confirmed Malpighi's account of these structures.
Malpighi refers to the problem, which almost simultaneously was
exercising the mind of Swammerdam, as to whether all the exuviae
of an insect coexist from the beginning, or whether they are formed
in succession as required. Swammerdam answered this question
emphatically in favour of coexistence, and based his well-known

---

[1] Jonston's academic and uninspired compilation (1653) covers 128 folio pages
and has 23 plates. There is scarcely any anatomy.

views on preformation in animal development largely on this unfortunate inference, but Malpighi is wiser and more cautious, and prefers to leave the matter open.

In this memoir Malpighi describes for the first time the spiracles of insects, and the system of vessels associated with them. There are, he says, nine pairs in the larva of *Bombyx*. Swammerdam wrongly gives ten pairs,[1] and accuses Malpighi of having overlooked one pair. There are, however, never more than eighteen spiracles in any Lepidopterous larva. Malpighi found that each spiracle had its own bundle of vessels, two of which anastomose with corresponding vessels in front and behind, and so assist in forming the longitudinal spiracular trunk on each side which stretches from head to tail. There are also transverse anastomoses *across* the body. He compares these vessels with the arterial system of the higher animals. Like the arteries they go on dividing and diminishing in size until they can hardly be seen even with the microscope. Also like the arteries they pervade all parts of the body without exception — even to its innermost and smallest recesses. He describes and figures the spiral thickenings or taenidia of their walls, which he interprets erroneously as *annular* rings, but rightly understands their function in maintaining the distension of the tubes. He names these " vessels " tracheae, and compares them with the lungs of vertebrates. This he confirms by finding in other insects, such as the cicada, stag-beetle, locust, wasp and bee, expansions of the tracheae which he identifies with the air chambers of the lungs — themselves nothing but appendages or extensions of the trachea. Hence these insects are provided with a pair of lungs in each segment, and furthermore every part of the body is permeated by them. Malpighi gives a good account of the structure of the spiracle, including the rim or peritreme, the fringed processes of the lips, which he compares functionally with the cilia of the eye, and the spiracular aperture. But what, he now asks himself, is the function of this curious apparatus ? His answer is admirably to the point. He stopped the spiracles with oil, and immediately the animal passed into convulsions and died " intra Dominicae orationis spatium ". If the anterior spiracles are blocked only the corresponding part of the body is affected, and the animal subse-

---

[1] His illustration is more correct than the text, and shows only nine spiracles on the one side figured.

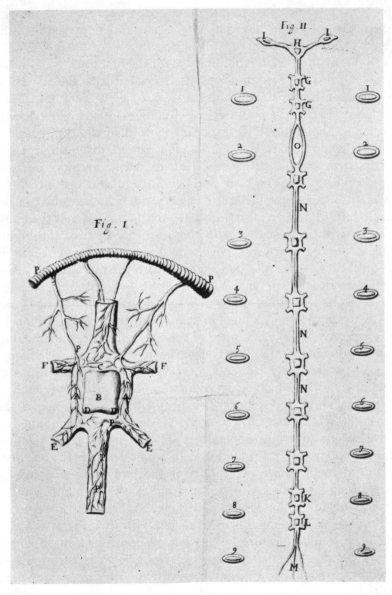

FIG. 79.—Malpighi, 1669. *Bombyx.* Nervous system, spiracles and tracheae of larva

quently recovers. Similarly if the posterior spiracles are obstructed only that section of the body reacted, and the beating of the heart in the same region was considerably slowed down. When the spiracles of one side were sealed, the larva remained motionless for a time, the anterior region recovering first, whilst the posterior remained languid. If on the other hand the whole body is painted with oil, but leaving the spiracles intact, the animal is not killed and even seems to be unaffected. Hence the effect of the oil is to throw the tracheae out of action, and the result is suffocation.

Many other experiments on a variety of species are described in confirmation of these conclusions. On the other hand insects kept under water for some hours do not die and the motion of the heart is maintained. In some experiments under water, bubbles of air were liberated which he suspected escaped from the spiracles, but he was not able to establish this point. Malpighi's final opinion is that only those substances which *completely* seal the spiracles are fatal, and that a constant in-and-out passage of air is just as essential to the insect as it is to the pulmonate animals. He suggests that to produce the aeration of the tracheal system abdominal movements must be necessary, and, although he is not prepared with a complete demonstration, he has noticed that the last three segments of the caterpillar are continuously expanding and contracting in a very striking manner. This he thinks may be due to the action of the heart, or to respiratory requirements, but that further research is necessary to decide the point. Severino had previously observed similar manifestations in *Gryllus domesticus* in 1645, and J. T. Schenck in 1665 had ingeniously detected thoracic and abdominal movements in " Scarabaeus " which he regarded as possibly respiratory, but it was not until 1860 that Rathke examined and described in detail the breathing mechanism of a number of insects. Malpighi's second surmise is, as we now know, the correct one.

The heart of the insect is identified for the first time by Malpighi. Libavius had undoubtedly seen the heart of the silk-worm in 1599, and described its pulsations, but he did not recognize it as the heart, expecting to find this organ in the same position as it occupies in vertebrates. In fact he says : " Whether they have any heart, let others seek out : yet there must be some such Principle ". The unexpected elongation of the heart in the silkworm deceived him into regarding it as an artery. Malpighi shows that the heart extends from the head to the tail, and in the living animal can be seen to pulsate along its entire length. He cannot assert positively that it is muscular, but assumes that contractile tissue must be present. The heart is a long tube, and there is no definite enlarged region in which pulsation originates. It exhibits, however, a series of segmental oval swellings or chambers which he regards as a series of " little hearts ". Swammerdam questioned this multiplicity of hearts, and it is true that in some insects the heart is unconstricted. Malpighi watched the pulsation of the heart, and correctly states that the wave of contraction proceeds normally from behind forwards,[1] the blood being pushed from heart to heart just as in the higher animals it passes from auricle to ventricle and from ventricle to artery. He did not succeed in finding any arteries arising from the heart, but he seems to have noticed the alary muscles in a pupa, and thought that they might be arterial trunks. He does not mention the ostia.

The elaborate fat body is described not only in the silkworm but also in other insects, and it is interpreted as a reservoir of fat comparable with the epiploon or omentum of the vertebrate. The gut, he says, is straight, without convolutions, and is plentifully supplied with tracheal tubes. He does not of course distinguish the three divisions of the gut of the modern morphologist, but he describes the oesophagus and crop of the fore-gut, with their circular and longitudinal musculature, and the stomach or mid-gut, separated by the pylorus from the hind-gut, which latter is recognized as consisting of an anterior or small intestine and the rectum. The relative extent of these parts is not accurately figured, the crop being exaggerated at the expense of the stomach. Malpighi discovered the glandular system now known as the Malpighian tubules, and noted that where it discharged into the hind-gut there was an enlargement — the excretory ampulla. He states correctly that the tubules associated with the small intestine and rectum in the silkworm have a smaller calibre than those attached to the stomach, and he is not quite certain that they belong to the same system, as they do. He found the tubules in other insects, and believed that they were subservient to digestion, transmitting highly assimilable food substances to the heart and

[1] Subject, however, to periodic reversal.

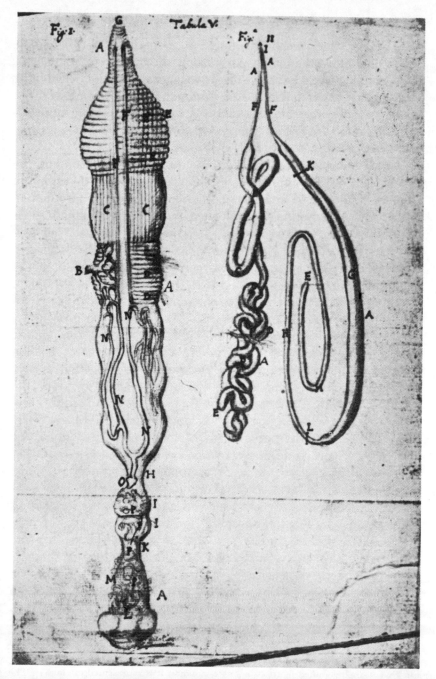

FIG. 80.—Malpighi, 1669. *Bombyx*. From the original drawings of gut and silk glands

skin, whence they were distributed to all parts of the body, the excess and non-assimilable parts being discharged into the gut. However, he concludes, these are only suggestions, and he willingly abandons the investigation of the function of the system to others. In the meantime he cannot decide whether they are caecal appendages similar to the pyloric caeca of fishes, or whether they are not analogous to the lacteal vessels of brutes.

Other discoveries by Malpighi include the nerve chain and the re-discovery of the silk glands. The latter were first described in 1599 by Libavius, who detected their relation to the production of silk, and Aldrovandus figured them in 1602, but Malpighi does not quote either of these works. It is possible that Libavius also saw the tracheae and the Malpighian tubules. His language, however, is too vague, and his observations are too casual, to justify any claim being made on his behalf to these discoveries. Malpighi on the other hand made a careful examination of the silk glands not only in *Bombyx* but in other insects. The description is accurate, but the figure wrongly shows the right and left portions of the gland to be asymmetrical. He was also not aware that each portion secreted its own filament, the two becoming stuck together as they traverse the single canal of the spinneret. The thread of the cocoon, therefore, is double throughout its whole length, as was first noted by Leeuwenhoek in 1700, but not published until two years later. As regards the nerve chain Malpighi saw the ganglia, the double nerve cord [1] and the segmental nerves. He recognized that the ventral ganglia consisted of peripheral and central sections, now known as the layer of ganglion cells and the medullary substance or neuropile. He examined in addition the nerve cord of the crayfish. In *Bombyx* he could find no difference in structure between the ventral ganglia and the brain, and thinks that the former may have arisen by the scattering of the latter. Malpighi's figure of the nervous system of the caterpillar has been often criticized. Swammerdam remarks that Malpighi had failed to discover, or at all events to mention, that the gut passes through an oesophageal nerve ring, and that he had overlooked one ganglion of the ventral chain *and even the brain*. Swammerdam's figure is

---

[1] It is strange that, in the general figure (p. 186), the cord is shown to be double, but that in the magnified figure of a single ganglion it is represented as single. The text, however, states clearly that the ventral cord is composed of two bundles enclosed in a common envelope.

assuredly the better of the two, and he also saw something of the sympathetic system, but it must not be forgotten that Malpighi was the pioneer — a consideration that Swammerdam himself justly acknowledges. In his notes published posthumously Malpighi indignantly denies having missed the brain, and it is not difficult to interpret his figure on that basis, if certain minor corrections are allowed. Thus the first spiracular opening is too far back and the second too far forwards. The brain has been severed in the median sagittal plane, and the connectives rotated right and left respectively. His silence on the oesophageal nerve ring is unaccountable, for he must have seen it. He did, however, overlook the first thoracic ganglion, which lies close behind the sub-oesophageal ganglion, and the divarication of the two halves of the ventral cord between the second and third thoracic ganglia is likewise not shown. On the other hand he apparently realized that the last abdominal ganglion was a composite structure supplying the eighth and succeeding segments. Therefore Malpighi's figure, if crudely drawn, is reasonably accurate, and the position of the ventral ganglia relative to the spiracular openings, with the exception of the first two, has been correctly represented.

Malpighi is somewhat less successful in his treatment of the pupa. He concluded that the spiracles were not operative during this phase, but he observed the reduction in the volume of the gut, although he was wrong in believing that the mid-gut or stomach became detached from the remainder of the intestine, and that the Malpighian tubules were fused at their extremities. He found the rectal caecum and discussed its function, and was aware that the ovaries and colleterial glands made their appearance in the early days of the pupa, as shown in the beautiful figures of the anatomy of the *Bombyx* chrysalis published by Pasteur in 1870.

The external features of the moth are carefully described, but the parts are not all correctly identified, his collum or neck being the prothorax, and his thorax the meso- and metathorax. His account and figure of the genital armature of the male reach a high standard of descriptive accuracy. In the ninth abdominal segment he distinguishes the tergum,[1] coxopodite, sternum or vinculum, claspers or harpes, and the gnathos — assuming the last mentioned to belong to the ninth and not to the tenth segment. In the latter

[1] This and the following terms are modern and not those employed by Malpighi.

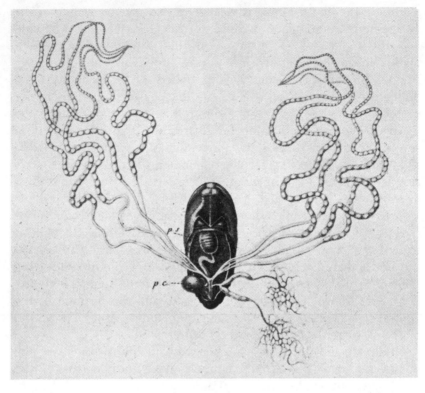

FIG. 81.—Pasteur, 1870. *Bombyx.* Anatomy of pupa, showing gut and female genitalia

segment he figures the tergum and its extension the uncus, and the anus. The penial apparatus is shown rotated upwards, and a description is given of the aedeagus and penis sheath or phallotheca, and the penis funnel or juxta. In the female, Malpighi erred in supposing that there was a kind of common cloacal space, " as in Birds ", in which the faeces and eggs were detained for a short time before emerging by the anus. It is true that the anus is close to the genital aperture, but there is no cloaca. Malpighi complains, as was inevitable, that the abundance of fatty connective tissue and the innumerable ramifications of the tracheae make the dissection of the moth very difficult. In his examination of the male genitalia he found that the testes had already appeared in the caterpillar. They are, he says, more developed in the pupal stage, and in the moth they swell up considerably until they become the most

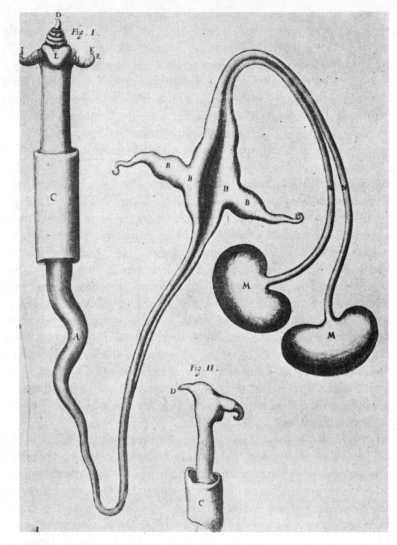

FIG. 82.—Malpighi, 1669. *Bombyx.* Genital organs of male moth

conspicuous features of the body, all other organs undergoing a corresponding degeneration. Malpighi describes and figures the two testes, vasa deferentia, accessory glands, seminal vesicles, and the ductus and bulbus ejaculatorius. This account was confirmed in detail by Swammerdam, writing shortly after receiving a copy of Malpighi's work.

The most impressive passage in the *Bombyx* memoir is the account of the successful unravelling of the complex female genitalia. Malpighi has, however, figured an unusual state, in which almost the whole of each ovariole has been converted into a greatly lengthened pedicel crowded with *full-grown* oocytes. Thus each of the eight egg-tubes presents a uniform structure throughout its entire length, instead of the normal graduated series of oocytes becoming progressively more mature from before backwards. Malpighi suspects that in pairing the male semen enters the vulva and passes at once into the bursa copulatrix. From this organ it is impelled into the common oviduct, where the eggs are fertilized one by one as they arrive from the egg-tubes. This inference, which falls short of complete accuracy in so far as it leaves out of account the receptaculum seminis, he endeavours to establish by dissecting females which had recently paired, and noting the exact position of the eggs which had been fertilized, but he did not succeed in fertilizing eggs removed from the egg-tubes with semen extracted from the bursa copulatrix. Again the receptaculum seminis is unluckily ignored. He was assisted in these experiments by observing that after fertilization the colour of the eggs changed from yellow to violet, whereas the sterile eggs retained the yellow colour, as Libavius had already pointed out in 1599. Malpighi rightly concludes that the function of the colleterial glands is to produce an adhesive secretion for the attachment of the eggs after deposition.

Malpighi discovered, described and figured all the parts of the female genitalia of *Bombyx*.[1] His only notable omission, assuming it to be an omission, is the extra duct mentioned by Ômura in 1938 by which the spermatozoa *leave* the spermatheca to re-enter the common oviduct prior to fertilization. Malpighi saw the copulatory opening or vulva, its connection (Fig. 83, *K*) with the bursa copulatrix (*I*), and the seminal duct (*M*) which extends from the bursa to the common oviduct. The eggs pass into the common oviduct or vestibulum (*A*), by which they leave the body, and into which open the spermatheca or receptaculum seminis (*F*), with its spermathecal gland (*G*), and the accessory cement or colleterial

---

[1] Malpighi's original drawing represents the parts as seen from above, but in the published version the sides are reversed in the printing owing to the engraver not reversing the figure on the plate.

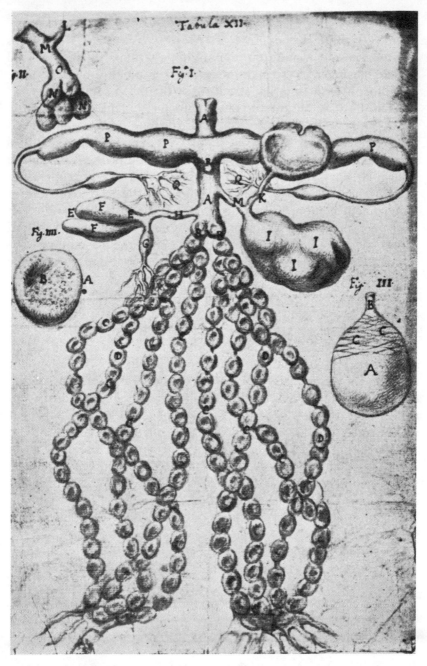

FIG. 83.—Malpighi, 1669. *Bombyx.* From the original drawing of female
genitalia

gland (*Q*), with its reservoir (*P*). The ovary consists of two groups of four ovarioles, each group having four terminal filaments [1] and an end chamber or germarium. A separate figure on this plate (II) illustrates the internal structure of the bursa copulatrix.

The heart of the moth, says Malpighi, resembles that of the larva, but its walls are thicker and more opaque, and its pronounced yellowish colouring makes it more conspicuous. He describes many interesting observations on the pulsation of the heart in the caterpillar, chrysalis and moth. These studies led to the conclusion that the heart of *Bombyx* was a string of little hearts capable of independent motion, but he does not appear to have examined it microscopically, since he mentions neither the ostia nor the ostial valves. He was, however, the first to describe the periodic reversal of the heart-beat which occurs in the late larval and subsequent stages of Lepidoptera.

Malpighi was evidently puzzled by the peculiar crop of the moth. This conspicuous lateral vesicular diverticulum of the oesophagus is, he says, full of air and opens into the mouth (Fig. III of the same plate). He believes its expansion and contraction assists in the expulsion of the genital products. He does not mention the oesophageal nerve ring of the moth, nor the striking changes undergone by the nerve chain during metamorphosis. In the letter of 1670 printed in the posthumous works, these changes are referred to, but what they are is not specified. Malpighi recognized the repetitive or segmental structure of the body of an insect, and contrasts it with what seems to him the unity or oneness which distinguishes the composition of the body of the higher animal. Each division of the larval body, he says, has its own heart, brain and lungs, but the possibility of independent existence is not thereby secured.

In the letter dated 1670, but not published in full until 1697, Malpighi corrects his earlier statement that the tracheae are strengthened by *annular* rings. He now recognizes that the " rings " are parts of a continuous *spiral* thickening, and claims that the tracheae themselves are strictly comparable with the tracheids of plants. He is also confident that the Malpighian tubules secrete the bile and

---

[1] In the original drawing on the right side eight are shown, presumably by mistake, as four of them have been partly erased. The latter do not appear in the published figure.

pancreatic juice, and he has found that they occur in all insects, and may be more numerous than they are in *Bombyx*.

From the foregoing analysis it will be apparent that Malpighi's monograph on *Bombyx* is an original contribution to zootomy of the first importance, and one of the earliest in point of time. Both in respect of observation and interpretation he has hardly an equal among his contemporaries, and his genius and tenacity are always employed in pressing a research home to a spacious conclusion, without at the same time sacrificing the attention due to detail. The work was too early to be used comparatively, except in so far as his own investigations permitted an approach to this method. He did in fact personally examine many other insect types in order to assist his understanding of *Bombyx*, and to ensure that he was not describing abnormal or specialized structural conditions. Nevertheless he missed the great opportunity of his career when he failed to perceive that the life cycle of his insect exemplified in a readily comprehensible form a great truth in animal development. The significance of the anatomy of metamorphosis did not completely take possession of his mind, or he would have looked more closely at the structure of the chrysalis and moth, and pondered on the dramatic internal transformations which discriminate the progress of the species from egg to imago. Not that he was wholly unconscious of his opportunity, since he did indeed touch the fringe of the problem. What he thus accomplished was good, but it was not enough. He ought to have convinced himself, and would doubtless have convinced Swammerdam, who was following close at his heels, that the preformation doctrine was fundamentally at variance with the facts of insect metamorphosis. Swammerdam himself thought otherwise, having, as he imagined, those very facts in view. Unhappily his own researches beguiled him from taking up the story where Malpighi had laid it down, and in consequence he failed not only to complete the penetration of the mystery, but even pushed the mastery of it further back than ever. Deprived, as it should have been, of the support of these two masters, and having to face their opposition, the luckless speculation of preformation, instead of hanging like a millstone round the neck of the embryologist for the best part of a century, would have achieved nothing more enduring than the casual notice of the historian.

XIX
TYSON

Tyson's more important contributions to comparative anatomy are as follows :

1680. Anatomy of a female porpoise.
1683. Anatomy of the male rattlesnake (*Crotalus*).
       Discourse on the tapeworm.
       Anatomy of the roundworm (*Ascaris*).
       Anatomy of the male " Mexico musk-hog ", the collared peccary (*Dicotyles tajacu*).[1]
1698, 1704. Anatomy of a female and male Virginian opossum (*Didelphis marsupialis*). In the second of these memoirs Tyson was associated with W. Cowper.
1699. Anatomy of a male pygmy (immature chimpanzee).

No student of Tyson's works could fail to perceive that he was a learned and accurate scholar and bibliographer, who knew and discussed the relevant literature as exhaustively as a modern author. His love of books and devotion to scientific research were crudely satirized by Garth in 1699, with, we may believe, little justice. It should be added here that in 1681 Tyson sponsored a translation from Low Dutch into English of Swammerdam's famous monograph on the anatomy of the may-fly.

The attitude of Tyson towards research may be deduced from his criticism of the scholastics. Natural history, he says, cannot be written without *zootomy* — science without experiment is a vain and unprofitable pursuit. He speaks of carrying out an ambitious scheme of comparative studies along four lines :

(1) Descriptive anatomy of selected types.

(2) Biochemistry of the more fluid constituents of the body.

(3) A *comparative* survey of the parts in order to obtain a clearer knowledge of them, and of ourselves.

(4) Embryotomia, or how the animal machine is built up. Nature, he says, more freely displays herself in the early stages " before she hath drawn over the veil of flesh, and obscured her first lines ".

In this research campaign he anticipates having the use of a

---

[1] This is the species suggested by Tyson's title, but his description tallies with the *larger* species, the white-lipped Peccary, *D. labiatus*.

" good microscope ". There is no doubt that he laboured to put these ideas into practice, and might have produced a comprehensive work on comparative anatomy had not a repressive diffidence and modesty induced him to consign the fruits of his toil to the " silent confinement of my study " ; hence the collections of unpublished manuscripts and drawings by him to be found in the British Museum and the Royal College of Physicians.

Tyson's treatise on the anatomy of the porpoise had worthy predecessors, which he had so carefully studied that he was able materially to improve upon them. They were : Rondelet, 1554; T. Bartholin, 1654; Ray, 1671 ; and Major, 1672. Belon, 1551, is not quoted by him, and it is an unfortunate omission, since Belon's work anticipates much of his own. Tyson opens his account with a comparison of the porpoise with a fish. When we view the porpoise externally, he remarks, there is nothing more like a fish, but when we look within there is nothing less like one. " It cannot abide upon the Land so much as the *Phoca*, yet is often drowned in its own Element, and hath a constant need of the reciprocal motion of Air in Respiration. It is viviparous, does give suck, and hath all its

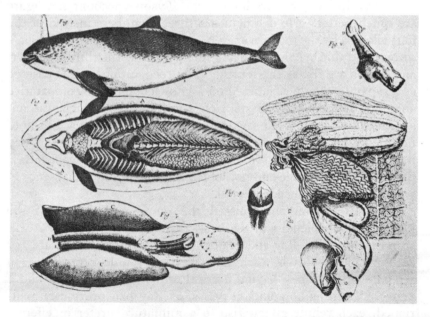

FIG. 84.—Tyson, 1680. Porpoise. Compound stomach, intranarial epiglottis and retia mirabilia

Organs so contrived according to the standard of them in Land-Quadrupeds ; that one would almost think it to be such, but that it lives in the Sea, and hath but two fore fins." Tyson and Ray point out that the caudal fins of the porpoise and fishes are not comparable in that they are orientated respectively in the horizontal and vertical planes, and the former adds an acute observation of his own that the pores of the lateral line system characteristic of fishes do not occur in the porpoise, nor does the latter have scales in its skin. Also the muscular system, viscera and all internal parts are exactly as in quadrupeds, the only difference being the external shape and the absence of hind limbs. He describes the subdivided spleen, one portion, however, being larger than the others, but Bartholin, Ray and Major had already seen these structures and correctly interpreted them as the spleen. Ray and Major were the first to discover the pancreas and its large duct, Bartholin having observed the gland but not the duct, and Tyson confirms their findings.

The remarkable compound stomach of the porpoise, overlooked by Belon, was observed by Bartholin, Ray and Major, but their descriptions of it are imperfect. Tyson's account and figure are much better. He distinguishes three chambers instead of the four allowed by the modern anatomist, the passage between Tyson's second and third chambers being interpreted as the third chamber, and Tyson's third as the fourth. The recognition of this passage as a separate chamber is based on a knowledge of the comparative anatomy of the Cetacean stomach — information which was not available to Tyson. He saw the curious small pylorus, but does not mention the bile duct. There is no caecum, as Belon and Ray had already noted. Tyson and Belon both state, as against Ray, that the liver is simple and not lobed, as in man, but the absence of a gall-bladder is conceded by all.

Belon unaccountably missed the compound nature of the kidney, so conspicuous in the bear, calf and sea otter, but it was seen by Bartholin, Ray and Major. Tyson says that there are 150 sections or renules in each kidney, each renule representing a separate little kidney having cortical and medullary zones and a pelvis. There is no common pelvis, as Ray had previously mentioned, each renule giving rise to a miniature ureter or efferent duct, the union of such ductules constituting the single common

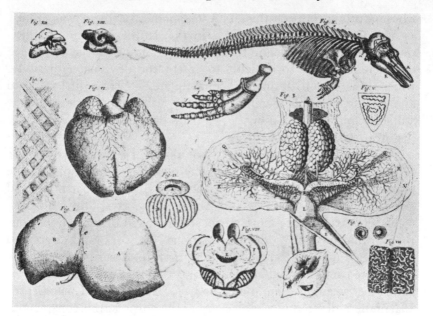

FIG. 85.—Tyson, 1680. Porpoise. 1, interlacing muscles of linea alba ; 2, liver ; 3, urogenital organs of female ; V, supra-renal ; VI, heart ; VII, rete mirabile of thoracic spinal region ; VIII, IX, structure of blow-hole ; X-XIII, skeleton

ureter. Tyson miscalculated the number of the renules, which in the porpoise is about 250. His account and figure of the genital system of the female with its highly elaborate vascular supply is excellent, and a marked advance on Major's version. He succeeded in finding all the essential parts, which are accurately identified, described and illustrated. The male genitalia had already been examined by Rondelet, Bartholin and Ray. " The *Organs of Generation* in this Animal," Tyson remarks, " which no less than the other parts did extremely imitate those of Quadrupeds ; and even in the whole dissection I could easilier imagine I was cutting up a Dog, a Swine, a Calf or any other terrestrial Brute, than an inhabitant of the watery Element." Nevertheless after carefully reviewing evidence which demonstrated conspicuously that the porpoise *must* be a mammal, Tyson nowhere has the courage to declare that it is *not* a fish, for once attaching more importance to habitat than to structure. He says he should *like* to think it was a mammal, but further than that he did not go. And yet he was writing more than a century and a quarter after Belon and Rondelet, whose attitude towards this static taxonomic problem was almost as advanced as his own. Why, in this stultifying and needless conflict between reason and tradition, it should be so easy to believe the incredible and so difficult to accept the obvious, is a riddle which it is not flattering to contemplate. " How little do we know," exclaims Clauder in 1686, " or rather how little does our stupidity permit us to know " — a truth deplorably exemplified by his own passion for alchemy.

One of Tyson's most important discoveries was that of the retia mirabilia of Cetacea, which, he observes, forgetting for the moment the well-known rete of ruminants, do not occur in land animals. " There is scarce any Animal ", he says, " in which the Veins and Arteries are more curiously branched or more numerous than in this . . . they formed a curious Net work, and afforded a very pleasant sight." He was the first to describe the highly remarkable rete mirabile of the thorax and spinal cord, which he styles a " seeming glandulous body ". It is indeed astonishing that in their memoirs on the anatomy of the porpoise this conspicuous and beautiful vascular mechanism should have been overlooked by Belon, Bartholin and Major. In modern times Breschet gave a detailed account of it with admirable figures in 1836, and Mackay added further details in 1886. Tyson is unable to offer any plausible explanation of the function of this rete, which he believes has not been seen in other animals, nor has he found any previous record of it in the porpoise.

The vestigial pelvic bones of Cetacea were found by Rondelet and Bartholin in the dolphin and porpoise respectively, and in the latter animal they are described again by Tyson. He says further that the diaphragm has no central tendinous portion but is muscular throughout, and that the heart is of the mammalian type, as Belon had asserted before him. The ventricles, however, are cleft at the apex or, as he puts it, " seemed a little divided ", and this condition is figured. The foramen ovale was *not* open, but its position could be recognized by a certain thinness and transparency. This was an important detail at the time on account of its theoretical bearings, the persistence of the foramen having been claimed as facilitating respiration under water in aquatic air-breathing species. Tyson did not look for the ductus arteriosus, which had, however, already been noted by Bartholin. The salivary and ductless glands are

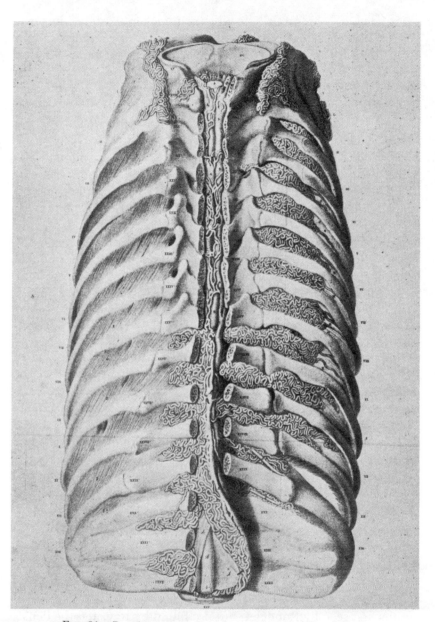

FIG. 86.—Breschet, 1836. Porpoise. Retia mirabilia of thorax

recognized as conforming to the mammalian pattern, but the peculiarities of the larynx are seized upon, and a good description with figures is given of this remarkable adaptation. Tyson observed that it was " somewhat inserted into the bottom of the fistula " [nasal passage], and that " it was very curious and different from other animals ", chiefly in the elongation of the cartilages, the epiglottis being two inches long. Tyson is concerned with the anatomy only of the larynx, and clearly failed to discover that he was handling an apparatus functioning as an intranarial epiglottis, which had been evolved to meet the needs of a terrestrial animal living in water, and had already been described with understanding by Belon nearly 130 years before. Tyson and also Ray, however, perceived that the fistula or blow-hole corresponds with the nostrils of other animals, and that it transmits air to the lungs, but Tyson misapprehends its office entirely in adopting the popular belief that, owing to the absence of gill slits, it also throws out water taken into the mouth when feeding — an error he shares with Major. Tyson, Ray and Major saw something of the nasal sacs and the valvular mechanism for closing the nasal passage, which latter Tyson considers to be glandular in nature.

The brain of the porpoise is a mammalian brain according to Belon, Ray and Tyson, and, although it is shorter and wider, the only important difference is the absence of the olfactory parts, as Ray also had noted. Tyson found " about 8 or 9 pair " of cranial nerves, including the eye muscle nerves, and discovered the corpora quadrigemina, which Ray had missed. The fact that the olfactory nerves are wanting involves also the absence of the sense organ with which they are associated, but Tyson is silent on the loss of olfactory sense in the porpoise. He claims that the structure of the eye and its muscles is the same as in man, with the addition, however, of the seventh or choanoid eye muscle " that is proper to Brutes, and which did inclose the Optick Nerve ". He noted that the sclerotic was much more dense and hard than in other animals — a condition now known to be characteristic of the Cetacea. The description of the ear is brief. It includes the very small external meatus first seen by Rondelet, but not found by Ray, and the dense bony tympano-periotic. He failed to observe the specialized auditory ossicles, as also did Major.

The skeleton is stated by Tyson to conform to the mammalian

plan — a conclusion previously reached by Belon. The description of the skull does not bring out any important point, but Tyson could scarcely be expected to unravel the peculiarities of the Cetacean skull without an extended comparative analysis. He fails also to describe correctly the cervical vertebrae, and it does not seem to have occurred to him that this region of the vertebral column in the Cetacea has undergone considerable reduction and fusion, but that nevertheless seven cervical vertebrae can still be distinguished. Major even says that, apart from the atlas, there are no cervical vertebrae, and therefore no true neck in the porpoise. Such a statement is inexcusable in view of the fact that six reduced cervical vertebrae are clearly visible behind the atlas in this species. Tyson, like his predecessors, recognized that the fore limb and pectoral girdle of *Phocaena* closely resembled the mammalian arm, but that the bones were " curiously articulated together ". Bartholin goes further, and holds that a comparison of the limbs in general terms is possible between the porpoise and man. The same distinguished author has described and figured a Cetacean type of fore limb in a grossly imaginative account of a " Syren " [manatee], of which a specimen " is also kept in the Anatomy School at Oxon ". Tyson corrects Major and agrees with Rondelet that there is no clavicle in the porpoise. The chief defects in Tyson's treatise on the porpoise are that he failed to examine the muscles and blood vessels, which, as a human anatomist, fell directly within his competence.

Tyson's memoir on the rattlesnake exhibits some advance in technique, but he could not have dissected under water, since he complains that so much time was spent on the drawings that the parts had dried before they could be thoroughly examined. This may perhaps explain why he failed to include the blood vessels, nervous system and sense organs in this paper. He acknowledges having received considerable assistance from Waller in the preparations and figures, and, apart from this, he was not the first to dissect a snake. His more noteworthy predecessors in this almost virgin field were Baldus Angclus Abbatius (1589), Severino (1645 and 1651), Vesling (1651–no figures) and Charas (1669), but his own description of the anatomy of the rattlesnake was the most workmanlike and accurate that had been published up to his time, and it might well have been better had his operations not been restricted to the single example available. He correctly, on anatomical

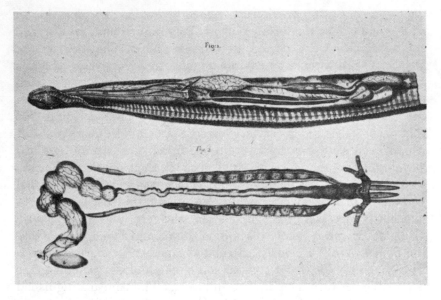

FIG. 87.—Tyson, 1683.   Rattlesnake.   Heart, lungs, gut and urogenital system
of male

grounds, assigned the specimen to the Viperidae, and deplores the neglect of anatomy by previous workers, because, he says, it is " the most principal Part in a *Natural History* of *Animals* ".

The sensory pit situated between the nose and the eye which is lodged in the cavity of the maxillary bone, and is the distinguishing feature of the Crotalinae, was discovered by Tyson. It was first examined carefully, and its sensory nature established, by Leydig in 1868, who wrongly compared it with the lateral line organs of fishes. Tyson at first, and in spite of its position, mistook it for the auditory meatus, but abandoned that interpretation when he found that it led into a blind cavity unconnected with any internal organ. Its structural features were investigated in 1900 by West, who was unable to throw any light on its function, and who erroneously attributed its discovery to Russell and Home in 1804. The absence of any external signs of an ear puzzled the earlier anatomists, who were disposed to conclude that this sense organ was absent in the snakes, and that audition had been taken over by the nostrils " as in Fishes ".

Locomotion in snakes is a process difficult to analyse, and it is remarkable that Tyson's solution should have been so close to the truth. He visualized how a combination of muscles, ribs and scales could produce a gliding motion, and how the elaborate articulation of the vertebrae contributed to this end by ensuring an extreme flexibility of the body. His description of the peculiarities of the respiratory system is also admirable. He found that the trachea, as soon as it entered the body cavity, developed air cells on its wall, and so assumed the structure of a lung, whilst the definitive lung extended far back behind the heart as a bladder-like organ, its walls gradually becoming smooth internally, with the result that the most posterior region is destitute of air cells. He points out correctly that such a pulmonary mechanism, which is unpaired in the rattlesnake, combines at once the smooth type of lung of the newt with the vesiculate type of the frog, and that the bladder-like portion of it is not respiratory, but acts as a reservoir of air. The latter point is indicated by the nature of its blood supply, of which, however, Tyson was ignorant.

Tyson's descriptions of the viscera are sufficiently detailed and accurate. He quotes Hodierna in Severino (1651), who thinks that the long protrusible bifid tongue is not used for catching flies and other small prey, but " rather for picking the Dirt out of their *Noses*, which would be apt else to stuff them, since they are always grovelling on the Ground, or in Caverns of the Earth ". The gut of the rattlesnake, says Tyson, is simple and almost straight, and he notes its adaptation to the form and feeding habits of the animal. He compares it with the unconvoluted intestine without specialized regions of other animals, and with the large sacculated crop of the leech, of which he gives a full description. According to Charas the heart of the viper has two ventricles. This Tyson denies, but is willing to concede two auricles — one at the entrance of the " vena cava " and the other for the " arteria aorta ".[1] He does not mention the sinus venosus, and has not succeeded in tracing the relations of the three arterial trunks to the ventricle.

The gall-bladder and its connections with the liver and gut are successfully treated by Tyson, and he found a gland which from its colour was regarded as the spleen, but which others had described as the pancreas. The colour of this organ is not, as he thinks, a sure guide to its identification in the snakes, and Tyson's gland

---

[1] A curious mistake. It is not a slip, as he proceeds to define the vessel by stating that it has one descending and two ascending branches.

was probably the pancreas, or possibly the opposed pancreas and spleen. He gives an admirable account and figure of the male urogenital organs, and says that the right kidney is longer than the left, but in the figure no difference is shown, the figure also confusing the relations of the ureter, renal portal vein and vas deferens — perhaps the error of the artist. The kidney is represented as having a segmental structure, with a bundle of " secretory vessels " in each section. Tyson considers this to be the fundamental structure of all kidneys ; and even the apparently single kidney of man, he says, consists of as many components as there are " corpora papillaria " [pyramids]. He describes and figures the scent glands of snakes, which open into the cloaca and produce a strongly odoriferous secretion. Vesling, who discovered these organs in 1644,[1] thought that they were the seminal vesicles, and Charas names them the epididymides, but, as Tyson points out, such interpretations are negatived by the fact that the glands are present in both sexes.

Tyson found that the testes of the rattlesnake were unequally developed on the two sides, the right being longer than the left. In the female viper there is the same disproportion in the sizes of the two ovaries. This vas deferens, although coursing in a straight line, is actually delicately convoluted, and can be unravelled to twice its apparent length. This he concludes must constitute a kind of epididymis. In the description that now follows Tyson is dealing with the viper, as the corresponding parts of the rattlesnake had dried before he was able to examine them. There are no seminal vesicles or accessory glands. The penis is paired, and each member of the pair bifurcates, thus producing four small penes, the vasa deferentia dividing so as to provide a seminal duct for each of the four organs. The penis of each side can be introverted into a pouch " as a Finger of a Glove may be by a Thread fastened to the End ", and is armed with very sharp prickles. The penes are withdrawn by a pair of retractores penum, which were mistaken for the penes by Abbatius, Aldrovandus (1640) and Charas.

The bone to which the large poison fang of the rattlesnake is attached is not recognized as the maxilla by Tyson, but " may be thought to be the earbone ", although what the ear-bone would be

[1] Published 1651.

doing in that position is not explained. The poison fang, he says, is tubular, its cavity being connected with the poison gland by a foramen at the base of the tooth, so that the poison can be induced by pressure to emerge from the slit-like opening near the apex of the tooth. Tyson gives a good description of the teeth, including an examination of the mechanism of their erection, but he did not investigate the poison glands themselves. The skull is considered only from the point of view of its adaptation to a considerable exaggeration of the gape when ingesting the large prey on which these reptiles feed.

Tyson's enquiries into the structure of certain parasitic worms are all the more important because they are among the earliest attempts in modern times to study the anatomy of invertebrates. The fact that these worms do not occur outside the body in which they are parasitic adds to the difficulty of explaining how they reproduce and how they pass from one host to another. In the roundworm (*Ascaris*) Tyson found that there is a perfect discrimination of sexes, but that in the earthworm, leech and snail the hermaphrodite condition prevails. He recognized that the narrow end of the tapeworm was not the tail but the head, and that his predecessors were " miserably mistaken " in holding the opposite view, and in looking for the head at the wrong end of the body. It says much for the impetuous but barren industry of scholasticism that Spigelius, a well-known and reputable anatomist, should have produced in 1618 at the mature age of forty an entire memoir of some ninety pages on the tapeworm without succeeding in discovering its head. Tyson, who based his knowledge of the parasite on observation rather than on books, was inevitably more fortunate. He found with others that, in some individuals, apertures [the genital pores] were situated " about the Middle of the Joints, on the *Edges* ; in others, about the Middle of the *Flat* of the Worm, near the Jointings ". He therefore distinguishes in this respect between the Taeniidae and the Bothriocephalidae.

The Cestode scolex was discovered by Tyson in a living worm which adhered to the gut of its host so firmly that it was not easily dislodged. The head end and scolex were highly contractile. After preserving the head in " spirit of wine ", he examined it under " some extraordinary good microscopes ", and found the scolex and its two series of hooks. There was nothing comparable with a mouth, the sole function of the scolex being to provide an effective attachment to the gut. No appearances of segmentation could be detected at the head end, and the jointing of the body appeared gradually behind the neck, the segments being at first narrow and increasing in width " as they descended towards the *Tail* ". Tulp and Spigelius (the latter at second hand) both print observations which may be translated as ambiguous references to the Cestode scolex, but in view of the clear and unequivocal description and figures of Tyson no claim to priority can be established on their behalf. Not having found a mouth in the scolex, and not suspecting the possibility of absorption through the body wall, Tyson was tempted to suggest that the genital openings were mouths, and that therefore the tapeworm had a mouth in every joint. He suspects that this statement will be received with scepticism, and his attempts at justification serve rather to encourage that scepticism.[1] His chief reason is that if these orifices are not mouths "I know not where in the *whole* Body to find them besides ". In his other memoir on *Ascaris* published at the same time Tyson had discovered and named the male and female genital pores, and it is therefore all the more astonishing that it did not inspire him to associate the openings in the tapeworm with reproduction. He omitted to study the internal structure of the tapeworm, but assumed that in addition to the " mouth " in each segment for receiving the food there were " no Doubt answerable Organs for the Digestion and Distribution of it ". In *Ascaris* he studied the contents of the uterus microscopically and discovered the eggs, and had he conducted a similar examination of the liquid which he expressed from the " mouths " of the tapeworm he could hardly have failed to correct his error, provided always that he was able to rid his mind of the belief that an animal can feed *only* through a mouth and gut.

Tyson's worm was unquestionably *Taenia pisiformis* (*serrata*) of the dog. There is no reason to believe that he ever saw the other common tapeworm of the dog — *Dipylidium caninum*. He gives figures of the scolex showing the two series of hooks, but *how* did he contrive to overlook the suckers ?

In the memoir on *Ascaris lumbricoides*, which worm he says is

---

[1] In 1688 (published 1691) Tyson established that the hydatids of medicine were " a Species of Worms, or imperfect Animals ", but Hartmann in 1685 had already reached a somewhat similar conclusion.

very common in children, Tyson points out that " nothing can be plainer " than the distinction of the sexes in this species, the male being shorter than the female. He found the mouth with its three lips, and the anus, and noted that there were no septa as in earthworms. He describes the pharynx and the straight gut, observing

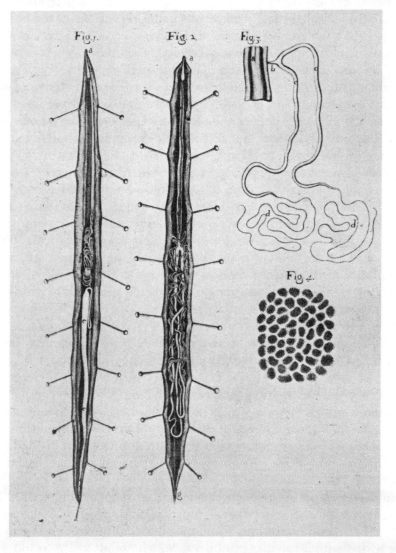

FIG. 88.—Tyson, 1683. *Ascaris.* 1, anatomy of male ; 2 and 3, anatomy of female ; 4, eggs examined microscopically

that the gut was not differentiated into stomach and intestines, and the description of the genitalia is particularly sound both as regards observation and interpretation. The recognition of the two sexes is based on anatomical criteria, and all the parts of the male and female are correctly identified and figured, the difference in position between the respective genital apertures being duly observed. The structure he called the " penis " was probably the sac of the penial setae, but after a microscopical examination of the contents of the uterus he correctly identified and figured the characteristic eggs of the species "covered with Abundance of small Asperities". Tyson emphatically disputes and explains away the opinion of other authors that *Ascaris* is viviparous, holding rightly that it is oviparous.

In the following year of 1684 Redi also published an account of the anatomy of *Ascaris*, but it does not compare favourably with that of his predecessor. The figures of the entire animal and of the dissected-out gut are inaccurate, and the two ovaries are represented as fusing proximally to form a single tube for the two sides. He found only four males, but failed to recognize them as such, and even seems to regard the male as a distinct species, arguing that the genitalia of the two sexes must be *superficially* identical. The fact that all the individuals dissected could be grouped into two structural types did not suggest to *his* mind that he was dealing with two sexes of the same species. Tyson's approach to precisely the same problem was not only more circumspect, but revealed a more instructed outlook.

The anatomy of the peccary, *Dicotyles*, had scarcely been broached before Tyson took it up, and hence his account is only slightly anticipated by some fragmentary observations. He gives a very good description and figures of the singular compound stomach of this species, which he describes as " three stomachs " — a statement which brought down upon him the mild censure of Daubenton. There is in fact but one stomach subdivided by partitions into three chambers — an expanded cardiac chamber on the left terminating in two large horns or culs-de-sac, a middle chamber into which the oesophagus opens, lined with an extension of the oesophageal mucosa, and a pyloric or right chamber communicating with the duodenum by an opening provided with a valvular protuberance, which latter Tyson did not see. The middle compartment is connected with the two others by conspicuous foramina, and the whole resembles the stomach of the true pig in an exaggerated form but approximating to the ruminant type. In the gut a short caecum is present, but there is no gall-bladder, in which respect the peccary differs from the pig.

Tyson, and also Daubenton, did not notice any inequality in the size of the testes, nor does the former remember whether there was an epididymis, which is, however, present and well developed. Tyson's figure of the male genitalia is good, but he does not appear to have found the small prostate, and is inclined to regard the large Cowper's glands as prostatic, in which he is followed by Daubenton. He compares the glands associated with the genital tract of Dicotyles with those of the boar and hedgehog.

The most remarkable and inscrutable anatomical feature peculiar to the peccary discovered by Tyson was the occurrence of one small and two large thick-walled " aneurisms " along the course of the dorsal aorta. This, he says, is what " surprised us most, and made us soon neglect the other Parts ". Tyson doubted, as well he might, whether these extraordinary structures were not pathological, but Daubenton, in his account of the other species, *D. tajacu*, found a similar but single enlargement of the aorta more

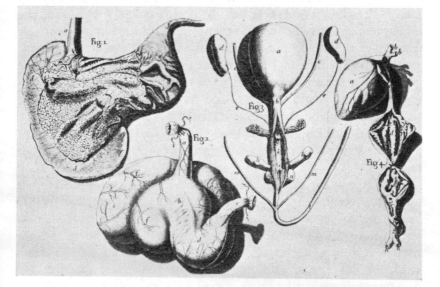

FIG. 89.—Tyson, 1683. Peccary, *Dicotyles*. Compound stomach, urogenital system, heart and pouched aorta

than 5½ inches long and 6⅓ inches in circumference which had substantial walls almost too tough to be cut with scissors. He was convinced that this growth, whatever it might be, was not an abnormal feature, but was present in all individuals.

The large and characteristic scent gland on the back is well described and figured by Tyson, who had already closely studied the scent glands in a number of invertebrates and vertebrates. Its aperture resembles, and corresponds in vertical position with, a dorsal second navel, and hence Cuvier's name of Dicotyles (1817) for the genus. Tyson exclaims that the gland constitutes the " greatest Wonder " of the species, " and differences it from any other Animal I know of in the World ". He shows that it is a dermal gland situated between the skin and the panniculus carnosus muscle, and hence comes away with the integument. Most of Tyson's predecessors identify the opening of the gland as a navel, situated irregularly on the back in this species, and one of them goes so far as to endow it with a long navel string, and to comment at some length on so extraordinary and baffling a phenomenon. The navel string, however, turned out to be nothing more genital than a reference line of an earlier figure introduced to direct the attention of the reader to the exact spot occupied by the " navel ". This perversion having been exposed by Tyson, the orifice was conceived to be the teat of a mamma, but when it was learnt that the young pigs " did suck the *Teats* under the *Belly* " another fiction had to be abandoned. The suggestion that the gland was a breathing mechanism was more than even the seventeenth century could swallow, and when F. Hernandez decided that it *was* a gland, he curiously failed to discover that it was a *scent* gland.[1] This was established by Tyson himself, who compared the structure with the scent glands of reptiles, birds and other mammals.

Tyson only briefly describes the skeleton of the peccary, but his figure compares very favourably with that published eighty years later by Daubenton. It is in fact more correct than his own text, where the fore-foot has three digits assigned to it and the hind-foot four — reversing the actual state of affairs. Tyson's work on this species did not cover the muscular, vascular and nervous systems, or the sense organs.

[1] Hernandez died in 1587 ; but this observation was not published until 1649.

We owe to Tyson the first account of the anatomy of a marsupial, and an excellent achievement it is, especially when it is remembered that a single female specimen was the only material at his disposal. He gives details of the pouch, its " ossa marsupialia " and their muscles, and how the latter operate to open and close the aperture. Incidentally he mentions that he had found the abdominal pores of fishes in a "dogfish", and discusses their significance. He did not discover these pores, which were recorded and figured in *Raia* in 1554 by Rondelet, who named them the " foramina vulvae ". Their connection with the abdominal cavity was known to P. Dent in 1675, this observation being published as his by Ray. Dent suggested to Ray, who accepted the suggestion, that they controlled the fluid content of the abdominal cavity, and in this way enabled the fish to vary its vertical position in the water.

The descriptions of the heart, gut, renal organs and skeleton of the marsupial call for little comment, beyond that he mentions the occurrence of chevron bones in the caudal vertebrae, the object of which, he rightly suspects, is to protect the caudal blood vessels in an animal which uses its tail as a prehensile organ, and requires, therefore, additional attachments for the extra tail musculature. Apart from the fact that Tyson refers to the inflected angle of the mandible, he fails to draw attention to the characteristic features of the marsupial skull, and his account of the teeth is inaccurate in attributing the same number of incisors to both jaws.

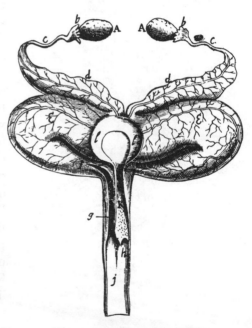

Fig. 90.—Tyson, 1698. Opossum. Genital system of female. *d, d,* " cornua uteri " ; *e, e,* the two folded "uteri"; *f,* slit in left uterus to show its passage into the left (opened) " vagina ", *g* ; *h,* opening of right " vagina " into the " common passage ", *j* [urogenital sinus], formed by the two vaginae and " urethra ", *K*

The greatest success achieved by Tyson in his dissection of the opossum is his admirable description of the puzzling female genitalia. He saw the ovaries, ostia and Fallopian tubes, and the highly vascular uteri, which latter he interpreted as parts only of the uteri, and named them the " cornua uteri ". The " cornua ", he says, open into the "uteri" — by what would now be termed a *double* os uteri. He thought at first that the " uteri " consisted of a single complete chamber brought about by the fusion of the " cornua uteri ", but afterwards discovered the median vertical vaginal septum of the modern anatomist, which ensures "that what is contained in the uterus on the right side, cannot pass into the uterus of the left side, on account of this partition ". The "uteri" become continuous behind with the paired " vaginae ", which unite with the urethra to form the urogenital sinus. Tyson's uteri + vaginae are the median vaginal culs-de-sac + the curved, lengthy, lateral vaginal canals of current terminology, but he correctly deduces that the presence of a double

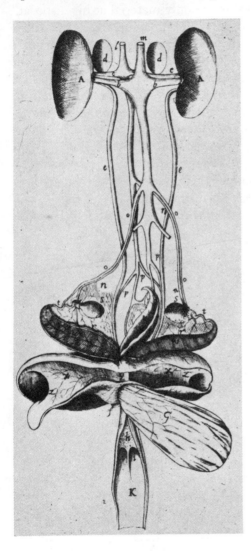

Fig. 91.—Tyson, 1698. Opossum. Urogenital system of female. The left " cornu uteri " and the " uteri " have been opened up ; *h, i, i,* junction of " urethra " with right and left " vaginae " to form the " common passage " *K*; *y,* the " diaphragm " which divides the two " uteri "

vagina presupposes a bifurcated penis. This acute inference makes it all the more surprising that when the opportunity of examining a male opossum came to him he left the publication of his own observation to Cowper. Further, Tyson did not note the interesting and significant fact that the clitoris is also bifurcated. This was established by Buffon and Daubenton in 1763, the latter assuming that the penis would exhibit a similar conformation, as Tyson and Cowper had in fact already discovered. Tyson points out that the urogenital sinus opens to the exterior so close to the anus as to justify the statement that " there was only one foramen for the exit of the faeces, the urine and the foetus ", but he does not mention the common sphincter of this foramen.

Some time later Tyson obtained a single male opossum, and another was dissected simultaneously by Cowper. Neither anatomist found a marsupium or its muscles in this sex, but the marsupial bones were present as in the female. Tyson exposed the brain and saw that the hemispheres were unconvoluted, but his assertion that the cerebellum is not furrowed is an under-statement of the facts. He noted the presence of the seventh or choanoid eye muscle. Tyson, however, does not pretend to have examined the male with much attention, surrendering his interests in the anatomy of the animal to Cowper, whose paper immediately follows his own in the *Philosophical Transactions* of 1704. Cowper succeeded in demonstrating in the opossum the urethral glands, since named after him, which he had described in man and other mammals in 1699, but not, as he believed, discovered. He was also able to establish the soundness of Tyson's prediction that the penis must be bifid, finding that the glans was deeply cleft and had two apertures, so that the seminal stream was divided and a portion directed into each vagina.

Tyson's monograph on the anatomy of a young chimpanzee, which he identifies with the orang-outang,[1] is his most considerable and imposing work, but it should be added that Cowper was responsible for all the figures and for the chapter on the muscles. It is also interesting to recall that the skeleton which Tyson prepared and described in this treatise has happily survived, and

---

[1] The name orang-outang was first used in a scientific work by Tulp in 1641, who, however, like Tyson, applied it to the chimpanzee. The latter therefore is, on priority, the true Orang!

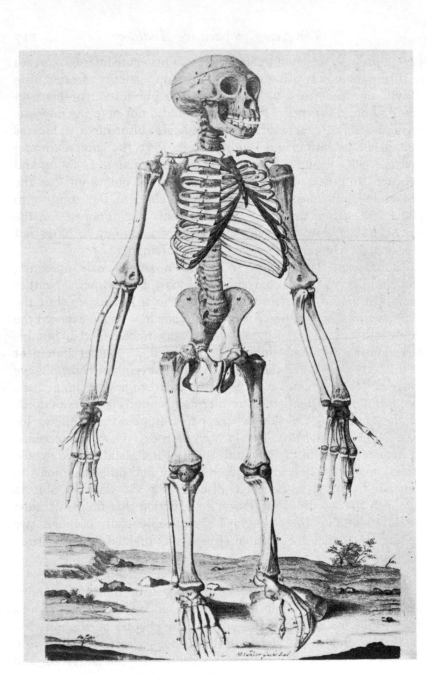

FIG. 92.—Tyson, 1699. Skeleton of young chimpanzee

the history of this unique historical relic is known. One of Tyson's relatives married a Cheltenham physician, and she brought as part of her dowry the skeleton of the "Pygmie". Her husband, viewing the intrusion of this osteological curiosity into his matrimonial affairs with some dubiety, made it over to the local hospital, the governors of which in 1894 presented it to the British Museum, where it may still be seen. It is surprising to learn that Tyson felt called upon to prove that his specimen, which he had examined in the living state, was neither a pygmy variety of the human species nor one of the common monkeys, but something between the two. That such a demonstration was necessary testifies to the state of knowledge at the time, and incidentally to the relevant and striking contribution which Tyson himself made to it. His detailed and skilful dissections of the osteological, muscular, alimentary, urogenital, respiratory and cerebral systems established that the animal was not a degraded or undeveloped human type, or one of the monkeys, or a hybrid, but a new and valid species. It is difficult for us to grasp the great significance of so obvious a conclusion, but it must be remembered that in Tyson's time the anatomy of the higher anthropoids was entirely unknown, and even their external form was familiar only to a few travellers and others. Hence material for investigation was almost beyond the reach of anatomists. The gorilla was not even discovered until 1847, and although the true orang appears for the first time in zoological literature in 1658, it was not until 1778 and 1782 that Pieter Camper put this species on the same anatomical footing as Tyson's pygmy. The chimpanzee emerges in 1625, and was examined but not dissected in 1641 by Tulp, who, in spite of his friend Rembrandt's celebrated " Anatomy Lesson ", was neither an anatomist nor a professor, and was not competent to determine the relationships of the animal. It was therefore Tyson who initiated the anatomical study of the man-like apes, and who was virtually responsible for instituting a new family of hominoid apes intermediate between man and monkeys. " What I shall most of all aim at ", he says, " will be to give as particular an Account as I can, of the formation and structure of all the Parts of this wonderful *Animal* ; and to make a *Comparative* Survey of them, with the same Parts in a *Humane Body*, as likewise in the *Ape* and *Monkey*-kind. For tho' I own it to be of the *Ape* kind, yet, as we shall observe, in the

Organization of abundance of its Parts, it more approaches to the Structure of the same in *Men* : But where it differs from a *Man*, there it resembles plainly the Common *Ape*, more than any other *Animal.*" " Our *Pygmie* is no *Man*, nor yet the *Common Ape* ; but a sort of *Animal* between both."

We must not expect to find in this work important or even novel contributions to anatomical knowledge. Its subject was too nearly related to man, whose structure had already been closely studied, to admit of further striking discoveries until new methods of attack had been developed, for which the time was not then ripe. The problem that *was* open to Tyson was a systematic one — the status of the higher Anthropoids, and it was a problem which had not even been considered. His handling of the evidence in this issue was to the point. He decides first that his ape is a true species, and then proceeds to discuss its lawful place in the animal series. His method is to compile a list of the more significant anatomical features, and from this list to assemble groups of characters emphasizing the possible and preponderating affinities of the animal. The modern zoologist follows much the same plan. It should be noted that Tyson's final alternative lay between the hominids on the one hand and the lower anthropoids or "monkeys" on the other. Was the pygmy to be placed in either of these groups, or must a new and intermediate family be constituted for its reception ? Some of his key characters may now be considered. The panniculus carnosus muscle, present in the monkeys, is represented in the pygmy and man only by remnants and the panniculus adiposus. There is some failure on Tyson's part to distinguish between the caecum and the appendix. This, however, is a matter of definition, and he clearly recognizes that the appendix is present in the pygmy and hominids, but is absent in the monkeys. In the latter the bile and pancreatic ducts open separately into the duodenum, whilst in the former they unite, and there is but a single aperture. In the pygmy the liver " was not divided into *Lobes* as it is in *Brutes* ; but intire as it is in *Man* ". Man and the pygmy have the left lung divided into two lobes and the right lung into three, whereas in the monkeys the number is greater — three on each side, and in addition there is a small mediastinal lobe. The monkeys have cheek pouches, but not the pygmy or man. When the brain of the pygmy was compared with the human brain

" with the greatest exactness, observing each Part in both ; it was very surprising to me to find so great a resemblance of the one to the other, that nothing could be more. So that when I am describing the *Brain* of our *Pygmie*, you may justly suspect I am describing that of a Man." [1] This statement is supported by a good figure of the base of the brain, in which all the parts, and also the roots of the cranial nerves, are correctly identified, except that his auditory nerve includes the facial, and his vagus the glosso-pharyngeal, his " ninth " nerve being the hypoglossal. These latter interpretations, however, were those current at the period, and were not Tyson's, who himself adopted Willis' classification of the cranial nerves (1664), which survived until Soemmerring initiated the modern system in 1778. Marshall, writing in 1861, says that Tyson's figures of the brain are " useless for modern science " — a criticism which could now be advanced against his own figures, as he should have anticipated before condemning a work already over 160 years old. As it happens, Tyson's figure of the base of the brain is more informative than Marshall's. Tyson's examination of the cranial nerves and of the internal parts of the brain confirmed his belief that the brain of the pygmy was unlike that of the monkey, but was " exactly as in a *Humane Body* ". In all the above features, selected from a list of forty-eight, Tyson found that the " *Pygmie* more resembles a *Man* than *Apes* and *Monkeys* do ; but where it differs, there 'tis like the *Ape-kind* ". The characters in which the " Pygmie differ'd from a *Man*, and resembled more the *Ape* and *Monkey-kind* " are hardly less numerous, but are of less weight. They include such considerations as that the great toe is set at a distance from the others like the thumb, that there is no pendulous scrotum, [2] that the os ilium is longer, narrower and less concave than in man, and so on. Tyson failed to examine the heart and vascular system of the pygmy, and there are other omissions, but the work was based on the dissection of a single immature specimen, and in the absence of more efficient methods of preservation could hardly have been more complete.

[1] It seems certain that the brain figured by Cowper was taken from Tyson's chimpanzee. Mr. M. A. Hinton, who specially examined the skeleton presented to the British Museum, reports that the brain *had* been removed before the bones were articulated, as indeed the figures of the skull in three of the published plates plainly indicate. The reason for this note is the suspicion that Cowper's drawing of the brain owes something to conjecture.

[2] It must not be forgotten that Tyson's specimen was immature.

XX
WILLIS

One of the dominating figures in the medical world of the seventeenth century was Thomas Willis, whose writings and personality inspired a " new sect of philosophers " called the Willisians. His name has also been attached to two cranial nerves,[1] and a vascular condition associated with the base of the brain was misnamed the circle of Willis.[2] He seems further to have been responsible for the introduction of the term neurology, which he defined as the doctrine of the nerves.

The works by him which include observations on comparative anatomy are the Anatomy of the Brain (1664) and the Soul of Brutes (1672). It seems certain that the observational basis of these researches was not supplied by Willis himself, and that his own part lay in providing the ideas and plans which animated the labours of his more objective collaborators, and in discussing the results with them and other friends before drafting the letterpress. He cannot be said to have achieved his ambition to make himself clear to the meanest capacity. His language is often involved, in spite of the elegant Latin in which it is cast, and it is not made more comprehensible by a tendency to practise the subtleties of speculative disputation born of the Schoolmen, which is what Willis calls " philosophising ". Neither has he the gift of clear and intelligible expression, but if we must admit the impossibility of discriminating the parts taken by the various members of the company in formulating the text, we may suspect that this task was mainly, if not entirely, the prerogative of their pontifical leader.

In the dedication and preface of his work on the brain Willis speaks of having slain in his anatomical sanctuary hecatombs of all kinds of animals. In these operations he had invoked the assistance of Richard Lower to make dissections and drawings, and Thomas Millington acted as adviser and critic. Christopher Wren was also called in as draughtsman, and most of the published figures are acknowledged to have been by him. The types examined, apart from *Homo*, are the calf, horse, dog, pig and sheep, the last named

---

[1] The nervus ophthalmicus Willisii (first division of the trigeminius) and the nervus accessorius Willisii (spinal accessory nerve).

[2] First figured in a posthumous work by Casserius in 1627.

being the focal point of their investigations. In 1672 Willis was still concentrating on the cerebral anatomy of this animal. His work on the brain, however, important as it is, has little *comparative* interest. He fails to homologize completely the parts of the bird's brain, but aptly contrasts the position of the lateral ventricles in bird and mammal. He apparently saw the foramen of Monro in the bird's brain, and institutes an astute comparison of the nates of the mammal with the optic lobes of the bird. In the fish (and by fish he means a bony fish) he found that the brain was small compared with the size of the head, that the optic nerves crossed without mingling, and that the auditory nerves were wanting.[1] Having discovered a pineal body in the fish brain, he concluded that this organ must be present in all vertebrates.

The treatise on the Soul of Brutes is more to our point, and we may be all the more grateful for the comparative anatomy included in this work since it is entirely surplus and irrelevant to the philosophical and medical issues which are its main concern. Here also experts are commissioned to make the dissections, and Edmund King and J. Master were selected for this important task. The types examined were the oyster, lobster and earthworm, and Willis therefore was one of the few anatomists of the period who displayed an interest in the structure of invertebrates, which, following Aristotle, he regarded as " bloodless creatures ", on account of their lack of *red* blood. If there are many points to criticize, as is inevitable in early researches, it must be admitted that important facts were brought to light, and that in this work Willis made a solid contribution to the armamentarium of the comparative anatomist.

The description of the anatomy of the oyster, of necessity incomplete, is yet admirable as far as it goes. He examined the single adductor muscle, and understood its action in closing the valves of the shell. He was the first to distinguish two parts in this muscle, and although it is certainly true that these two components have not the same physiological value, Willis is wrong in concluding that the anterior member *opens* the shell. Both in fact are adductor in function, the difference between them having relation to the speed and manner of the closure, and not to the final result. Willis looks upon the mantle as a pair of " circular

[1] A surprising error, but one not confined to Willis.

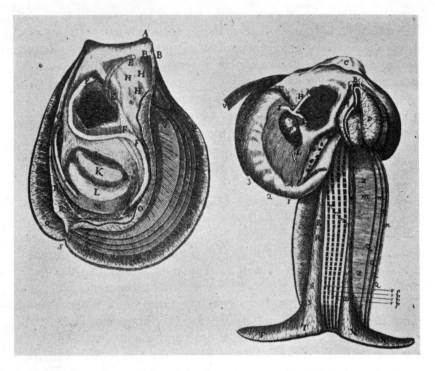

FIG. 93.—Willis, 1672.   Anatomy of oyster

muscles ", and the labial palps, which he noted enclose the mouth, are his " superior gills ".   The gut is well described, although he unfortunately missed the crystalline style caecum and the style itself.   He found, however, the mouth, oesophagus and stomach, into which opened " by little foramina " the digestive gland or " representative of the liver ", which, he thinks, acts as an absorbent body.   From the stomach he traced the winding course of the gut up to the anus.   He saw the typhlosole, which is compared legitimately with the similar structure in the earthworm, but which does *not* " seem like a spinal cord ".   The position of the pericardium and heart, which in the oyster are situated well below the rectum, is correctly described and figured, and the pulsation of the heart was observed.   This organ, he says, consists of a single auricle and ventricle — a statement which is only technically inaccurate.   The double nature of the auricle in this species has been questioned, and in any event it would not be reasonable to expect a pioneer, unfamiliar with the comparative anatomy of the Lamellibranch heart, which normally has two auricles, to recognize in the auricle of the oyster indications of its descent from a paired ancestor. Willis describes a " vena cava " or efferent branchial vein entering the auricle, and three arteries arising from the ventricle, which are apparently the anterior aorta and the rectal and reno-genital arteries. He also traced the arterial supply of the head, gut, adductor muscle and the gills, and for the time gives a competent description of the structure of the latter organs. They are, he says, permeated everywhere by water, and extract " nitrous particles " from it. Having no knowledge of ciliary motion, which was first seen in the gills of a bivalve by de Heide in 1680,[1] he is compelled to fall back upon muscular action when attempting to explain the production of the inhalent and exhalent respiratory currents which he found entering and leaving the oyster at its posterior region. Willis was aware of the open as well as the closed portions of the vascular system, although without recognizing that they were both parts of one and the same circulation. The nervous system is not mentioned.

The dissection of the lobster has also been conducted with much skill, and is illustrated by excellent figures. This animal, says Willis, moves backwards, and, as if to be in accord, all its parts exhibit the same topsy-turvy relations. Thus its " bones are not covered with flesh, but the flesh with bones ", the heart is dorsal, the spinal cord ventral and so on. The brain and nerve chain are described, but, although he saw the circum-oesophageal connectives passing into the ventral cord, he says nothing of their relation to the oesophagus. Also he attempts to trace in the cerebral ganglia and adjacent tissues the parts of the vertebrate brain, distinguishing two hemispheres, a medulla oblongata [the cerebral ganglia], and a cerebellum [? the bladder of the green gland]. Even the optic nerves are mysteriously wrong. The description of the gut is sound, and includes the oesophagus, the gastric mill and its three teeth, and the digestive gland, "commonly called the liver ", with its paired opening into the gut, from which the gland can be inflated. The genitalia are satisfactorily treated, except that the position of the external opening in the male is inaccurately given, although the figure would suggest the correct

[1] Published 1683.

appendage.   In the female he found the ovary [" uterus "], and the short oviduct [" neck of the uterus "] ;   and in the male the testis and the three parts of the vas deferens, of which the terminal section is misinterpreted as a penis.   The motion of the muscular heart was studied in the living state, and he noted that it was moored to the wall of the pericardial sinus by the alae cordis, but

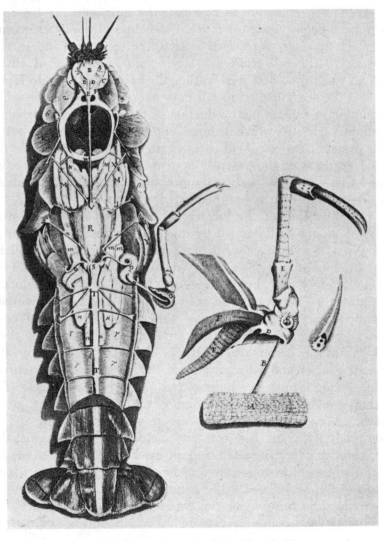

FIG. 94.—Willis, 1672.   Anatomy of female lobster

was under the mistaken impression that the heart was enclosed in a definite pericardial sac as in a vertebrate. He mentions the colourless blood, but missed the ostia of the heart, and did not understand the nature of the circulation. The anterior median ophthalmic and antennary arteries were seen, although their distribution is incorrectly given. Similarly he describes the dorsal abdominal and sternal arteries, but regarded them as veins or " venae cavae " transmitting blood *to* the " auricle " of the heart, which blood then left it by the three anterior arteries. There is, however, no auricle in the heart of the lobster. Willis attempted a detailed account of the gills, finding the afferent and efferent blood channels, and also a third vessel, the existence of which has not been confirmed. He is again unsound in stating that if a black liquor be injected into the *heart* it passes to the gills and traverses both afferent and efferent branchial channels, which, indeed, it might do if the injection were introduced into the pericardial *sinus*. Willis saw the beating of the scaphognathite, and considered that this appendage represented the most anterior gill. He also examined the ventral blood sinus, and found that by injecting or inflating it the black liquor or air reached the gills, limbs and body spaces. His ideas on respiration in the lobster do not readily commend themselves to the understanding, or call for exposition.

The earthworm receives the same careful treatment as the other invertebrate types, and many novel and interesting observations are recorded. Willis knew and figured the four series of setae or " little feet " which extend throughout the entire length of the worm, and he understood how they were of assistance in locomotion. The prostomium or " proboscis ", small as it is, did not escape him. The worm was dissected from the ventral surface, which exposed the nerve cord and a part of the circum-oesophageal connectives, both of which are figured, although there is no mention of the latter in the text. The brain would not appear in this view, Willis' "brain" being the pharynx, and the nerve cord is similarly mistaken for an artery. He saw the hearts, or at least some of them, and records their pulsations, and must therefore have noticed the red colour of the blood. Seeing that Willis accepted the Aristotelian pronouncement that the earthworm was a " bloodless " animal, *i.e.* without red blood, his misinterpretation of the nerve cord was to that extent in order, but it is puzzling that he should

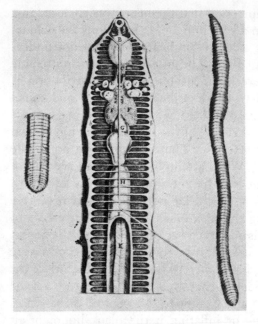

FIG. 95.—Willis, 1672. Anatomy of earthworm

have failed to comment on the fact that such undoubted blood vessels as the hearts *did* contain red blood.[1] Another equally baffling surprise is that he discovered and figured the dorsal pores, which are far from conspicuous, and yet overlooked the openings of the vasa deferentia, which are always so obvious. His comparison of the dorsal pores with the spiracles of insects, and the assumption that they transmit air to the internal parts as the tracheae do, are inferences which the knowledge of the time seemed to justify. The well-known oesophageal glands and pouches, which proved of such interest to Charles Darwin, were discovered by Willis, and he saw also the two pairs of spermathecae and the seminal vesicles, the posterior pair of the latter being identified as ovaries, in the belief that they contained " very many eggs " [? cysts of *Monocystis*]. The oesophagus, stomach [= crop + gizzard] and intestine are described and figured, and attention is drawn to the typhlosole or " liver " — " an intestine within an intestine " as he calls it, also found later by Redi in 1684. Willis noted that the body cavity was divided by " partitions of ringlike muscles " [septa] into a series of compartments — " like the colon of the higher animals ".

In addition to his dissections of invertebrates, Willis discusses respiration and circulation in a fish — cognate problems which had already been somewhat clarified by W. Needham in 1667, his material being the pike and the salmon. Fishes, says Willis, breathe by their gills, which, when they expand, withdraw " some-

[1] Swammerdam mentions the red blood of the earthworm in 1675. It is always possible that Willis was describing what he had not himself observed. Sir Thomas Browne describes red blood in worms in 1646.

OSSERVAZIONI
D I
FRANCESCO REDI
· *ACCADEMICO DELLA CRVSCA.*
INTORNO
AGLI ANIMALI VIVENTI
CHE SI TROVANO
NEGLI ANIMALI VIVENTI.

IL PIV BEL FIOR NE COGLIE

IN FIRENZE, MDCLXXXIV.
Per Piero Matini , all' infegna del Lion d' Oro .
Con licenza de' Superiori .

FIG. 96.—Redi, 1684.   Title-page of the work containing his observations on
comparative anatomy

thing nitrous " from the water.   This substance having, as far as
possible, been turned to account in the system, the gills contract
and expel the waste products.   The heart of the fish, he adds, has
one ventricle only, which, in the absence of lungs, is all that is
necessary.   The auricle and sinus venosus are not mentioned,
although the former was known to Needham.   In the sturgeon,

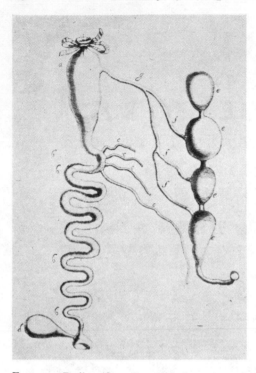

FIG. 97.—Redi, 1684. Gut, pyloric caeca and moniliform swim-bladder with pneumatic duct of ? *Pagrus auratus*

salmon and cod the cardiac aorta dispatches arteries to the gills, which break up in the substance of all parts of these organs. From such branchings another vessel is constituted by union, and this branchial " vein " courses upwards to assist in forming the dorsal aorta. A black liquor injected into the cardiac aorta emerges by the gill " veins ", but his statement that a part of the injection passes straight into the "veins" *without* circulating through the gills is erroneous. Nevertheless Willis gives us an early defensible explanation of the circulation of the blood in the gills of a fish.

The current doctrine that the presence of lungs presupposed a heart with two ventricles was supported by Willis, and he cites as examples snakes, lizards and some amphibious animals such as the frog. He raises the question of those amphibious species [aquatic mammals] which have lungs but no gills, and cannot therefore respire under water. Bartholin, he says, easily solves this difficulty by assuming that the foramen ovale of the embryo persists throughout life, with the result that the venous blood passes straight from the right auricle into the left, and so reaches the aorta without traversing the lungs. This solution was questioned by Needham, who failed to discover a persisting foramen ovale in the species examined by him, and, apart from this, the explanation cannot be applied to the lower vertebrates, where the foramen does not occur. Nor is its presence consistent with respiration *out* of the water, unless some kind of closing mechanism can be demonstrated.

Willis was familiar both with the swim-bladder of fishes, closely studied by Redi in 1684, and the air sacs of birds, and explains how air reaches the latter via the lungs. He believes that the object of these structures is to increase the buoyancy of the animal.

XXI
MARTIN LISTER

Martin Lister's original contributions to comparative anatomy are restricted in their scope, but reveal a sound, if imperfect, appreciation of the morphological method. They include some notes on the anatomy of spiders (1678), the anatomical appendix to the History of Conchology (1692), separate monographs on snails and slugs (1694), marine and freshwater snails (1695), and bivalves and cuttlefish (1696). A short paper on the anatomy of the scallop appeared in 1697. His attitude towards these researches is well represented by a quotation from Adrien Auzout which is printed on the title-page of the 1694 monograph, and may be freely translated as follows : " We must dissect not only the human body but also Birds, Fishes, Quadrupeds, Insects and Amphibia, in order that the function of the various parts may be discovered. For indeed some structures may be wanting in certain animals, or represented by analogues, or perchance variants may occur, which events are often of the greatest significance."

By general acclaim Lister is the originator of British conchology. He started as a naturalist and speciographer, but finally settled down to plan an ambitious introduction to the comparative anatomy of the Mollusca, of which he completed several parts. Admitted that so cramped a field compels the operator to work in miniature, and therefore to concentrate on craftsmanship, Lister was yet one of the first to attempt, and partly to establish, a system of comparative anatomy on an imposing scale. He justifies his interest in the smaller animals by quoting with approval the opinion of Bacon that the small and lowly often throw more light on the great than the great on the small.

In the 1694 preface Lister complains that unless one is a second Harvey, Malpighi or Redi, studies in comparative anatomy provoke only contempt and laughter, but he claims with some ostentation that fame and applause were to him matters of indifference, even if they were not beyond his reach. This work, he adds, had been the privilege and delight of his youth, and was now the solace of his old age — but here speaks the valetudinarian who was then but fifty-six years of age. He goes on to lament his defective sight, which, however, when supplemented by the microscope, still enables him to pursue and enjoy his scientific recreations. He explains also, and this is an interesting bibliographical point, why his molluscan researches are issued in conjunction with purely medical dissertations, the reason being that his medical works, though less valuable than his anatomical writings, command a larger sale. Nevertheless, he concludes, in minatory vein, unless medicine is based on a knowledge of Nature it is a thing altogether vain and worthless.

Lister was at first interested in the structure of Arthropods, and published in 1678 some observations on spiders, which include a few anatomical notes on the gut and genitalia of no special interest and importance. Moreover, in a later section of this work he gives no support to Ray's views on the organic nature of fossils.

The great folio work on conchology occupied Lister continuously for ten years, and cost him £2000 to produce.[1] The collation of this volume, which was originally issued in parts, is a bibliographical nightmare, for no two copies appear to be alike. The figures are unaccompanied by letterpress, and many of them are not even named. The descriptions of the anatomical plates, and the text belonging to them, appeared some years afterwards as expanded sections of the parent treatise. De Bure's elaborate collation of 1764, which runs to thirty-five pages of print, concludes with the remark that the first edition of this work is very difficult to find and still more difficult to certify.

In 1684 Redi described and figured reciprocal union in the slug *Limax maximus*, but owing to the engraver failing to reverse the drawing on the copperplate, the figure, when printed, showed the penes and pulmonary apertures on the left side of the animals instead of the right.[2] In 1685 Lister's engraver reproduced this

---

[1] The copperplates are now the property of the Oxford University Press. This work was derided by the wits of the age as an abuse of time and a waste of money, but the importance of conchology in the interpretation of fossil remains had not then emerged.

[2] The craftsman's reason is obvious. It is easier to engrave the figure *as drawn* than to work to a looking-glass image of it. The practice was therefore not uncommon

unreliable figure and followed the same procedure, in consequence of which the sides are again reversed, to the advantage of truth. Thus two wrongs may occasionally make a right. Ray observed the mutual fertilization of snails in 1660 in the gardens of Cambridge, as Lister and Swammerdam admit, and the latter figured it in 1667.[1] Lister says that Swammerdam's figure is " fictitious ", and that the penes of the copulating snails pass straight into the vaginae and are not intertwined as represented by Swammerdam. Lister knew that there were dioecious as well as monoecious " snails ", although he sometimes failed to distinguish between them.

The only *original* figures of Molluscan anatomy in Lister's *Historia* are to be found in the Appendix to Lib. IV, dated 1692, but one plate is inscribed " Anna Lister delineavit 1693 ",[2] and it has been stated that the last part may have appeared as late as 1697. Lister was thus preceded in this work by H. Power and Redi, whose dissections, if less numerous and representative, rival his own. J. J. Harder's figures of 1679 of the anatomy of the snail, so often reproduced, are certainly less complete and accurate. It is unfortunate that Lister often fails to define the species which appear in his anatomical plates, and although many of them can be identified with some confidence, nothing short of a tedious investigation would succeed in running down the remainder.

Lister's account of the anatomy of *Helix pomatia* is representative of his best work. He describes the pulmonary chamber with its " valved " aperture, and the gut, interpreting the salivary glands as the " omentum " and their ducts as the " suspensory tendons of the omentum ". The heart has a cleft or bifid single ventricle. Harvey, he says, is " false and assuredly contemptible " in holding that the heart of the snail is an auricle without a ventricle, and that it is a primitive or imperfect heart, whereas even in those *Vertebrates* which have a single ventricle only, the structure of the heart

with the early engravers in line, and produced many anomalies which are not often understood or even suspected by writers on the history of zoology, who are apt to put the blame on the author. As an example, Swammerdam's figure of the reciprocal union of snails as first published in 1667 is, as far as the sides are concerned, correct, but the version in the *Biblia Naturae* shows the reversed (and wrong) position.

[1] " In the *Snail* each is both Male and Female together ; which I doubt whether it is so in any other Animal " (Swammerdam, 1675).

[2] In the Bibliothèque Nationale there is a copy of the anatomical plates with the four leaves of explanation dated 1693.

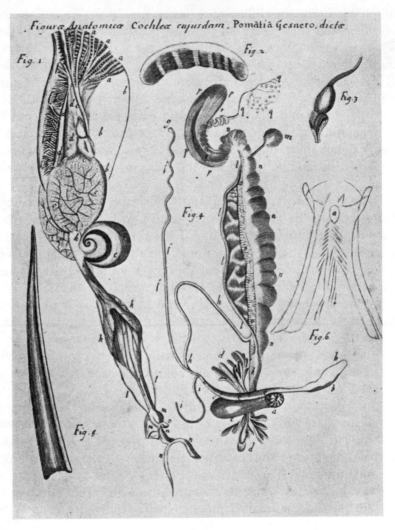

FIG. 98.—Lister, 1692.   Anatomy of snail, *Helix pomatia*

is not more perfect and elegant than it is in the snail.   In the hibernating snail Lister established the important fact that the heart exhibited a slow and feeble pulsation which was scarcely perceptible.   He contradicts Grew who had maintained that on account of its beat and constant motion the heart in all animals was red in colour, Lister having no difficulty in demonstrating that in Invertebrates generally the heart was colourless.   We note with surprise that Lister rarely mentions the auricle of the Molluscan heart, and that with him the " heart " is the ventricle only. As regards the question of definition, however, he is but conforming to the tradition of his time, although it was *not* the practice of his contemporaries entirely to ignore the existence of auricles. He found the kidney, and apparently also the ureter, and suggested that the former might be the true liver, but later abandoned this view. The genitalia were carefully examined and all parts are well shown. Even the identification is good for the time, and he did at least recognize that he was handling an *hermaphrodite* set of reproductive organs. According to Lister the ovo-testis is the ovary, the spermatheca and its duct are the testis and vas deferens, the albumen gland is a large uterine gland, the oviduct is the uterus, the vas deferens is a uterine ligament, and the mucous glands are the seminal vesicles, but the last of these suggestions is removed from his later publications. The funda jaculatoria or dart sac and the spiculum veneris are also figured, and he understands the use of the spiculum in copulation, even if he could not decide where it was formed. One of his misinterpretations might well have been avoided. The figure in question clearly and accurately shows the vas deferens being detached from the compound genital duct to connect up with the penis, whereas *his* vas deferens has *no* connection with the penis. This should have convinced him that his ligamentum uteri was the true vas deferens. Lister hardly looked at the nervous system in any of his dissections of Mollusca — in fact he at first doubted the existence of a brain in the snail, but afterwards admitted that there was a very small one. In *Limax* he saw the cerebral ganglia, to the nervous character of which, however, he hesitates to commit himself. It is remarkable that Lister should have employed the microscope in his researches and yet not have used it to assist him in identifying the ovary and the testis. Leeuwenhoek had described the spermatozoa in the *Philosophical Transactions* of 1679, and Lister could scarcely have been unaware of so astonishing a discovery, but it was not until 1695 that Leeuwenhoek mentions the spermatozoa of a mollusc. On the other hand such an examination might have given Lister some justification for his identification of the spermatheca as the testis.

Plate II is specially interesting as including a microscopic

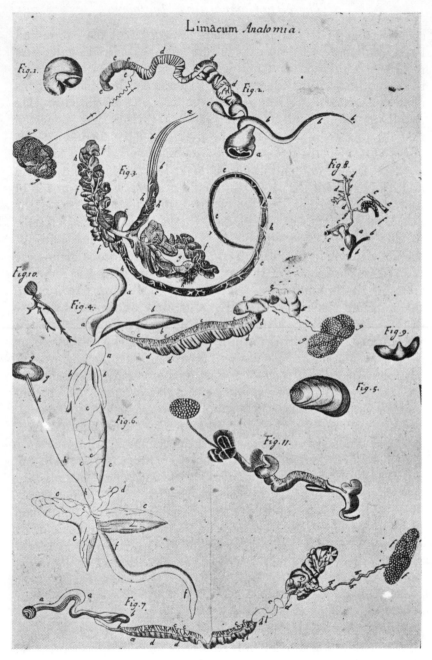

FIG. 99.—Lister, 1692.   Anatomy of slug, *Limax maximus*

figure of the spermatophore of *Helix hortensis*, and also a good diagram of the pulmonate *Xerophila itala* — a species which Lister himself discovered in 1674, and the genitalia of which are now figured without the omission of any important part. He failed, however, to discover that the bifid dart sac contained twin spicula. Of the slugs, Lister examined more particularly *Arion ater, Limax flavus* and *Limax maximus*, but confined his attention to the gut and genitalia, with the addition of a few notes on the heart. He discusses the internal shell, the salivary glands (" omentum "), enlarged oesophagus, and caecal stomach, indulging his gift of gentle irony at the expense of Jacobaeus (1676), who, says Lister, claims that there are four stomachs in the slug, " but to us it is not given to be so meticulous ". The liver and its ducts are closely examined, and he finds that if the gut is inflated the liver also swells up, which suggests that it arises as an outgrowth from the gut. He then proceeds to compare the gastropod liver with the enteric caeca of the insect and the pyloric caeca of the Teleostean fish. In his investigation of the gut he established that the rectum opened to the exterior, not posteriorly, but in the neighbourhood of the mouth. The details given of the heart and circulation include the ventricle and arteries only, and Lister wrongly concludes from experiments that the motion of the heart is voluntary, and not wholly " innate " as in man. He deduced that the mantle cavity was primarily a respiratory organ — a conclusion which explained its superabundant vascular supply. It should be noted that Redi figured the two-chambered heart of the slug before Lister, and his figure is the better of the two, but the text is not informative beyond the statement that the pulsation of the heart is visible on the external surface of the living animal. Lister's description of the genitalia is less comprehensive than that of the gut. He saw the ovo-testis (" ovary "), the short spermatheca (" testis ") and the compound genital duct, although he does not represent it as compound in all species, nor does he show its relations to the penis sheath and atrium, notwithstanding the fact that these points are readily establishable.

The Pectinibranch *Viviparus* was the subject of some of Lister's most successful dissections. He distinguishes between the sexes, figures the embryo-laden uterus, and shows that the male may be recognized by its very curious shorter, thicker and deformed right tentacle, having at its extremity the male sexual aperture. In the same sex he saw the eyes, the coiling of the gut, the heart, and the vas deferens with the enveloping prostate (" penis maris "). He found and correctly identified the testis, but does not figure it. In the female he describes and figures the ovary, heart, ctenidium, the uterus crowded with fully developed young, and the female genital opening. The kidney is also shown, but he was not aware of its nature.

If Lister's dissections of snails and slugs be now compared with those previously carried out by Redi, it will be found that the latter interprets the albumen gland of the snail as the testis. Hence the oviducal portion of the common genital duct is the vas deferens, and he has no explanation to offer of the vas deferens properly so-called. The ovo-testis is just " a gland ", and he found the spermatheca equally puzzling. Mucous glands, flagellum and retractor penis are ignored. His worst mistake was in severing the connection of the vas deferens with the penis and apparently not realizing that he had done so, nor does he suspect that he is dealing with an *hermaphrodite* set of genitalia. Redi's figure is good but not equal to Lister's, and he is also less sound in his reading of the facts. On the slug he is equally at fault. The penis sheath is represented as ending blindly, the retractor muscle and vas deferens

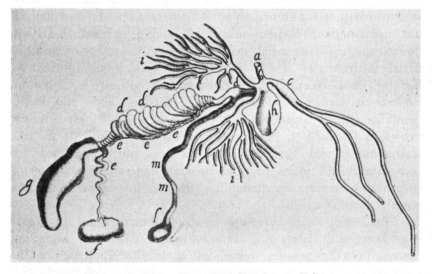

Fig. 100.—Redi, 1684.   Genitalia of snail, *Helix* sp.

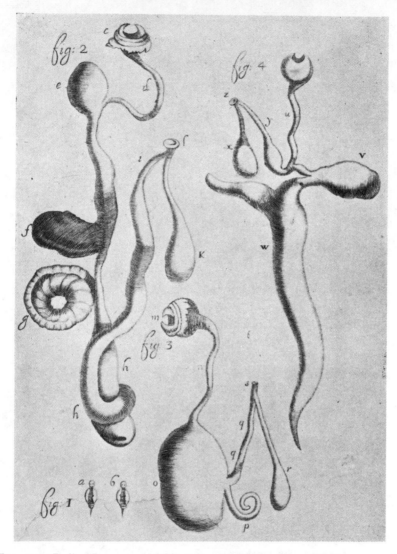

FIG. 101.—Redi, 1684. 2, gut of *Octopus* ; 3, of *Sepia* ; and 4, of *Loligo*.   1 *a* and *b* are larval cestode parasites formerly known as *Scolex polymorphus*

having been removed and not even noted, whilst the spermatheca is figured but otherwise ignored.

The observations on the anatomy of *Loligo, Octopus* and *Sepia* published by Lister are taken mostly from Redi, but he has himself dissected *Loligo*, although not very successfully.   This is partly

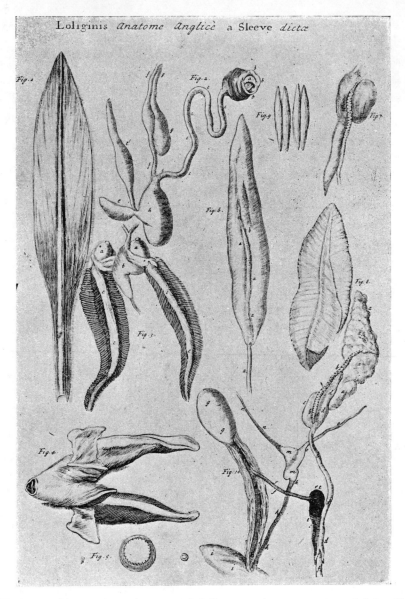

FIG. 102.—Lister, 1692. Anatomy of *Loligo*. 2, *e'*, caecum, not spiral in this species ; 3, *a*, *a*, branchial hearts and pericardial glands—below them are the auricles, *b*, ventricle ; 7, vas deferens, seminal vesicle and Needham's pocket ; 8, genital gland (? ovary) = Lister's albumen gland ; 10, *a*, testis, *b*, vas deferens, *c*, seminal vesicle, *d*, Needham's pocket with *very long* efferent duct, *e*, ink sac, *f*, rectum—the two sections of which marked *f* should be continuous and have no connection with the ink sac, *g*, stomach, *h*, oesophagus, *i*, liver, *k*, cephalic artery, *m*, ventricle, *n*, *o*, ? efferent branchial vessels — but associated auricles not shown. In this figure Lister has mistaken the sex, which is obviously a male, and hence most of his interpretations are wrong

explained by the fact that the only material available had been preserved in brine, and that he had not even seen fresh or living individuals. His figure of the gut does not accurately represent the *relations* of the parts, and, in addition to confusing oesophagus and rectum, shows *two* intestinal caeca, one of which must have been the elongated outgrowth of the caecum already known to Aristotle. This outgrowth is spirally twisted in some species, and since Lister figures it as straight his species was probably *Loligo vulgaris*. In another figure he shows the auricles and ventricle of the systemic heart, on the discovery of which he modestly congratulates himself, but his vena cava is the cephalic artery, and his branchial arteries are the *efferent* branchial vessels. He saw also the pericardial glands and branchial hearts, which, he says, on the approach of winter are so contracted as to be almost invisible. These structures are regarded as the ovaries, although he apparently saw the true ovary also, which he compares with the gastropod albumen gland. The specimen described by him as a female was undoubtedly a male, showing, as it does, the testis, vas deferens, seminal vesicle and Needham's pocket. Owing therefore to this mistake all his identifications are invalid, except those which have no relation to the genital organs, viz. ventricle of heart, stomach, liver and ink sac. He, however, properly contested Redi's belief that the Cephalopod ink sac was its gall-bladder. Lister's general neglect of the nervous system is relieved by some attempt to explore this system in *Loligo*, and he succeeded in finding a " small cerebrum " contained in a " small cartilage ", but he could discover no trace of a " spinal cord " although he carefully searched for it.

Lister's researches into the anatomy of the freshwater mussel *Anodonta* are confined mainly to the parts exposed when the mantle is reflected. He noted that the outer gill may act as a " uterus " and contain embryos, but he says nothing of the structure of these embryos, and therefore cannot be regarded as the discoverer of the glochidium larva. Leeuwenhoek also saw this larva in 1695, but again did not observe its distinctive characters, which were first described in 1797 by J. Rathke, who considered it to be a parasite on the mussel and named it *Glochidium parasiticum*. Lister established the dioecious state of the mussel, the males and females being almost equal in number, and easily distinguishable in the sexually mature animals by the diverse nature of the genital

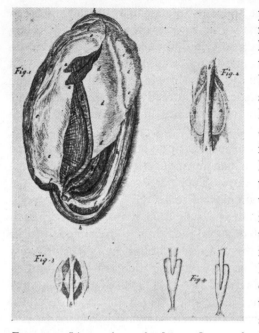

FIG. 103.—Lister, 1692. *Anodonta.* Organs of mantle cavity and heart

products. He found the mouth, coils of the gut, liver, ovary and testis — the organs which form the bulk of the visceral mass, and also the kidneys, the last named, however, eluding his attempts at identification. The exhalent aperture is called the "tracheae orificium", which is not altogether illogical, since this aperture is regarded as both inhalent and exhalent by Lister, whose experiments on the flow and strength of the respiratory current unaccountably failed to result in complete enlightenment. Nevertheless, he is not far from the truth in maintaining that the stream produced by the gills is not solely, or even mainly, respiratory, but that it plays an important part in bringing food material into the mantle cavity, and in nourishing the young before they leave the parent. This, he says, is particularly important in a stationary species like the oyster, which cannot feed unless water mixed with food is directed towards the mouth by the action of the gills. Lister figures the ventricle of the heart traversed by the rectum, and at first he looked upon it as two auricles, and the rectum as the aorta. Later he realized that there were serious objections to these conclusions, his second thoughts, however, being not much sounder than his first, although he now describes the same structure as the "*ventricle* or if you like the auricles". The anterior aorta is labelled the "vena cava". In another variety of freshwater mussel he got a better view of the heart, and gives a good figure of it, which introduces the two auricles. Having already concluded that the ventricle was the auricles, he had no choice but to miscall the auricles the ventricles. In his subsequent notes on the anatomy of the oyster he is still playing with the idea of the existence of two ventricles in bivalves. Having perforce abandoned his first idea that the rectum was the aorta, the intrusion of the rectum into the heart still continued to perplex him, unless, he conjectures, its object is to give stability to the heart. He did not therefore grasp the nature of the Lamellibranch heart, which was first correctly described in *Anodonta* by Mery in 1710. Lister sums up his work on the freshwater mussels by pointing out that the position of the principal organs in these animals is the reverse of that which obtains in vertebrates, " as Aristotle rightly demonstrated centuries ago ".

The marine bivalves next claimed Lister's attention, and again he concentrated on the organs of the mantle cavity, particularly in the cockle *Cardium*. The two siphons which are not conspicuous in *Anodonta* are strikingly evident in the cockle, where Lister figures them widely separated, and in the relaxed condition. Nevertheless, he did not suspect that they served different functions, and the only difference he recognizes is that the inhalent siphon, being the larger, will *eject* water more swiftly and copiously than the exhalent siphon. He understood, however, the relation of the anus to the latter siphon. Writing later on *Pholas* he states that the larger [inhalent] siphon *discharges* water mixed with air, producing a great eddy in the surrounding medium, and if the siphon be stimulated a discharge of great violence is produced. It will be noted that Lister makes no provision for an inhalent current, and yet his own views on the method of feeding in bivalves presume such a current to exist. The two adductor muscles, " by which the valves of the shell are opened and closed ", are described, although how they could discharge the former office is a question which neither Willis nor Lister appears to have considered. The labial palps, gills, heart with rectum, foot (appropriately called the vomer or ploughshare), and the visceral mass with its included genital organs and other viscera, were all carefully examined by Lister. He also figures the "stylus crystallinus", which was discovered and named by de Heide in 1683 in *Mytilus*. Lister found it again in *Pholas*, and looked upon it as a part of the generative system. He did not see the vestigial byssus in the cockle, but figured it in *Tellina* and *Tapes*, remarking that it looks " just like a penis ".

Lister's final contribution to the anatomy of the Mollusca is his brief paper on the scallop *Pecten*. He describes the " head " and

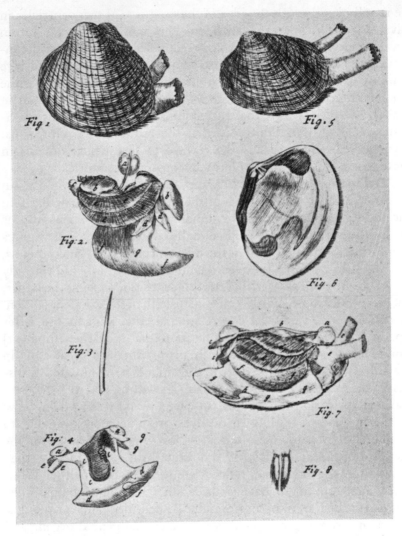

FIG. 104.—Lister, 1692. *Cardium, Tellina* and *Tapes*. Organs of mantle cavity, adductor muscles, heart, rectum, crystalline style and vestigial byssus

mouth, the digestive gland, pericardium (" perhaps the urinary vesicle "), and the rectum passing *over* the ventricle of the heart, which has no auricles. Lister is wrong here — the heart has two auricles and the rectum passes through the ventricle as in most Lamellibranchs. He correctly describes, however, the adductor muscle and its two parts, which he says are distinguished by differences of colour and texture, and the gonad or " uterus " is rightly " suspected " to be hermaphrodite, having a dorsal white [male] and a ventral yellow [female] portion. Since the foot has the appearance of being a process of the gonad, Lister regards it as an " extension of the uterus ", and in the byssal groove he professes to recognize the male and female genital apertures. He describes also the gills, the muscular mantle or " outer gill ", and the hinge mechanism with its two ligaments. Nothing is said of the pallial eyes, and he seems to have missed the labial palps, or rather to have confused them with the mantle, nor does he understand how the animal feeds, but believes in a kind of mechanical ingestion brought about by the mantle.

XXII
GREW

In 1681 Nehemiah Grew produced a work bearing the frank and unassuming title of *The Comparative Anatomy of Stomachs and Guts Begun*. It was Grew's practice to take a modest view of his labours, and he had previously, in 1672, published *The Anatomy of Vegetables Begun*. We have seen that this was not the first use of the expression " Comparative Anatomy " — in fact Grew himself had already employed it in 1672 and again in 1675, but it is the first occasion that the term appears on the title-page of a zoological book, and also the book itself represents the first attempt to deal with *one system of*

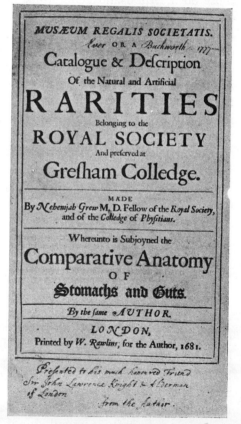

FIG. 105.—Grew, 1681. Title-page of his work on the comparative anatomy of the gut

organs only by the comparative method. In 1685 Collins discussed animal types organ by organ, and to some extent anticipated John Hunter, but Grew confined himself to the alimentary canal to the exclusion of other features. Although his comparisons have all the defects of tentative knowledge, it must be allowed that his attempt to understand the nature and homologies of the different regions of the gut by directly comparing a number of types *inter se* gives this work considerable historical importance.

Grew explains that he does not propose to repeat what is already known, but to confine himself to his own observations, adding in justification the following quotation from Seneca : " There is still much work to be done, and much will always remain, nor yet will those born a thousand generations hereafter, lack the opportunity of adding something further ". Grew's ideas on comparative anatomy, however, revolve round the human type — the common denominator to which all quadrupeds must in the last resort conform. Thus, he says, of all such animals dissected by him only the pig, horse and rabbit have a true colon " if that of a Man be the standard for the Definition of it ", and he evidently declines to recognize any part of the gut as a colon *unless it is sacculated,* as in man.

On the basis of his somewhat numerous dissections of " quadrupeds " Grew suggests the following classification of these animals based on anatomical and physiological criteria drawn from the gut :

Carnivorous.   Weasel, polecat, stoat, cat, dog, fox.
Insectivorous.   Mole.
Frugivorous.   Hedgehog, shrew, squirrel, rat, mouse.
Frugivorous and Graminivorous.   Rabbit, pig, horse.
Graminivorous.   Sheep, calf.

It will at once be obvious that this classification does not harmonize with the facts which Grew himself discovered, due consideration of which should have enabled him materially to better his scheme. For example, the *identity* of gut structure in the mole and hedgehog, which he had observed and described, is rendered meaningless if these animals are assigned to different groups, and *diversity* is equally ignored if the hedgehog is made to lie down with the Rodents. He is, however, less erratic in physiological considerations. Thus, he points out, if the food is rich and easily digestible the gut is very short, but in an animal like the sheep, which feeds on grass, it is necessary that the gut should be able to contain a large quantity of food (hence the large stomach), and also that it should be long and slender so as to ensure a longer voyage to a type of food from which it is difficult to extract the nutritive substance. Therefore he deduces the principle that the morphological significance of any organ cannot be grasped *until* its function has been considered.

Peyer's glands of the gut, which were discovered by Severino in 1645 some thirty years before Peyer, were closely examined by Grew, who noted their variable development in a number of mammals. Apart from these structures, Grew's observations on the gut of mammals, birds and fishes were confined to the more obvious macroscopic characters. In the weasel he found no caecum or sacculated colon, the gut having the same width from stomach to anus, and being simpler and more uniform than any other mammalian intestine he had examined. There were a pair of anal scent glands embraced by the external sphincter muscle of the anus. In the cat he wrongly concluded that the caecum was absent, although he noticed its very small appendix, and there were also a pair of anal scent glands. The caecum, however, is not well defined in the cat. In the dog Grew describes a conspicuous caecum six inches long forming a short spiral or bent tube of three turns, which he thought opened into the rectum [ascending colon].[1] The fox was found to have a large spiral caecum, and, as in the dog, Peyer's patches were numerous and conspicuous, being easily seen even when the gut was inflated and dried. The gut of the mole and hedgehog agreed in having few or no Peyer's glands, and also in the absence of a caecum and anal glands.

According to Grew the gut of the [black] rat is distinguished by two peculiarities. First, there is the large and striking caecum, having the appearance and size of an extra stomach ; and, second, immediately following the caecum, there is a curious region about two inches long, the cavity of which is provided with about twenty-five pairs of oblique plates meeting in the middle line, and forming a conical plumose figure as in the abomasum of the sheep. The comparison with the sheep has reference to the lamellar folds of the

---

[1] The caecum of the dog varies considerably in length and appearance.

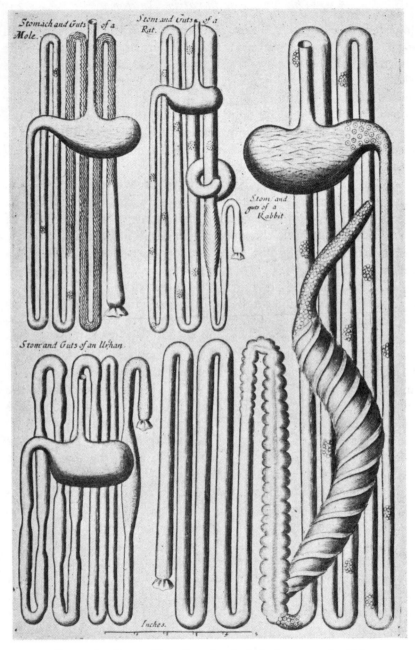

FIG. 106.—Grew, 1681.   Gut of mole, hedgehog, rat and rabbit

mucous membrane of the abomasum, which cross the great axis of the chamber very obliquely, so as to produce a kind of spiral arrangement. Grew calls this part of the gut in the rat the aboma-sideum, without realizing that any community of structure between two such diverse regions of the gut can have no comparative value. He has, however, raised quite an interesting point. In the part of the gut immediately following the caecum for about two inches the mucosa is raised into prominent oblique lamellar folds, which in the grey rat are so disposed as to form two conical systems, the posterior of which is the more loosely spaced out and less regular, and the apices of the cones pointing backwards or forwards according to whether the gut has been opened along the median dorsal or ventral line. Behind this region, and separated from it by a prominent smooth area, the folds of the mucosa are parallel and longitudinal. Grew apparently saw only the anterior of the two cones, but there may be some differences in the two species of rat, his observations having been restricted to the black rat, since the grey species was unknown in England before 1728. Blasius figured, but did not describe, the " abomasideum " in 1673.

The figure of the gut of the rabbit is good for the time, and shows the pyloric glands of the stomach and the distribution of Peyer's glands. The sacculus rotundus is described as the pancreas intestinale, but is not included in the figure. Grew found the spiral lamellar fold of the caecum, and witnessed the peristaltic motion of the gut in the freshly killed animal. He examined the villi of the mucosa, which he says form a kind of pile and stand up " round and high, like an infinite number of *Papillæ* : the Mouths of each visibly open ; from whence a Mucus may easily be express'd ". The belief that the villi open *directly* into the cavity of the gut survived even the careful observations of Lieberkühn, Hewson and Cruikshank, and as late as 1849 was still accepted by some anatomists, although it had been exploded by John Goodsir in 1842. Grew explains that he " had not time to observe the inside " of the stomach of the horse, and was therefore not acquainted with its highly peculiar character, which had, however, already been discovered by Severino. He criticizes Glisson, who had professed to find two caeca in this animal, and his own statement that the two limbs of the loop of the colon are connected by a narrow non-sacculated stretch of gut over four inches long is also correct.

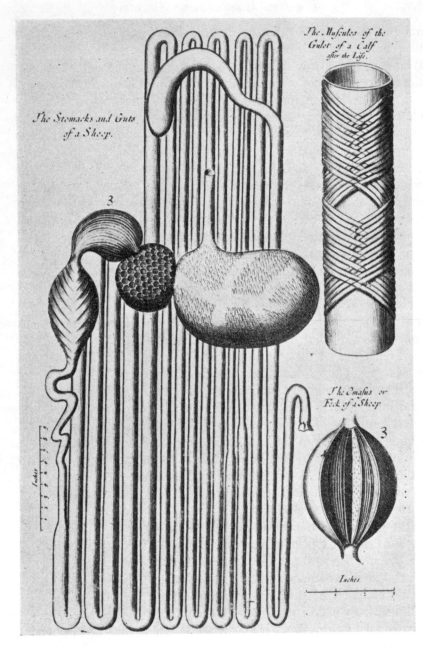

The Muscules of the
Gulet of a Calf
after the Life.

The Stomacks and Guts
of a Sheep.

The Omasus or
Feck of a Sheep

FIG. 107.—Grew, 1681.   Gut of sheep and oesophageal muscles of calf

The oesophagus of ruminants is provided with two opposing systems of spiral muscle bundles which at intervals cross each other almost at right angles, and are at those points interlaced. Grew was one of the earliest to establish this condition, which he saw first in the sheep. Steno (1664) and Willis (1673) had previously noticed spiral decussating muscle fibres in the oesophagus, without, however, stating the species they had dissected, and a similar condition was recorded in the dog and reindeer by the Amsterdammers (1667) and T. Bartholin (1672). These fibres are well developed in the oesophagus of ruminants, where they form a unique spectacle which was better illustrated by Grew in 1681 than by Peyer in 1685.

Aristotle is responsible for the first account of the compound stomach of ruminants, which was described by several modern naturalists before Grew, including Coiter (1572), Aldrovandus (1613), Fabricius (1618), Ambrosinus (1642),[1] Severino (1645), Jonston (1650), Blasius (1667, 1674), Perrault (1669, 1671, 1676, 1680), Glisson (1677) and Charleton (1680). After Grew, Peyer's monograph (1685) is the most important of the earlier contributions to our knowledge of this organ. Grew investigated with reasonable accuracy the compound stomach of the sheep and calf, but he did not study the precise relations of the paunch, honeycomb and manyplies, nor is his figure correct on this point. Moreover, the cells of the honeycomb are represented conventionally by *circles*, and the disposition and form of the chambers are too diagrammatic. A separate figure of the dissected omasum is added. Grew did not perceive that the stomach and caecum may represent one functional whole. Although he was fully alive to the fact that the structure, length and relative importance of the various parts of the gut were conditioned by the nature and digestibility of the food, he was none the less puzzled to find that the same type of food was consistent with a large complex stomach in one animal and a small simple one in another. Had he noticed that the caecum was simple in the former case and complex in the latter, his doubts might not have appeared so formidable, although there is evidently some remote suspicion of the truth in his remark that in the horse " Nature hath here transfer'd the greater part of the *Alimental Luggage* into the *Caecum* ", or " second stomach ". It has already been pointed

---

[1] This author gives a figure of the compound stomach of a lamb.

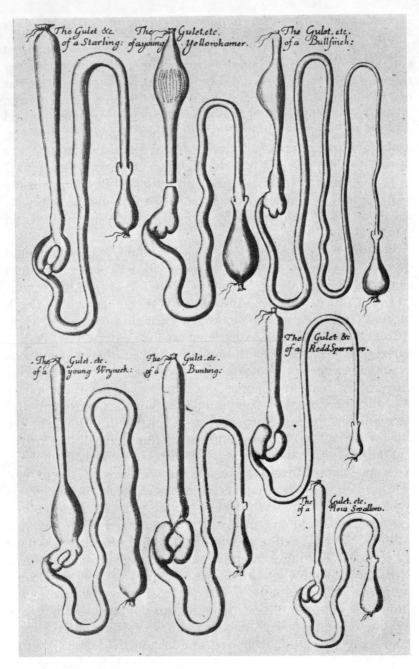

The Gulet &c. of a Starling:

The Gulet. etc. of a young Yellowhamer.

The Gulet. etc. of a Bullfinch:

The Gulet. etc. of a young Wryneck:

The Gulet. etc. of a Bunting:

The Gulet &c of a Redd Sparrow.

The Gulet. etc. of a Hors Swallow.

FIG. 108.—Grew, 1681. Gut of starling, yellow-hammer, bullfinch, wryneck, two buntings and swallow, illustrating reduction of rectal caeca

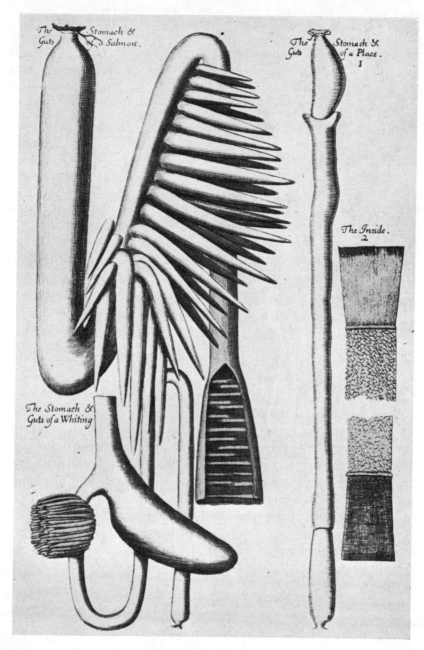

The Guts · Stomach & of a Salmon.

The Guts · Stomach & of a Place. *1*

The Inside. *2*

The Stomach & Guts of a Whiting

FIG. 109.—Grew, 1681. Gut of salmon, whiting and plaice, illustrating reduction of pyloric caeca

out that he correctly homologized the fourth chamber or abomasum with the undivided stomach of other mammals, and his conclusions that the first three chambers constitute the organ of taste, and that the function of the gastric villi is mainly gustatory, are not so misguided as they appear, since he confuses with taste the process of judging the state of comminution of the food.

Grew investigated the gut of some forty species of birds, but he describes only the cassowary, fowl, pigeon, swallow, bullfinch, " a Bunting ", reed-bunting, yellowhammer, starling, jackdaw, wryneck, cuckoo and owl, all of which are figured except the cuckoo. In the anterior region of the gut he distinguishes the crop, a glandular " Echinus " [proventriculus], and the gizzard, but of these the second only was present in *all* the birds dissected by him. He enters into some detail on the structure of the gizzard of the fowl, and notes the extremely variable size of the rectal caeca in birds — long and club-shaped in the owl, small in the jackdaw and starling, vestigial in the swallow, and absent in the wryneck. He failed to observe that the caeca of the cuckoo are asymmetrical.

The fishes are briefly dismissed in the text, although he had examined the perch, barbel, tench, bream, pike, salmon, trout, cod, whiting, gurnard and plaice. He doubts whether the stomach is present in all fishes, by which he means a stomach of the obvious expanded type, and the pyloric caeca of the Teleostean fishes are compared with the caecum of the mammal, although they vary in number from a few in the perch to more than eighty in the salmon. In agreement with some modern authors, Grew found only two caeca in the plaice, instead of the four normally present. The figure of the gut of the whiting exaggerates the blind end of the stomach, and has been badly displayed for the draughtsman, with the result that the duodenum does not, as it should, form a U-shaped figure with the stomach, but continues the backward direction of the oesophagus.

VI

THE DUTCH SCHOOL

XXIII
LEEUWENHOEK

IT is often stated that, whatever may be said of other spheres of human activity, science is of necessity international in scope and character. Unlike the arts and music, which must and do reflect the environment in which they flourish, and the mentality of the people who enjoy and cultivate them, science is by contrast impersonal and universal, and scientific principles and phenomena capture the imagination only in so far as they are the embodiment of doctrines and truths which transcend all racial boundaries and are common to all peoples. Thus it is true that in times of national unrest and peril scientific research seems to proceed almost as usual — unmoved by the distresses of mankind, and it is none the less true that when the national fortunes are in the ascendant science follows soberly in the wake of the general expansion and prosperity, and takes its place in human progress as if unaffected by the prevailing passion. The history of comparative anatomy in Holland provides us with a significant example of these tendencies.

The seventeenth century has been appropriately described as the Golden Age of the Dutch nation. With the foundation of the Dutch Republic the dominance of Spain had at last been broken, although peace and stability were still far from assured. The Dutch Reformed Church regulated the morals of the people, and bent all the resources of austerity and discipline to the task of moulding the national character. The incorporation of the wealthy and powerful Dutch East India Company was the outcome of a period of commercial inflation and private enterprise, and it was accompanied by a sudden and almost unparalleled outburst of national genius. Their great admirals swept the seas, their merchant marine carried Dutch traders, collectors and naturalists to all parts of the world, and in art, literature and science men of the highest distinction emerged in extraordinary profusion.

Rembrandt, Hals and Vermeer had their analogues in science in Leeuwenhoek, Swammerdam and Christiaan Huygens. The microscope itself, without which even the bare identification of the parts of the body cannot be established, owes much to Dutch science of this period, and Cornelis Drebbel, born in the delightful town of Alkmaar in 1572, and Zacharias Janssen, of ancient and placid Middelburg, have each in turn been acclaimed as the inventor of the compound microscope incorporating *two* convex lenses. But if the truth on this point is perhaps beyond the reach of the historian, it cannot be denied that Drebbel played a great part in making the microscope known in Western Europe.

The history of comparative anatomy in Holland in the seventeenth century, when science in its modern form may be said to have begun, centres round the great figures of Leeuwenhoek, Swammerdam and the youthful de Graaf — worthy successors of Coiter of the sixteenth century, and followed in the eighteenth century by B. S. Albinus,[1] Lyonet [1] and Camper. Lesser, though still important figures in this Olympian period, were van Horne, Blasius and Ruysch. Many others like Haller, not born in Holland, were trained or studied there in the Universities of Leyden (established in 1575), Franeker (1585), Groningen (1614) and Utrecht (1634), but the genius of Leeuwenhoek and Swammerdam overshadows all else. The methods and outlook of these two observers of Nature had little in common. Swammerdam was a trained man of science. He knew how to prepare and how to state a case. A glance at his works reveals at once how far he had got, and the degree of accuracy he had succeeded in reaching. Even in his perversely verbose treatise on the may-fly, we find that directly he abandons his crazy theology and applies himself to the objective aspects of the research he steps into his proper place among the big men of his century. But the untrained, irresponsible amateur Leeuwenhoek stands by himself in a morass of difficulties. This is due partly to bibliographical problems which are still in process of liquidation, but particularly to the fact that his discoveries were written up for publication whilst they were still fermenting in his mind, nor did he even clearly define the various species on which he worked. But whatever the explanation may be, it must be admitted that even now the works of Leeuwenhoek have not been

---

[1] Adopted sons of Holland.

studied sufficiently to enable us to assess the value of his contribution to the fabric of biological science, and not until the whole of the letters have been studied with the same devotion and scholarship that Dobell brought to bear on the protozoological correspondence will it be possible to do so.

Leeuwenhoek's impetuous enthusiasm made it difficult for him to digest the facts and to clarify his ideas before appealing to the arbitrament of print. Thus it happens that a fresh paragraph is often the only indication we have that the subject has been abruptly changed, and that we must adjust our minds to the consideration of a new problem. Neither does he coin scientific terms himself nor use those devised by others. He spoke, and understood, no language but his own Low Dutch. The word *dierken* (animalcule) is applied indiscriminately not only to microscopic animals but to many larger ones. Even the crab *Portumnus* is an animalcule ! Hence the doubts that arise when, without warning, he switches over from one species to another with no further definition or explanation beyond that they are both " animalcules ". All this raises difficulties of interpretation which are often considerable and sometimes insuperable. The time that can be lavished in disentangling this confused and baffling series of observations can be realized only by those who have found it necessary to study in detail the writings of this remarkable and unmethodical man. These difficulties, however, have one considerable advantage. They enable us to follow, stage by stage, not only the complete evolution of his discoveries and theoretical views, but even to track them in his own snow. The interest and significance of this point is particularly evident in his attack on the problem of the parasites of Aphidae, where we can follow the gradual emergence of the solution, as observation and inference expanded under his scrutiny. It would indeed be difficult to cite another pioneer investigation of the period, outside his own writings, which resulted in so many new and important discoveries.[1]

But if Leeuwenhoek unduly exercises our patience and speculative faculties, he provides also ample compensations. There is something very modest and human in these unpretentious letters of his, and doubtless on this account he is held in greater affection by his own countrymen than his Olympian contemporary Swam-

[1] For details cf. Cole, *Ann. Sci.* (1937) ; *J. Quekett Micro. Club* (1938).

merdam. They like his ingenuous and indignant surprise at the
scepticism with which his discoveries were received, and especially
his occasional lapses into broad humour, such as we find in his
famous letter on the body louse, which he winds up with the
remark — " so much for this lousy chat ".[1]

The extant scientific correspondence of Leeuwenhoek comprising some three hundred letters addressed to friends or to
the Royal Society, is not yet completely available to the historian.
In the Dutch edition of the *Opera Omnia*, however, prepared
and issued a few years before Leeuwenhoek's death, we have
available in printed form, and in his own language, the bulk of the
work which the author himself considered to be of value. Some of
the letters have never been published in any form, or have only
been abstracted. The Royal Society partly or wholly translated
their own letters into Latin or English, and published the translations in the *Philosophical Transactions*, but 118 of the letters
were printed for the first time in the collected Dutch and Latin
editions of his works which appeared later. The Leeuwenhoek
Commission, still, let us hope, sitting in Holland, has undertaken
to collect, annotate and publish in Dutch and English the *whole* of
the correspondence which has survived ; but so far only the first
two volumes have been issued, and at the present rate of progress
it will be almost a generation before this monumental task is
accomplished.

Leeuwenhoek examined at least 214 animal types from In-
fusoria to whales, and employed more than four hundred microscopes made with his own hands, nearly all of which have vanished
beyond recall. Even the twenty-six microscopes which he
bequeathed to the Royal Society have long since been lost, but it is
known that they magnified from 50 to 200 diameters, and the
Utrecht example, said by van Cittert to give a magnification of 275
diameters, resolves structural details the size of which is somewhat
more than 1μ, in spite of the fact that the lens is badly scratched on
the side of the object. Leeuwenhoek rightly preferred a single
biconvex lens of very short focus to the imperfectly corrected
compound microscopes then available. Nevertheless, with this

---

[1] Even the Latin translator rose to this, and rendered it *Sed hic finem imponam
pediculoso huic sermoni* ; nor, in the English translation, was it too much for the
fastidious Hoole.

simple apparatus results which have become classic were obtained. A notable feature of his work is that having made an important observation on one animal, he proceeds to search for the same feature in others. He was a firm believer in the fixity of species, but that did not prevent him from following out a structure in a whole series of types and comparing one with another. He was in fact the first to practise this fundamental method on a big scale, and the persistence and resource which he displayed in these quests are presumably unknown to those of his critics who profess to see in him nothing but a superficial amateur. On the other hand he was not always successful — he made his mistakes, although they were surprisingly few in number, and as a rule not important in their consequences. His intuition and judgment in interpreting his results and in devising his experiments stamp the unassuming draper of Delft as one of the foremost men of genius of the time. The compliment justly paid to Pasteur that " no one has walked so surely through the circles of elemental nature " applies with equal force to the Dutch biologist, whose ambition, as he himself put it, was to free the world from its errors and to teach himself.

Most of Leeuwenhoek's discoveries are outside the scope of the present work, but in one respect his importance to comparative anatomy can scarcely be over-estimated. As the pioneer of micro-dissection he not only initiated a type of research of the utmost value, but he succeeded in obtaining results transcending what could reasonably be expected from the exploitation of a new and difficult method.[1] His manipulative skill must have been astonishing, for his apparatus was of the simplest. We can only regret that he deliberately omitted to record how his work was carried out, and some of it is sufficiently intricate to tax a modern observer armed with all the refinements of binocular dissecting microscopes. He warns his contemporaries that they must not expect, without repeated effort, to imitate his achievements in dissection, and they, on their part, finding this warning amply justified by the event, retaliated by declining to believe that there was anything to imitate. This phase of Leeuwenhoek's activities covers the period 1680-1701, between the ages of forty-eight and sixty-eight, but he was attempting micro-dissections even after passing his eightieth

---

[1] It is unfortunate that, unlike Swammerdam, Leeuwenhoek did not illustrate his micro-dissections with adequate figures.

year. It is, however, apparent that he was some years in acquiring the necessary skill, and that in the last twenty years of his working life as a biologist he found it necessary to confine himself to microscopical work of a less exacting nature.

Leeuwenhoek's first attempt at micro-dissection was undertaken in 1680 [1] when he selected as his subjects the mite *Aleurobius* (*Tyroglyphus*) *farinae* and the flea, but in 1683, and again in 1693, he carried out more detailed examinations of the latter animal. He describes tracheae, the male genital organs [? part of the gut], ovary and mouth parts. In another letter, also written in 1683, he found that *Pediculus* develops within the egg-shell, and has all its limbs when still within the body of the parent. It is thus ovoviviparous. In *Pulex*, he says, this is not so. In 1685 he dissected a sheep foetus of about seventeen days which was no larger than the eighth part of a pea, and professes to have found in it *all* the parts of a sheep. Another sheep embryo of " three days " (as he was told), which was about the size of a large grain of sand, was nevertheless a little sheep fully formed. He could see in the head the eyes and jaws, the backbone of the body, and so on. His next subjects were the shrimp and the lobster (1686), the embryos of which, when dissected out of the egg-shell, were complete miniature shrimps and lobsters, in which he could distinguish all the parts, including the head, eyes, mouth parts, body, tail fin, feet and chelae. In another letter written a month later in the same year, similar conditions were described in the egg of the prawn, *Palaemon*. Leeuwenhoek now proceeds to follow up this line of investigation, and in 1687 he reports on the results of his studies on the development of *Bombyx*, which process he found could be accelerated by artificial heat. He saw the micropyle of the ovum, first noticed by Malpighi in 1669, and opened eggs at intervals in order to examine the development of the ovum, and the progress of the larva within the egg-shell throughout the winter months. He describes the embryo surrounded by the yolk sac, and traces the growth of the external parts and appendages of the larva. He mentions dissecting the gut in an unhatched larva, and observed the food-yolk mass being taken into the interior of the embryo. He includes also an

---

[1] The dates quoted in this section are those which appear on the letters themselves and are not those of publication. For the identification of the letters see the writer's memoir on *Leeuwenhoek's Zoological Researches*, Part II (1937).

account of the structure of the egg and its membranes. In the same year he extended his researches to the small granary weevil *Calandra*. He was the first to understand the nature of the rostrum, since he observed the mouth and mouth parts *at its extremity*, and saw how this apparatus could be used to bore a small hole in the husk of the seed. He distinguished the sexes, observed copulation, examined the spermatozoa and the structure of the " penis " [male genital tube], and found the ovary in a dissection of the female. In this letter he also records having dissected a louse and an ant. In his next letter of the same year he mentions having repeated the dissections of *Formica*, and having found the " stomach " and " intestines " of the larva. He describes a sting in the red ant which he did not find in the black and yellow ants, but the sting is not the simple feature Leeuwenhoek supposed it to be. In 1688 he gives a description, remarkable for the time, of the mouth parts of *Culex*, in which he justly criticizes Swammerdam's first account of these structures published in 1669, but Swammerdam's second attempt in the *Biblia Naturae* is on quite another plane. Leeuwenhoek found both piercing and sucking parts, and showed how they all fit together and are put away when not in use. He noted that there were four piercing organs, some of which were barbed, and that they differed in length on the two sides. He is rather pleased with this success, and emphasizes its difficulties. Others will not be able to confirm his results, he says, without the careful and skilful examination of a large number of gnats. He wrongly states that the labium is grooved laterally instead of dorsally in order to lodge the piercing parts, and he evidently saw the maxillae, mandibles, labrum + epipharynx and the hypopharynx, but he does not mention the maxillary palps. His figure shows the labium with the labella.

In 1692 Leeuwenhoek examined the rectractile telescopic ovipositor of the grain moth *Tinea granella*, and dissected some females, in each of which he found 50-70 eggs, and in the following year, and also in 1696, he essayed the difficult task of dissecting the louse *Pediculus*, with results that are indeed remarkable.[1] He saw the oesophagus and the imbibed blood running swiftly through it. In the stomach and gut there is, he says, an incessant " to and fro "

---

[1] Not, however, as remarkable as the masterly dissections and figures of the same animal by Swammerdam, published in the *Biblia Naturae*.

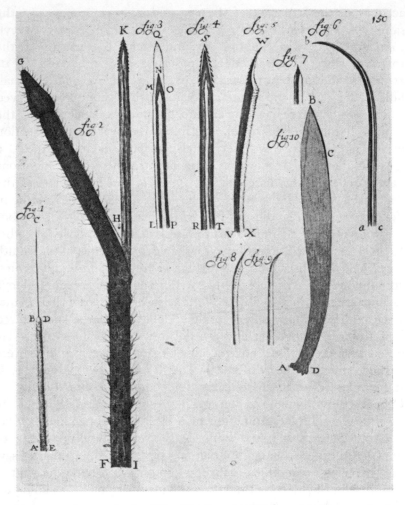

FIG. 110.—Leeuwenhoek, 1688. Mouth parts of *Culex*. 1 is Swammerdam's *first* figure of these mouth parts. 10, mandible of *Tabanus*

movement of the blood. He describes and figures the eversible mouth parts as an exceedingly fine piercer or " sting " enclosed in a sheath. He thought he saw the blood circulating in the foot, and having exposed the brain he found a " multiplicity of vessels " in the head. He dissected, figured and described the structure of the male copulatory organ or penis, which he regarded as a posterior " sting ", but noted that it was found only in the male. Leeuwenhoek was wrong in stating that there was only one penial rod, but he correctly describes the ovary, the two pairs of testes, the vesiculae seminales and the vasa deferentia. He removed eggs from the ovary and found that they would not develop [because not fertilized], but in one case the louse was shown to be ovo-viviparous, since an egg taken from the body of a female had a little louse inside it.

Like Swammerdam, Leeuwenhoek at first thought that the louse was hermaphrodite and that eggs were to be found in every individual, but he corrected himself when he discovered that the

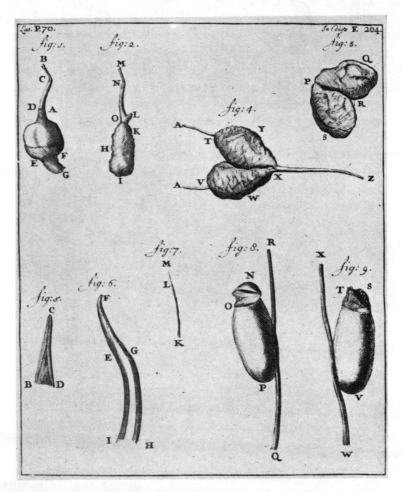

FIG. 111.—Leeuwenhoek, 1696. Body louse, *Pediculus*. Mouth parts, male genitalia, eggs

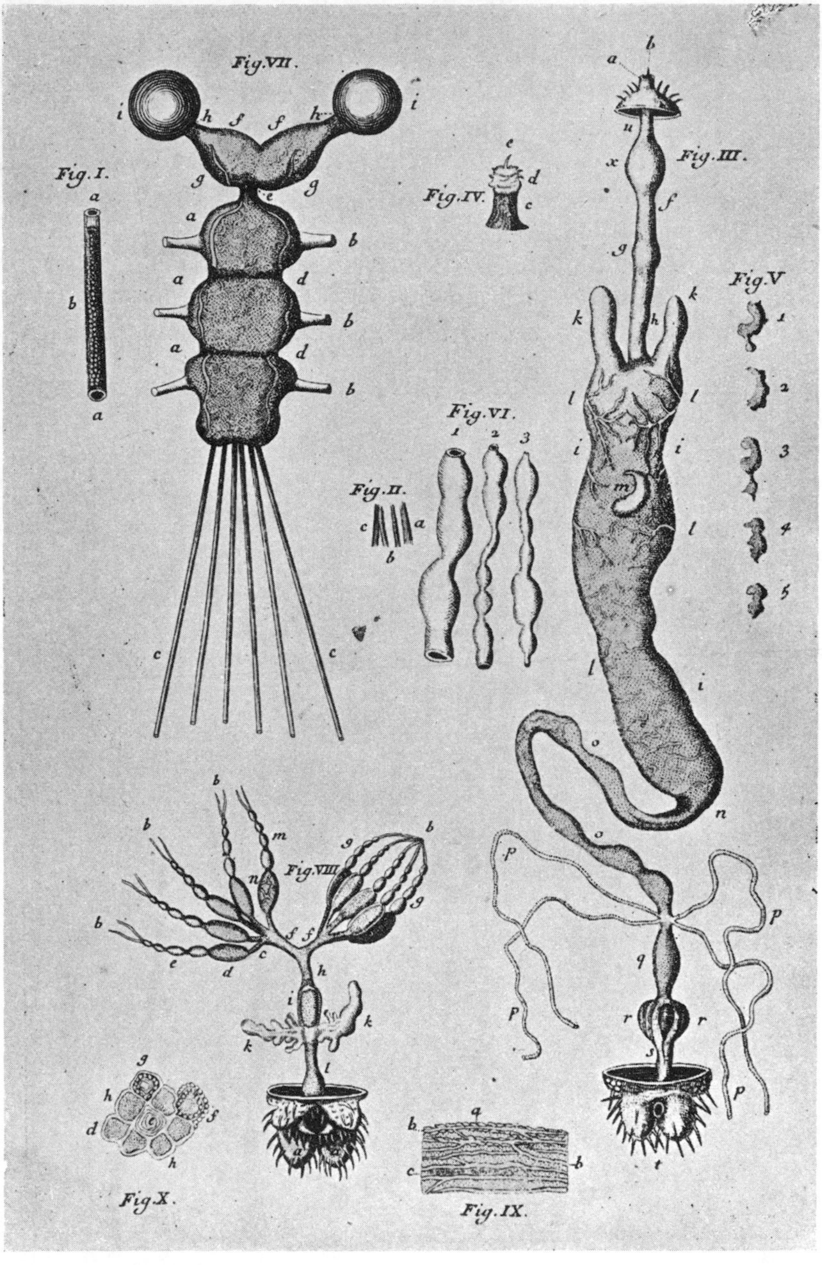

FIG. 112.—Swammerdam. *Pediculus*. Gut, female genitalia, nervous system. In III the oesophagus is much too wide for a blood-sucking animal, and the salivary glands have been overlooked

" eggs " of certain specimens were in fact the testes. This enabled him to establish the existence of the two sexes in this species. Swammerdam on the other hand left this problem unsolved, but we cannot sufficiently admire his detailed and accurate dissections of the human body louse, and especially those of the gut, nervous system and female genitalia.

In 1693 Leeuwenhoek also dissected females of the cheese mite *Tyroglyphus*, and found eggs within their shells in the abdomen. He probably saw the testes of the male, and states correctly that the young mite has only six feet [hexapod larva]. One of the few serious mistakes made by Leeuwenhoek was his identification of the Polyzoan *Membranipora* incrusting the shells of *Mytilus* as the spawn of that mollusc (1694). He dissected these supposed eggs and described the zooecium with its orifice and the contained polypide. Some of the polypides were removed and figured, and he regards the tentacles as the byssus of the " young mussel ". The crystalline style of *Mytilus*, he says, is a protrusible ovipositor which arranges the eggs on the shell, but in the postscript of the same letter this function is attributed to the foot. His comment that if we are to come by the truth we must submit to the discipline of many arduous trials and experiments only reminds us that by so doing we can still fall into error. In the same letter Leeuwenhoek mentions having dissected a *Balanus*, and describes the mantle cavity with its contained eggs [which appear to have been segmenting], the cirri and the long penis.[1] Individuals without eggs he erroneously identified as males, and did not therefore discover that the animal was hermaphrodite.

Leeuwenhoek's famous dissections of aphides, which resulted in the discovery of parthenogenesis in those animals, were accomplished in 1695, 1696 and 1700. His chief species at first was the currant aphis *Myzus ribis*, and the year 1694, but his first letter on the aphis is dated July 10, 1695. He searched carefully for eggs, without, however, any success. He next attempted the delicate task of *dissecting* some individuals in the hope of discovering the eggs, but found instead, in one specimen, four young aphides *just like the parent*. This was the first demonstration that the aphis was viviparous, and that there was no metamorphosis through a larva as then understood in other insects. In the same summer he

---

[1] Leeuwenhoek was not aware of the nature of this organ.

dissected some more examples, and in one of them found as many as seventy young. These embryos were not all of the same size or state of development, which made it possible to work out the life-history by examining the stages found in the body of a single female. He confirmed his belief that the species was viviparous by observing the act of parturition in two individuals, which produced respectively six and nine young in twenty-four hours. He studied the phenomenon of ecdysis, and noted that *all* external cuticles were shed, including the cornea of the compound eyes and the mouth parts. He compared the alate and apterous forms and established that they agreed essentially in structure, and therefore belonged to the same species. He established also that what we now know to be the pupal stages with their lateral wing rudiments moulted to produce the alate individuals, which he showed were themselves viviparous females, although he erred in concluding that *all* members of the society became alate in course of time. He describes the piercing and sucking action of the mouth parts, and even observed the peristaltic action of the gut in the unborn young.

In this letter Leeuwenhoek was already greatly puzzled by the fact that all the aphides he opened were *females*, and the smallest ones, which might have been males, could not be identified as such. Even the youngest aphis, only one twenty-fourth the size of the parent, contained round particles of different sizes which Leeuwenhoek concluded were early embryos or eggs. He does not in this letter propound any theory of parthenogenesis, but in the following year, 1696, he repeats that the male aphis had eluded him, and after an interval of three years, in 1699, he is beginning to suspect that *males may not exist*. It was not until 1700 that, in a reference to his first letter on the subject, he quotes himself as having stated that aphides " bring forth their young without coming together " or " without the male kind " — a phenomenon to him unique in the animal kingdom. In the 1695 letter, however, he does not specifically say this, but simply that he had not *found* any males.

Leeuwenhoek was a fervent supporter of the animalculist interpretation of the generative process, as was natural in the discoverer of the spermatozoa ; but his further discovery of parthenogenesis should have suggested to his mind the expediency of revising his opinions on that doctrine. Instead, it led him even further astray, and he was induced to advance the proposition that the aphis was a spermatic animalcule, and it was thus the *female* sex that was wanting in those deceptive insects. He admitted that owing to the minuteness of the spermatic animalcules it was not at the moment possible to demonstrate that they were foetuses, so that in the meantime it must suffice, he says, that in the aphis this phenomenon *can* be observed. He opened some young aphis and saw *immature* embryos in them. On dissecting older specimens he found the embryos proportionately advanced, and so on until in the mature aphis the young were ready to be brought forth. He therefore attempts to establish a parallel between the production of parthenogenetic aphides and the multiplication of spermatozoa in the testis.

In 1698 Leeuwenhoek attempted two very difficult operations on a non-bloodsucking gnat [? *Culex*]. He cut off the head and removed the brain " with difficulty ". He gives a drawing of the brain, which might be a mass of tracheae, but is certainly not the brain. From the same insect he extracted what he identified as the extensor and flexor muscles from a joint of one of the legs. This must have been an exacting operation if successfully carried out — more so in fact than the removal of the " brain ". In 1699 Leeuwenhoek dissected a *Daphnia*, and from the brood-chamber removed the embryos, which he found were provided with limbs for swimming. In the following year he reported on some " worms " taken from a corrupt tooth which had been sent to him by the Royal Society. These worms were almost certainly the larvae of the cheese-fly *Piophila casei*. He dissected out the gut and evidently saw tracheae, but did not realize that they were respiratory organs. In the same year he refers to his first letter of 1673, and gives us an account of his anatomical researches on bees. In this work he had been anticipated by Swammerdam, although the latter's results had not then been published. Leeuwenhoek describes the structure of the sting, which he says consists of two parts or stings provided with teeth, and a third part or " sheath ". He dissected some bees and looked for eggs, without, however, finding any, and he holds that there is only one female in the society — the so-called " King " [Queen]. He dissected a queen and extracted from her an incredible number of oblong eggs, and in another queen which had died he found only very small eggs, from which he concluded that she had recently discharged her ripe eggs. He refers also to the colon with its " rectal glands ", and saw a great number of vessels and organs.

In the same letter he mentions having dissected some spiders, and noted the circulation of the blood and the wonderful organs which produce the web. In 1701 he returns to the spiders and gives us an excellent account of the garden spider *Araneus diadematus*. He found on dissection that there were four hundred silk glands arranged in eight groups [an under-estimate], which he " thinks " would produce eight threads, each composed of a great number of smaller ones. He describes the eggs, which he opened, and assumes that the " globules " in the egg serve as food for the spermatic animalcule when it enters the egg, and so enable it to grow into a spider.

In 1704, at the age of seventy-one, Leeuwenhoek carried out some of his most remarkable micro-dissections on the cochineal insect *Coccus cacti*, a subject which had first attracted his attention in 1687. He did not discover the animal nature of cochineal, which was fully recognised by Hartsoeker in 1694, but he was the first to establish that the cochineal particle was the remains of an insect *allied to the aphis*, and having the same type of generation. He now found on dissection that *all the dried insects were ovigerous females, and that there were young ones in the eggs*. He counted two hundred well-developed oval eggs in one large individual. An embryo was removed from its egg-shell and he discovered that it was not a " worm " [that is, a worm-like larva] *but a young segmented insect*, its three long pairs of legs, jointed antennae and rostrum being folded against the body. He describes the rostrum with its setae, and compares it correctly with the beak of an aphis. He also dissected out the tubular ovary, and saw eggs even in some of the small individuals one-eighth the size of the large ones. He mentions extracting the embryos from the eggs only " with a great deal of pains ".

In 1711, when Leeuwenhoek was seventy-eight years of age, he dissected some female mites,[1] removed the eggs and watched the growth of the larva within the egg-shell, but he does not in this letter mention its hexapod character. Leeuwenhoek's last attempts at micro-dissection were made in 1713 and 1714 on the mite *Tyroglyphus*. He extracted the larva from the deposited egg, and

---

[1] *Glycyphagus domesticus.*

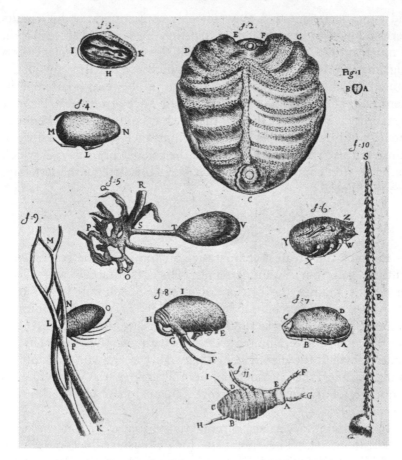

FIG. 113.—Leeuwenhoek, 1704. Dissections of cochineal particles. 2, single grain
magnified ; 4, egg removed from parent and larval insect dissected out ; 5, part of
ovary including one egg—*T-V* ; 6, another hexapod larva taken out of the egg
shell ; 8, larva showing the rostrum—*H-G*, " wherewith it gets its Nourishment
out of the Leaves " ; 11, dissected unhatched larva illustrating segmented body,
three pairs of legs and antennae

even dissected the ovary. It was full of eggs of varying sizes,
and, he says, looked like the ovary of a fowl seen without the
microscope.

So much for the researches of our worthy Dutch draper and
chamberlain of the ancient town of Delft. In estimating the
importance of his work it is essential to view it against the background of his own time. To us it is inevitable that the researches
of the naturalists of the seventeenth century should seem crude

and superficial, but to Leeuwenhoek's contemporaries his results were so novel and revolutionary that they hesitated to accept them. It must, moreover, be recognized that the progress of science is a pilgrimage towards an equilibrium which may exist only in the human mind, and so elude us when we think we have attained it. Hence our activities find expression in advance or retreat, but never in stability. Wherefore the fact that we have outlived Leeuwenhoek should not blind us to his *historical* importance, for in the nature of things we also shall be outlived, and the science of to-day, in which we have such confidence, will then be as obsolete as the letters of Leeuwenhoek.

XXIV

SWAMMERDAM

Perhaps the greatest comparative anatomist of the seventeenth century was Jan Swammerdam. His grandfather was Jacob Dirksz.[1] (= James the son of Dirk, *ang.* Theodore or Richard), who was a wood merchant living in the Dutch village of Swammerdam or Swadenburgerdam (now Zwammerdam) on the Old Rhine, and it was he who emigrated to Amsterdam and there adopted the name by which the distinguished naturalist is known. Swammerdam's father, the prosperous apothecary, with whom he is frequently confused, was therefore Jan or Johan Jacobsz. (= Jan the son of James).

In addition to various Parerga to be found in the works of T. Bartholin, sen., Thévenot, van Horne, Blasius, Boccone, Steno, the *Philosophical Transactions* and the works of the Amsterdam " College ", Swammerdam published four major works during his lifetime : (1) an Inaugural Dissertation on Respiration (1667) ; (2) a History of Insects in Dutch (1669) ; (3) on the Structure of the Uterus (1672) ; and (4) on the Life-Cycle of the Ephemera (1675). To mollify an angry parent Swammerdam wasted much valuable time in cataloguing the family museum, of which the animal section formed about one-sixth of the whole, and this Catalogue,[2] which was published anonymously in 1679, shows that

---

[1] An abbreviation of the genitive Dirkszoon. The stop which indicates the contraction is in modern usage often omitted.

[2] Miall states that there is a note at the end of the Catalogue to the effect that the collections of both father and son were offered for sale at the same time. In the only copy of it which the writer has seen, a copy which includes manuscript corrections and

F<small>IG</small>. 114.—Swammerdam. From the material for a revised edition of his *History of Insects* of 1669 which was never published the collection was not comparable with many others which existed in Holland at the time. The early death of Swammerdam was not entirely responsible for the fact that the bulk of his work, the *Biblia Naturae*, remained unpublished, for to his own indifference were added the apathy and neglect of his successors. The manuscripts passed first into the hands of Thévenot, who designed to print them, but he died without discharging this important trust. After some vicissitudes, the papers were acquired for publication by G. J. Duverney, the celebrated Parisian anatomist, who had already undertaken other charges of a similar nature, but he lacked the energy or inclination to liquidate any of them. Finally Boerhaave was moved to intervene, and in the subsequent proceedings played an important and honourable part. He purchased the documents from Duverney, and published them under his own guarantee as two imposing folio volumes in Dutch and Latin in 1737–8, exactly a century after the birth of the author, but nevertheless not yet out of date ! [1] These volumes include the substance of the treatises on insects and the Ephemera, together with a considerable body of additional matter, and it was probably the industrious editor, Boerhaave, who gave to the work the title of " The Bible of Nature ".[2] It is no exaggeration to say that this book is one of the outstanding classics of the literature of zoology, and is an enduring monument to the genius of its neglected and unhappy author.

Since Leeuwenhoek's first letter is dated 1673 and Swammerdam's last important work was published in 1675 they can scarcely be described as *working* contemporaries, although Swammerdam was born only five years after Leeuwenhoek, and it is known that he visited Leeuwenhoek in 1674, when the oracle of Delft demonstrated to him some of his microscopical observations. By that time, however, Swammerdam's career as a naturalist was almost over, and it had lasted but twelve years, from 1663 until 1675. In 1673 he succumbed to the distracting influence of Anthoinette

---

additions in Swammerdam's own handwriting, no such note is to be found.   Thévenot printed a short list of the Cabinet of Jan Swammerdam himself in his *Recueil* of 1681, a very few separate issues of which still exist, one of them in the Library of the British Museum.

[1] The manuscript of the *Biblia Naturae* is now the property of the University of Leyden, but twenty of the plates are either wholly or partly missing.

[2] Mistranslated " The Book of Nature " in the English version.

Bourignon, and thenceforward his interest in science began to decline. It flared up from time to time, but the peril of extinction could no longer be averted. " I am now ", he says in 1675, " resolved to addict my thoughts more to love the Creator of these things, than to admire him in his creatures." The rigours of five years' incessant and exacting toil on the *Commentarium de Apibus* (1668–73) had so damaged his health that the abandonment of his scientific career was doubtless suggested. He recovered sufficiently to publish the monograph on the may-fly in 1675, but it proved to be his last anatomical recreation. It was, moreover, never completely finished, and lacks those amplifying and corrective revisions which give such completeness and accuracy to his work on bees. He was now convinced that the pursuit of natural knowledge was vain and even impious, and that man must not aspire to rival the Almighty in the perfection of his works. The closing years of his life, therefore, were devoted wholly to religious exercises, and he died in an atmosphere darkened by the turmoils of an unbalanced mind. Nevertheless, even during this period of strife he was occasionally found working, as the rhetorical Boerhaave tells us, exposed bareheaded to the scorching heat of the sun and " dissolving into sweat under the irresistible ardours of that powerful luminary ". When he died he had already been almost forgotten, and it was only the belated publication of the *Biblia Naturae* that revealed to the world what a great naturalist he had been — a conclusion which a comparison of his work with that of his contemporaries establishes beyond question.

Swammerdam's commendation of the comparative method has already been quoted, but there are times when he fails to apply it, or he must have hesitated before making the pulmonary veins of the frog open into the precavals. He was, however, one of the first anatomists to develop the *technique* of research, as we learn from Boerhaave. He early acquired a remarkable facility in the dissection of insects, the structure of which he demonstrated in public, and " by his silent skill effectually suppressed the talkative ignorance of others ". For the examination of very small objects he designed a simple dissecting microscope made by Samuel J. van Musschenbroek which had two arms — one for holding the object and the other for the lens. These arms were provided with coarse and fine adjustments. He employed a great variety of microscopes and very delicate dissecting instruments of the finest workmanship, and his practice was to begin with the lowest powers and proceed by degrees to the highest.   He preferred very fine scissors to knives and lancets, because the latter disturbed the natural relations of the parts, and all his instruments were sharpened under the microscope.   By these means he was able to dissect the viscera of a very small insect, such as the nymph of the may-fly, with the same precision and thoroughness that others could achieve only with the larger animals.   He was particularly dexterous with finely drawn-out tubes of glass, with which the smallest vessels could be inflated, or injected with a tenuous coloured liquid. Swammerdam was possibly the first to dissect under water, and to remove obstructions, such as the fat body of insects, by solution with turpentine or other reagents, and one of the first after Boyle to recognize the efficiency of alcohol as a preservative of animal tissues.   He mounted insect larvae by squeezing out the viscera through a puncture in the skin and substituting wax — the modern process of mummifying these desiccated larvae being not in all respects an improvement on Swammerdam's procedure.   He also exploited the method of inflating and drying portions of the viscera, and of preserving biological material in "balsam". Swammerdam, the superlative craftsman, occupies a place in Dutch anatomy similar to that of Gerard Dou in Dutch art.

In 1651 Harvey demonstrated the possibility of complete injection by throwing warm water into the pulmonary artery and finding that it returned via the pulmonary vein to the heart, but this crucial experiment was not published until 1687, and then by Ent.   In the meantime, in 1660,[1] Malpighi, and in 1667, Gayant, had independently carried out precisely similar operations, and in 1679 Ent had repeated Harvey's experiment, using the serum of milk as the medium.   Swammerdam is usually regarded as the inventor of the *solidifying* injection mass.   This claim cannot strictly be maintained, since Boyle in 1663 was recommending three different types of such masses,[2] and there are even earlier aspirants to the honour, but it is certain that Swammerdam demonstrated the practicability of the method, and was responsible for its general

[1] Published 1661.

[2] Whether Boyle ever tried out any of his numerous suggestions, including this one, is unknown, but it is probable that, as he says himself, they were " little more than bare designs ".

adoption by anatomists after his time. The first *wax* injection was carried out by Swammerdam in van Horne's house at Leyden in January 1666. It was on July 5, 1672, that Swammerdam sent to the Royal Society, accompanied by descriptions and figures (published in 1672), his famous preparation of the human female genitalia, in which the blood vessels had been injected with soft yellow and red wax, and the uterus blown up and dried.[1] This preparation was still in the Museum of the Society in 1681, when it was catalogued by Grew, but in 1781 the Society's collection of " rarities " was handed over to the British Museum, where, as the official history of the Museum assures us, we may search in vain for Swammerdam's historic specimen.[2] He gives a brief description of his injection method. Pure white wax was the material selected, to which, when liquid, the red, yellow, green or other colour as required was added. Arteries and veins were distinguished by different colours, and the medium was injected with a syringe as a hot liquid, the blood having been first pressed out of the larger vessels so that they might easily and quickly fill. When cool the wax solidified in the vessels, and the result was a permanent preparation which could not bleed should it be necessary to dissect it. In the *Biblia Naturae* he refers to the injection of the sturgeon, whiting and other fishes with wax of several colours by a " method peculiar to myself and of my own invention ". The History of Bees, completed in 1673 but not published until 1738, includes an interesting description of how he injected the blood vessels of a lepidopterous larva through the heart by means of a glass capillary tube. Unless the vessels of insect larvae are injected, he says, they are so very delicate and transparent that they cannot usually be seen.

Further (anonymous) experiments, almost certainly to be attributed to Swammerdam, include a successful attempt in 1672 [3] to inject mercury into the bronchial artery of a calf. The result was said to be a marvellous spectacle, for all the arteries stood out *as if made of silver*. In the *Biblia Naturae* Swammerdam mentions having filled the veins and arteries of a frog with quicksilver. He

---

[1] Also possibly at the same time he sent the red wax injections of the gall-bladder and spleen which the Society at one time possessed.

[2] Camper saw the remains of the Royal Society's collection at the British Museum in 1785.   [3] Published 1673.

was thus one of the first to inject blood vessels with mercury — a method which produced such beautiful preparations that it became very popular with his successors ; but in this he was anticipated by Malpighi (1661) and Bellini (1662), and after him Leeuwenhoek was independently experimenting with mercury injections in 1689 and 1692. Before graduation Swammerdam had visited France, where he was befriended by Melchisedech Thévenot. It was here that he publicly demonstrated the valves of the lymphatic vessels on June 19, 1664 (published 1667), but again he was too late, since another Hollander, Frederik Ruysch, had produced a detailed description of these valves in 1665.

The Swammerdams, father and son, following the example of many wealthy merchants of the period, made private collections of natural " rarities ", but that of the son, judging from the brief description we have of it, differed from the father's in having a specific morphological outlook and purpose, and was probably the first zoological museum to be so distinguished. It included three thousand species of insects, many of which he had carefully dissected and compared in their various stages. It was on this secure foundation that he based his famous classification of these animals. During his lifetime several attempts were made to dispose of the museum, but without success, and his heirs having failed to obtain five thousand florins for the specimens and instruments, the collection was broken up and dispersed.

The inaugural dissertation on Respiration (1667), which Swammerdam submitted for his doctor's degree, was a work of considerable achievement and even greater promise, and it has become one of the classics of the history of physiology. The discovery of the part played by air in respiration may here be recognized in one of its earliest phases, and the mechanism of mammalian respiration is ingeniously demonstrated by means of a pair of bellows containing the windpipe and lungs, in which the action of the bellows on the lungs can be observed through a window cut in the side. In the interests of priority it should be recorded that this experiment had been published for thirteen years before Borelli issued the first volume of his *De Motu Animalium*. Swammerdam was among the earliest to test the effects of injecting the blood vessels of *living* animals, and his discovery that the mammalian lung sinks in water before it has functioned, but floats

FIG. 115.—Swammerdam, 1679. Engraved title-page of his Inaugural Dissertation, first published 1667, which includes his earliest studies in comparative anatomy and physiology

afterwards, has important medico-legal bearings in cases of suspected infanticide. Significant observations on clinical and chemical thermometers are also to be found in this remarkable thesis, and even topics foreign to its express purpose, such as the hermaphroditism and reciprocal union and fertilization of snails, are not excluded.

Passing over the first work on insects (1669), which deals only with external morphology and metamorphosis, and also the memoir on the uterus — important from the point of view of anatomical injections and the Preformation Doctrine, we come to the more relevant monograph on the may-fly, first published in Dutch in 1675, six years after the appearance of Malpighi's treatise on *Bombyx*. To Anthoinette Bourignon the may-fly was a " little beast which lives for only a single day, and throughout that time endures many miseries ". She gave, however, a grudging consent to the printing of the book, but warned Swammerdam to devote himself in the future to the more serious interests of eternity. The statement so often made that Swammerdam's emotional obsessions were wholly due to the influence of his spiritual directress seems to be an exaggeration. His friendship with her was the result, and not the occasion, of his deep interest in religion. The Dutch edition of the *Ephemeri Vita* testifies to his own innate and meticulous piety. Never was a sermon more heavily underlined or more laboriously annotated. A scriptural basis for almost every word proclaims at once a profound knowledge of the Bible, and an ingenuous belief in the literal interpretation of isolated texts. He himself says that he wrote the ephemera " to give us wretched mortals a lively image of the shortness of this present life, and thereby to induce us to aspire to a better ". In this expansive and verbose work he was following independently the lead of Servetus, and assuming the difficult task of establishing an ethical system which reconciled the diverse attributes of biology and religion. In spite of the fact that it had a numerous following, among whom may be mentioned Lesser in Germany and Kirby and Spence in England, Insecto-Theology, as it was called, fell an inevitable victim to the strife of its own incongruous elements. The biology of the *Ephemeri Vita*, however, greatly diluted as it is, shows us Swammerdam at his best. He started work on the may-fly as early as 1667, and mentions dissecting the nymphs in 1670, and making

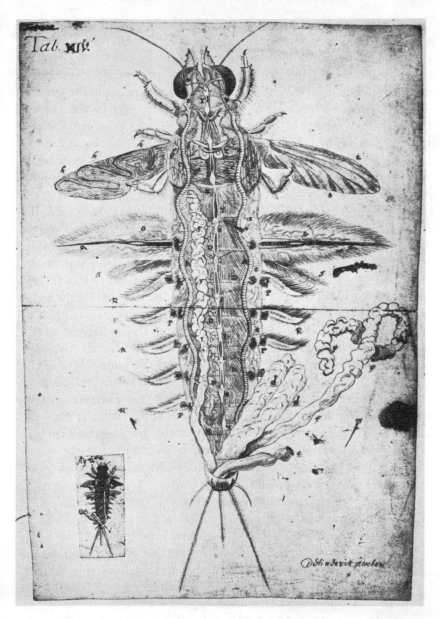

FIG. 116.—Swammerdam.   May-fly.   From the original drawing of his dissection
of the male nymph, *delineavit auctor*

notes on the metamorphosis in 1671. That some of his observations were completed several years before the memoir was written is indicated by the remark that many records had been lost, and the " contents of them are wholly out of my memory ". He is even puzzled to describe certain features which appear in the figures, having " entirely forgotten " what they were, and in the case of the heart he was unable to name the region of the body from which the portion figured had been taken. Notwithstanding this, the anatomy of the small nymph is described from beautiful dissections, and in this respect Swammerdam is clearly superior to Malpighi. Moreover, the zoological works of the Italian anatomist cover a much smaller canvas, and he has nothing to compare with the formidable series of researches assembled in the *Biblia Naturae*. In addition to the anatomy of the species the astonishing life-cycle, in which a momentary adult existence closes, with the savage ruthlessness of Nature, a prolonged and active larval life, is laid bare for the first time. The anatomical section of this work is included in the *Bybel der Natuure*.

Swammerdam was the first to study the anatomy of an Ephemerid, but apparently he did not dissect the adult fly, confining his attention largely if not entirely to the later stages of the nymph. An account is given of the method to be followed in examining so small an animal. His species was *Palingenia longicauda*, the *Ephemera Swammerdiana* (*sic*) of Latreille (1805), the imago of which, according to Swammerdam, lives at most for five hours. The metamorphosis, he says, is so rapid that the wings can only be expanded in the time by a sudden injection of blood from the heart and of air into the air tubes. Details are given of the changes which occur at metamorphosis, and the sub-imaginal stage, found only in the may-flies, which is terminated by an extra moult *after* the insect has left the water and reached the adult winged form, was first described by Swammerdam. Irrelevant observations are by no means excluded from this discursive volume, as when he tells us how deeply he was impressed by the " scaly feathers " on the wing of a butterfly, the structure of which, he says, " deserves an entire Treatise ". Some three years later, in 1678, Leeuwenhoek was equally astonished by the beauty of lepidopterous scales, but he did not publish his notes on them until 1693. Swammerdam correctly states that the body of the nymph

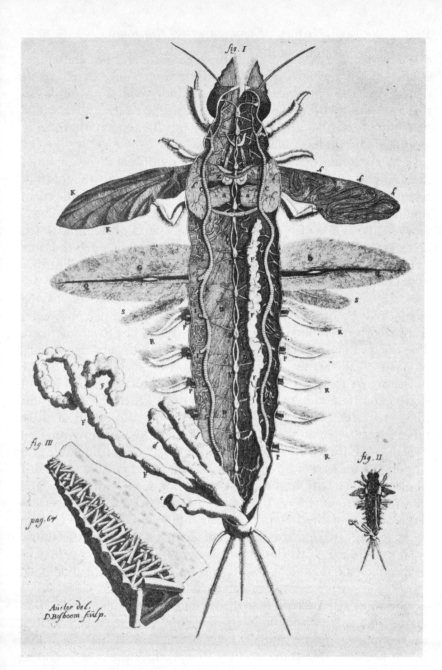

FIG. 117.—Swammerdam, 1675. May-fly. Dissection of the male nymph as engraved in the first edition of the published work

of *Palingenia* consists of a head, three thoracic and ten abdominal segments, but the mouth and mouth parts are only barely mentioned, and the figure shows the mandibles rotated outwards, so that the biting surfaces are turned away from each other. The remaining external characters, however, are more thoroughly examined and understood.

The six biramose tracheal gills of the nymph and the air tubes ramifying in them are accurately described, although they do not resemble in disposition the gills of Crustacea, as Swammerdam would have us believe. He mentions their incessant " trembling " movements, and is of opinion that the two parts of the gill, which he calls the " gill " and the " fin ", serve different purposes, one " cooling the blood in this Worm, as is done in Fish ", and the other functioning as a swimming organ. Both these surmises are incorrect. Vayssière (1882) accuses Swammerdam of having overlooked the rudimentary first pair of gills, which pair, however, is apparently not found in the species investigated by Swammerdam. The tracheal system, he says, transmits air by its very fine ramifications to " all the outer and inner parts whatsoever ", just as the tissues are irrigated by the blood vessels, and it consists of a pair of conspicuous lateral longitudinal trunks from which the rest of this complex system originates. He is aware that the lining of these tubes is cast when an insect moults, although he has not witnessed it in the may-fly itself. This phenomenon is " so considerable that all humane understanding must stand amazed thereat . . . and it would seem incredible if I myself had not seen it distinctly and shown it also to others ". In another passage, however, he describes the moulting of the lining of the air tubes as if he *had* seen it in the may-fly. Swammerdam did not discover that the respiratory organs become a *closed* system in the nymph, and professes to have seen spiracles in the thorax, but the ten open thoracic and abdominal spiracles occur only in the imaginal state.

Swammerdam asserts that the nymph of his Ephemerid fed on the organic matter contained in mud, but that it ceased to feed when approaching the time of metamorphosis, with the result that its gut became quite transparent. He describes the oesophagus as a " thin thread ", which is the imaginal condition, the oesophagus of the nymph being a wide tube, as required by the nature of the food. He points out correctly that the adult may-fly does not feed,

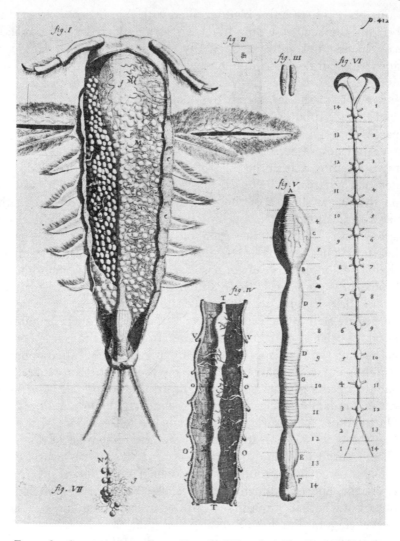

FIG. 118.—Swammerdam, 1675. Nymph of may-fly. Female genitalia, gut, nervous system and heart

and that the now empty gut functions no longer as a digestive tract, but he makes no reference to the degeneration of the mouth parts. He does, however, suspect that the gut of the winged fly is modified for aerostatic purposes, and that the stomach functions as a balloon for the storage of air. He says : " There is yet another reason that our Flie thus lightly driveth on the water, which is, that in its body it hath a fine bladder filled with Air, except it be said to be the Stomach, now only filled with Air, which I cannot strictly say, having not fully satisfied myself therein ". Swammerdam gives an accurate description of the straight unconvoluted gut of the nymph, and describes the varied lining of its different parts. It is surprising that he completely overlooked the numerous and striking Malpighian tubules, nor does he mention that the oesophagus passes through a nerve ring.

After describing parts of the thoracic and abdominal musculature, Swammerdam proceeds to an examination of the blood and nervous systems. The heart, he says, is long and tubular, and lies on the dorsal surface of the gut, as Malpighi described it in the silkworm. It exhibits a swelling in each segment, but he does not agree with Malpighi that each of these swellings is a separate heart. He has seen only irregular motions of the heart in the mayfly. The double nerve chain possesses eleven ganglia, including the brain, suboesophageal, three thoracic and six abdominal ganglia. From these ganglia, which are well supplied with " air vessels ", nerves are sent to the various parts of the body in their neighbourhood, the large optic nerves arising from the brain. Vayssière charges Swammerdam with having missed the suboesophageal ganglia, but this charge is based on the assumption that there should be the normal seven ganglia in the abdomen, whereas in Swammerdam's species it has been stated that there are only six. On the other hand Swammerdam failed to take account of the fact that the brain was dorsal to the gut and the remainder of the nerve chain ventral to it, and that therefore the two halves of the chain must diverge to surround the oesophagus. This was a capital oversight, and one difficult to explain in a skilled and meticulous observer who had himself seen the oesophagus passing through a nerve ring in the hermit crab, the snail, a coleopterous larva (*Oryctes nasicornis*), two lepidopterous larvae (*Vanessa* and *Bombyx*), and two dipterous larvae (*Stratiomys* and *Piophila*), and who believed that this condition occurred in all insects except the louse. His lack of success in the last-mentioned case is readily excusable, but a similar failure in the cuttlefish *Sepia* is beyond comprehension. Nowhere, however, does he appear to suspect that he has discovered one of the fundamental characters in the morphology of the invertebrate animals.

The characters which distinguish the sexes in the nymph are recorded by Swammerdam, and he noted that the female is much the larger " as in all Insects " [?], because she has to carry so many eggs. He was mistaken in believing that the eggs were fertilized *in the water*, whereas fertilization is internal and takes place during flight, as was first suggested by Collinson in 1746 and by De Geer in 1755. Swammerdam was here possibly influenced by his own discovery of external fertilization in an animal, which he had established in the case of the frog. He found two large ovaries in the nymph which contained small eggs, and were supplied with innumerable air tubes, but he does not identify the genital ducts in either sex. He seems to think, however, that they have a common opening with the gut. He therefore failed to observe that the genital ducts of the may-flies are unique [1] among insects in having paired external openings, and hence there is no question of a cloaca. In the male he describes and figures a pair of large tubes which must be the testes + vasa deferentia, and he found also what appear to be paired vesiculae seminales.

The posthumous " Bible of Nature ", handsomely sponsored by the enthusiasm and opulence of Boerhaave, is famous for the detail and accuracy of its plates. They were the work of an unknown artist employed and directed by the author, but the most striking of the ephemera plates is signed " Delineavit auctor " in the original drawing, and " Auctor del." in the engraved plate. In all, eight editions were published, including German, French and English translations. The plan of the treatise, based on selected types, is monographic, comparative and experimental, and it is indubitably the foundation of our modern knowledge of the structure, metamorphosis and classification of insects. In addition, there are equally valuable observations on the Crustacea (for example, the hermit crab), Mollusca (Gastropoda and cuttlefish), and the life-history and anatomy of the frog in both larval and adult stages. There is almost no end to the discoveries which were buried in the *Biblia Naturae* and forgotten for more than a century, and it is possible here to deal only with two examples selected from this stupendous work, but they are representative of the quality of the remainder. Omitting, but with regret, the numerous and important researches on the anatomy of the Mollusca, one of our

---

[1] Not quite. A few earwigs exhibit the same feature.

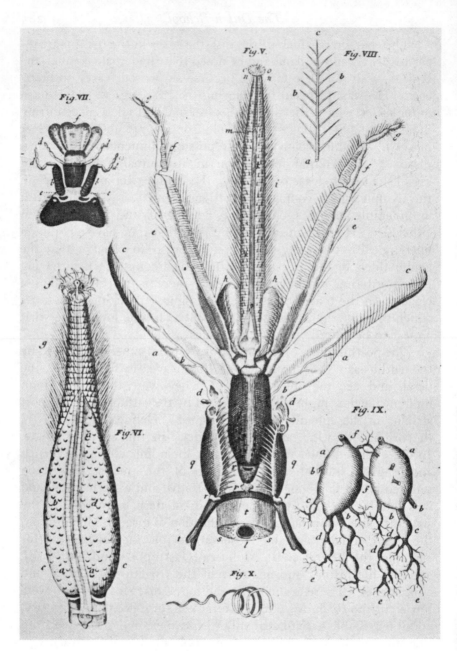

FIG. 119.—Swammerdam. Mouth parts of honey-bee and wasp. Tracheal
sacs or " lungs " of the bee

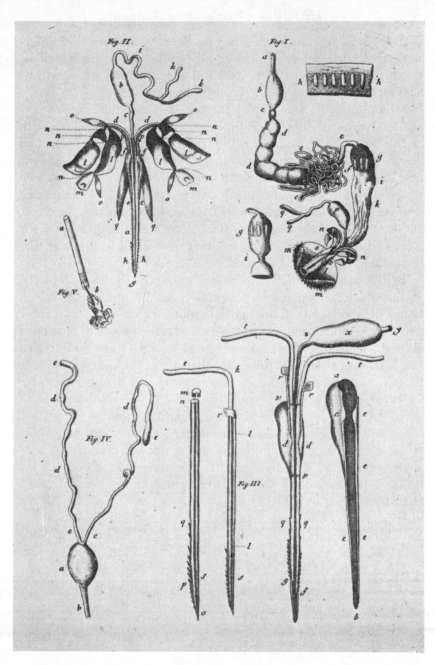

FIG. 120.—Swammerdam. Honey-bee. Gut, sting and acid poison glands

examples must be the Commentary on Bees, by consent Swammerdam's most detailed and accurate anatomical work, on which he
was closely engaged between 1668 and 1673, and the other will be
the Tractatus on the Frog, written probably in 1668.

Swammerdam's anatomy of the honey-bee is the first *comprehensive* account of that much-investigated animal. He deals with
the life-history, the anatomy of the male, female and neuter in all
their stages, and the general economy of the hive. Its accuracy is
such that it would be difficult to cite any other pioneer research
which left so little for subsequent workers to criticize. He proved
by *dissection*, as also independently did Leeuwenhoek in 1700,
that the queen [1] is the mother of the colony, that the drones are the
males and that the workers are neuters, but failed to discover that
the workers are arrested females.

In the *proboscis* all the parts are correctly shown except the
lorum, which is represented accurately as in one piece, but it
should be V-shaped. This figure is more trustworthy than those
appearing in many modern textbooks, where the cardo is usually
omitted. Swammerdam, however, was wrong in concluding that
the glossa was a *complete* tube, and that it was the only entrance to
the gut. He therefore missed the mouth, which was afterwards
described by Réaumur in 1740. In the bee the solid food is in-
gested by the mouth, and the nectar and honey by the glossa.
Pollen would obviously block the glossa. The account and figure
of the *sting* are again admirable, and compare favourably with
current versions. He showed that the sting was absent in the
drone, but overlooked the unpaired or alkaline poison gland,
which, however, is inconspicuous and not easily found. All the
other parts were known to him. He illustrates the stylet sheath and
the barbed stylets or lancets with their diverging arms, and
describes how the lancets move *alternately* along guiding rails
provided by the sheath. The palp-like appendages were observed,
and so also were the three levers — the fulcral, quadrate and oblong
plates. The poison sac and the acid poison glands are accurately
described. The physiology of the mechanism is carefully explored,
the main facts being fully recognized, and he shows that it is the
poison which produces the pain.

The structure of the *compound eye* was investigated by

---

[1] At that time called the king.

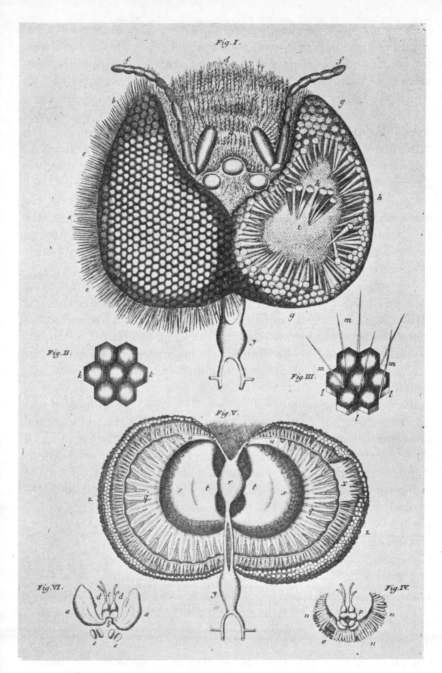

FIG. 121.—Swammerdam. Honey-bee. Structure of compound eyes

Swammerdam before it was taken up by Leeuwenhoek, but owing to the delayed appearance of the *Biblia Naturae*, Leeuwenhoek has priority of publication. The two researches, however, only slightly overlap. Swammerdam was also the first to describe the three simple eyes, and to indicate how they differed from the faceted eyes. He painted over the eyes and produced blindness. His attempt to work out the microscopic structure of the compound eye was very successful considering the crudity of the methods then available. He found that minute tracheae actually penetrated into the substance of the eye. He described the corneal facets, the elongated crystalline cones characteristic of the honey-bee, the rhabdomes + retinulae, and the packing of pigment. His attempt to explain the mode of action of the compound eye was purely speculative, and in consequence a failure.

Swammerdam's description and figure of the alimentary tract include all the important features except the cephalic and thoracic salivary glands, which apparently escaped him. This and other omissions suggest that he dissected the head and thorax much less carefully than he did the abdomen. He traces the course of the very narrow oesophagus through the thorax into the honey sac, which he identifies as the stomach, but recognizes that it is employed for the storage of honey. He saw the mouth of the proventriculus, or stomach-mouth, which he calls the pylorus, and he seems to have found the proventricular valve at the entrance of the stomach, but not the ventricular valve at its exit, or the peritrophic membranes of the stomach which are thrown off by the lining when food enters this part of the gut. His " small intestine " is the true stomach, and the vasa crocea or " yellow gut vessels " are the Malpighian tubules, the " infinite number " of which he was unable to estimate owing to the confusing intrusion of tracheae. There are in fact a hundred or more of them in the bee. The small intestine is his " narrow part of the intestine ", and he divides the colon or rectum into three regions, although its unity is demonstrated when it becomes distended with faeces during the winter months — a condition with which he was familiar. Swammerdam discovered the six problematical rectal glands or glands of Chun, the function of which may be to resorb water from the undischarged faeces and thus to regulate the water-content of the body. He did not devote particular attention to the respiratory system of the imago. He saw the ten spiracles of the larva, but neglected to study them in the adult. He confined his efforts to the abdominal part of the system, where he discovered the two large " lungs of the bee " [1] [abdominal air sacs] and their commissures, of which latter only two out of the six present were seen. He notes that the sacs do not exhibit the spiral thickenings characteristic of the tracheae, and that therefore, unlike them, they collapse when deflated. He regards the sacs as dilatations of the main tracheal vessels, and from them, he says, the small air tubes which proceed to all parts of the body arise. He does not mention the spiracles, nor any connection between the air sacs and the exterior. As regards the blood system he refers only to the dorsal position of the heart, and remarks briefly that it resembles the heart of the silkworm and many other insects.

Coming to the *female genitalia* Swammerdam shows that the queen lays all the eggs. He describes and figures the ovarian tubes, the paired common oviduct, the vagina and the bursa copulatrix. Opening into the vagina is the spermatheca with its associated coiled gland. The poison sac and acid gland are shown, but he again fails to include the alkaline poison gland. This is the only important difference between his beautiful figure and that published by Snodgrass in 1925. In the *male genitalia* Swammerdam found the testis, vas deferens and seminal vesicle. Each of the vesicles communicates with the large accessory mucous gland of its side. From the junction of the mucous glands the single long ejaculatory duct, which opens into the penis, arises. The proximal region of the penis is large, and bears two dark chitinous plates. Further back there is a median dorsal gill-like projection, and ventrally in the same region a series of five transverse plates may be seen. The terminal section of the penis consists of a thin walled chamber which gives rise to two large copulatory pouches ending in points. A very remarkable piece of minute dissection. Here Swammerdam has overlooked nothing, and his figure is again very similar to that in the modern monograph by Snodgrass. Swammerdam was less successful in his investigation of the *nervous system*. The number of the ganglia given is correct, but he has three in the thorax instead of two, and four in the abdomen instead of five. His third

---

[1] This expression seems to imply that Swammerdam was not familiar with the *thoracic* air sacs.

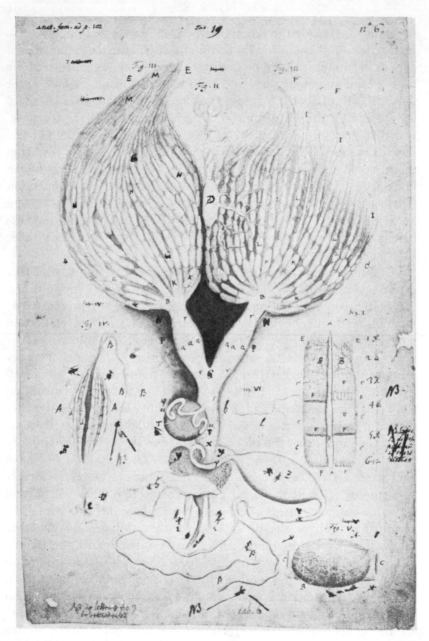

FIG. 122.—Swammerdam. Honey-bee. From the original drawing (much faded)
of the female genitalia

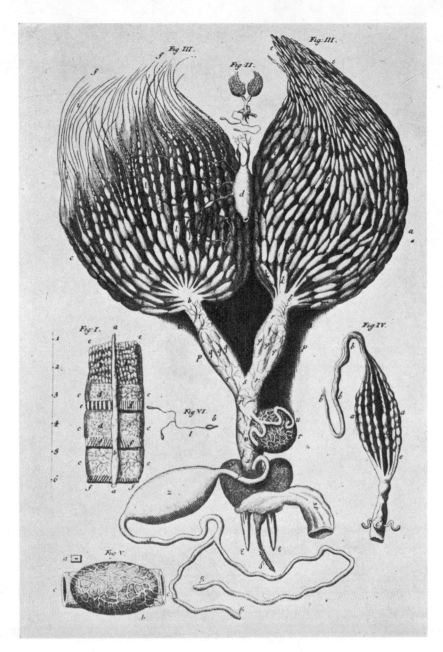

FIG. 123.—Swammerdam. Honey-bee. Female genitalia, sting and acid poison gland as engraved in the first edition of the *Biblia Naturae* of 1738

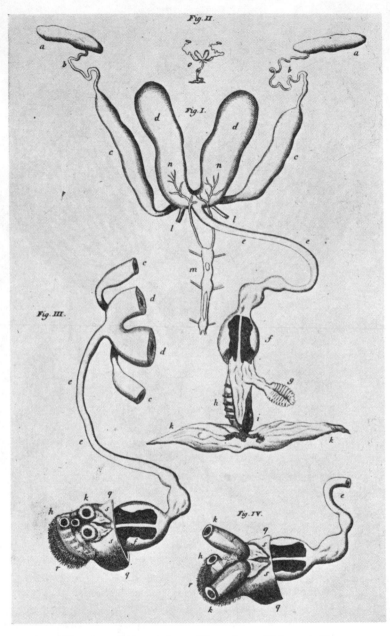

FIG. 124.—Swammerdam. Honey-bee. Male genitalia

ganglion is obviously the composite second thoracic, and his first is apparently not a ganglion at all. He did not find the "first" abdominal, and again failed to observe the nerve ring through which the oesophagus passes. This, however, is not an easy point to establish in the bee, although it was by no means beyond his skill and equipment.

Such are some of the discoveries to be found in this memorable treatise on bees. One of Swammerdam's Amsterdam contemporaries, Dr. Peter Hotton (1648–1709), makes the following minor but welcome contribution to the history of the manuscript: "I am truly grieved that the *Commentarium de Apibus* of our friend was not published, since of everything that he worked at it was the most detailed. It was written as a Dutch discourse, and illustrated with many plates. I well remember to have seen it more than once at his house, but where it is now hidden I am quite ignorant."[1]

Swammerdam's work on

[1] Original in Latin.

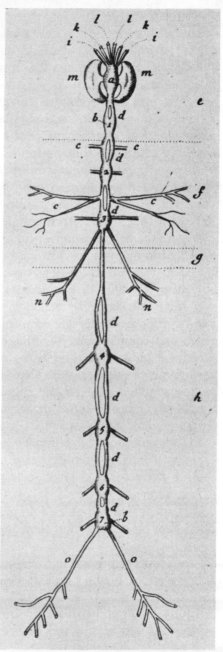

FIG. 125.—Swammerdam. Nervous system of honey-bee

the frog [1] may be considered under three heads : (1) generation ; (2) blood system ; (3) muscle physiology.

(1) *Generation.*—A detailed comparison is instituted between the development of man, the insect and the frog. In all cases *moulting* is assumed, the new parts developing under the old skin, which is then thrown off, as for example in the fore limbs of the tadpole. The vocal sac and copulatory pad on the thumb, both of which are figured, are found in the male only, and are an indication of sexual maturity. The figure of the male genitalia is so good that with slight modifications it might be used in a modern textbook. He discovered and figured the seminiferous tubules and the vasa efferentia, which latter he thinks discharge *directly* into the vas deferens of their side. In the male the ureter is also a vas deferens, but in the female a ureter only. Swammerdam could scarcely be expected to discover or surmise how the semen is conveyed across the kidney in the frog. He describes the adrenals as " singular and strange bodies ", but admits that he did not examine them carefully. He claims that each ovary consists of *separate* compartments without the least communication with each other, so that when one compartment is inflated the others are unaffected. He finds ripe and unripe eggs in the ovary, and points out that in the spring only the ripe eggs are shed, the others being reserved for the next spawning. But *all* the pigmented eggs are not discharged, those remaining behind undergoing resorption in the ovary. Swammerdam discovered the ostia of the oviducts, and mentions this discovery during his lifetime in 1672 in his work on the uterus. He now describes and figures them, and indulges in a polemic against Bartholin and Jacobaeus, the former of whom is accused of extolling the work of Jacobaeus on the frog " in the most disgustful manner ", and of being more interested in the reproduction of his own portrait than in plain and intelligible anatomical figures.

Swammerdam does not distinguish a cloaca in the frog, and states that the bladder and uterine portions of the oviducts open into the rectum. How the eggs get into the oviducts is to him " utterly incomprehensible by human understanding and hid in mysterious darkness ", and has remained so until our own times.

[1] Said to have been written in 1668. Cf. Ent's dissections of the frog of 1677, and Jacobaeus of 1696.

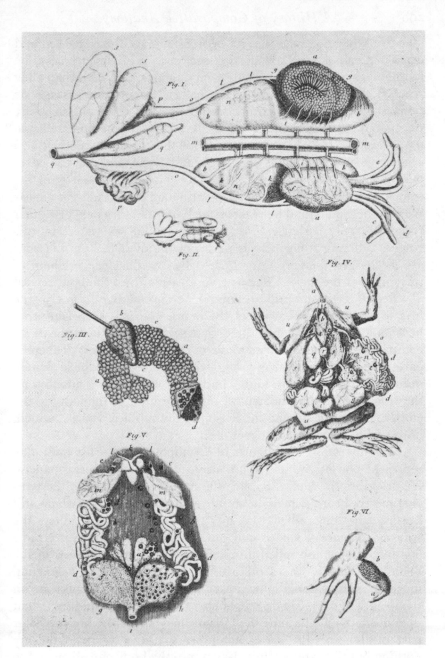

FIG. 126.—Swammerdam, *c.* 1668. Frog. Urogenital system of male and female

The ovary itself, he says, has no opening and can be distended and dried without any escape of the enclosed air. Nevertheless, he found eggs in the abdominal cavity, and concluded that they reached the oviducts through the ostia. In another passage he remarks that the eggs rupture the ovary and roll into the body cavity, but again he cannot conceive how they are manipulated into the oviduct. It is interesting to note that Swammerdam found Distomids and Ascarids in the lungs, and discovered that the latter parasites were ovo-viviparous. He recognized that the jelly of the spawn was secreted by the oviduct, although he did not get a clear view of the glands, and he regards the " uterus " as a non-glandular receptacle only. Hence it does not swell up in water, in this respect differing from the remainder of the oviduct. As the eggs are discharged they are impregnated by an effusion over them of the sperm of the male. This is a very early, if not the first, record of *external* fertilization in an animal. Swammerdam was also the first to observe the cleavage of the ovum, but it is surprising that he should have seen only the first furrow, which he appears to confuse with the medullary groove of the later embryos. He compares the developing frog's egg with that of the chick, which induces him to look for, and " find ", an amnion and allantois in the frog, and he can understand the absorption of the jelly by the embryo only by assuming the existence of umbilical vessels, which, however, he owns he has never seen.

The dissection and figure of a tadpole of the fifteenth day represent one of Swammerdam's most remarkable performances. He describes the four gills, the afferent branchial system, the two-chambered heart, which he removed from the body when still beating, the lungs filled with air, the liver and gall-bladder, the stomach, pancreas, spleen and gut, which latter, he says, is now five inches long and rolled into coils owing to the shortness of the body, and finally the structure of the tail with its segmental muscles. The external gills, he points out, begin to disappear on the twentieth day. He describes how this happens, and thinks that they may possibly become the gills of the young frog, but in a later part of the work he states definitely that the internal gills are nothing but the external gills drawn into the body and advanced to a more important office. He believes that the tadpole breathes by gills and lungs together, and that air mixes with and alters the

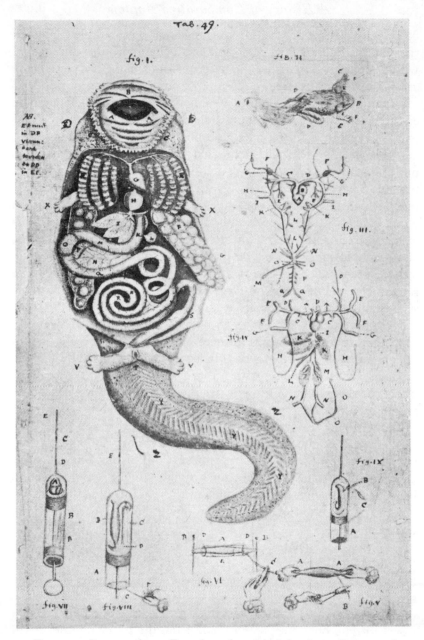

FIG. 127.—Swammerdam. Frog, from the original drawings. Anatomy of tadpole, arterial and venous systems of frog, nerve-muscle experiments

blood in *both* these organs. It is strange that he should have missed the external opening of the branchial chamber, the absence of which he says distinguishes the tadpole from a fish. His statements are not always consistent, and exhibit some vacillation of mind, as for example when he doubts whether the blood of the tadpole circulates through *both* gills and lungs. He thinks that the greater part is dispatched to the gills, and only a small part to the lungs, and then only to nourish them. So far he is on the right track, but unhappily he proceeds to conclude that in the adult frog, when the gills have been " moulted " with the last larval skin, an occurrence which he admits he has not observed, the bulk of the blood goes *to the body*, and only a minor fraction of it to the lungs. This rather fundamental error doubtless explains his most important observational lapse — that of making the pulmonary veins open into the precavals. This he could not have seen, and hence we must regard it as one of those *a priori* "facts" of which the literature of science affords only too many examples.

Swammerdam's success in explaining the relation of moulting to metamorphosis in insects induces him to study the sloughing of the epidermis in the frog, which he carefully describes and figures. He compares the tadpole with an insect because it hides its limbs under the skin, and moulting brings them suddenly into view. The frog, he now says, moults its larval mouth and tail, having previously, and also later, stated that the tail shrivels and dries up. " We see ", he adds, " the tail contracted more and more every day, until at last no vestige of it appears." This moulting " forms another out of one and the same animal, which though different in appearance, yet remains one and the same creature. May not the resurrection of the dead be exemplified in this illustrious instance."

(2) *Blood System.*—The opinion of Malpighi and Needham that the blood is perfected in the gills and lungs, which organs are comparable with each other and answer the same purpose, is not accepted by Swammerdam. He disputes it on the erroneous basis that in the frog and other Amphibia and Reptiles the lungs receive very little and only nourishing blood from the heart, and hence the liver must be restored to its former dignity as the laboratory of the blood, especially as the frog has no lacteal vessels and therefore all matters absorbed from the gut must pass to the liver. In two letters

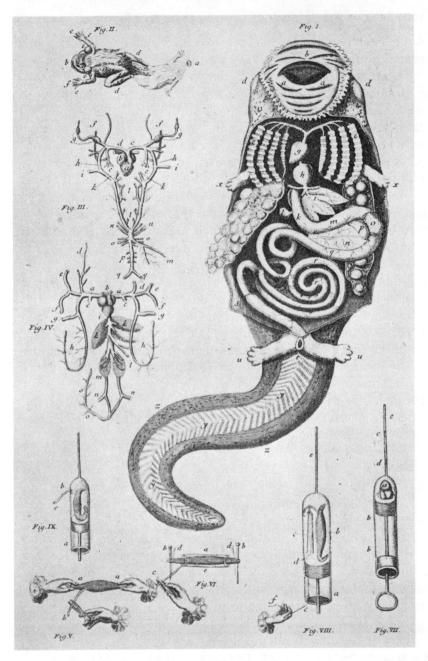

FIG. 128.—Swammerdam. Frog. Preceding plate as engraved in the first edition of the *Biblia Naturae* of 1738

dated January and March 1673 Swammerdam claims that the frog has no pulmonary artery, " so that the blood is sent straight from the heart to every part of the body, without previously undergoing a circulation through the lungs ". This passage must not be misunderstood. Swammerdam saw and correctly figured the pulmonary artery of the frog, but as it "*arose from the aorta*", and not separately and directly from the heart, it had the significance only of any other branch of the aorta, which was concerned with supplying *all* the tissues of the body indifferently, including the lungs.

In the heart of the frog, Swammerdam distinguishes a ventricle and only one auricle, but he mentions that the latter is divided by a membrane not unlike a valve, which looks as if he had seen the inter-auricular septum and had failed to realize what it was. This mistake was bound to nullify any attempt he might make to understand the circulation in the frog. He discovered the very muscular dilated bulbus cordis, but did not observe the spiral valve. The arteries are very accurately described for the time, and he demonstrated the anastomosis of pulmonary artery and vein in the lung by injections of quicksilver. He mentions *two* pairs of carotid glands, one of which was doubtless the thyroid gland, and he fails to trace the internal carotid to the head, wrongly interpreting as the carotid a branch of the systemic arch. The circulation of the frog, he says, differs from that of fish and mammal where *all* the blood passes through either the gills or lungs.

Swammerdam's account of the venous system is somewhat confused. If, as he says, the pulmonary veins are nourishing veins only, and open separately into the precavals, they cannot be compared with the efferent branchials of the fish or the pulmonary veins of the mammal, and, further, if the pulmonary arteries are also nourishing vessels, and transmit very little blood to the lungs, it is necessary to explain how the adult frog respires. He even claims that a true pulmonary vein is wanting in the frog, since *all* the blood does not pass through the lungs. It is curious that Swammerdam made no attempt to extricate himself from the toils in which he became involved when he failed to distinguish two auricles in the heart and to observe correctly the course of the pulmonary veins. Another error, this time more venial, is that the renal portal veins are said to be directly continuous with, and to form a part of, the postcaval. Otherwise he was familiar with the general scheme of the renal portal system, but did not of course recognize it as such. He says that all the veins of the body can be inflated through the anterior abdominal, and he briefly mentions the oval corpuscles of the blood, which he was the first to observe after Malpighi and before Leeuwenhoek. In spite of its defects Swammerdam's figure of the venous system is remarkably good compared with others belonging to his period, but it is not as accurate as his figure of the arteries. It should be noted that he was the first to describe the periganglionic glands of the spinal nerves, now known as the " glands of Swammerdam ", and he states that their contents ferment very strongly when mixed with acid, the otoliths of the skate and other fish, and the gastroliths of Crustacea, being formed of the same " alkaline substance ". Jacobaeus (1686) was the first to *publish* an account of these glands.

(3) *Muscle Physiology.*—It is only in modern times that Swammerdam's classic experiments on the frog, during which he invented a form of plethysmograph, have been brought to the notice of physiologists. His ingenious muscle-nerve preparations, by which he studied the relations of these two tissues, and the nature of muscle contraction and nerve impulse, established the fact that when the cut nerve was mechanically stimulated the muscle contracted, and this reaction could be evoked as often as he willed.[1] He controverts the " idle and absurd opinion " [of Erasistratus], accepted in his time, that muscle contraction was accompanied by an increase in *bulk* — a result supposed to be due to the addition of a subtle spirituous liquor reaching the muscle from the central nervous system via the nerves. Swammerdam interpreted his experiments as proving that there was on the contrary a *discharge* by the muscle, whose bulk was thereby decreased. It must be admitted that Swammerdam here failed to accept the testimony of his own researches, which indicated that there was practically no alteration in bulk when the isolated muscle contracted. He should also have realized that on this point the entire excised heart was not a relevant subject, and, employed as it was in his experiments, proved nothing either way. He must, however, be credited with recognizing the status of afferent and efferent nerves, and he clearly understood the difference between reflex and voluntary

---

[1] Successful experiments on muscle contraction in the frog, obviously by Swammerdam, were conducted in Amsterdam as early as 1665.

actions, but his ingenuous statement that he has seen a muscle contract when boiled in the same balsam in which it had been preserved for several years assumes the identity of two entirely different reactions. His discussion of muscular contraction in the frog is long and argumentative, and lacks that complete experimental backing without which no hypothesis can expect, or deserve, to survive. Swammerdam was very pleased with the results of his physiological ventures, which he demonstrated to the Grand Duke of Tuscany in " 1658 " [1668], and despite their manifest imperfections they justify us in acclaiming him as one of the founders of experimental biology.

Swammerdam was the first to throw any light on the *anatomy* of metamorphosis in insects. In 1668 he had already discovered by skilful dissection, and had publicly demonstrated, that an insect larva, pupa and imago may *at one stage* of the life-cycle exist simultaneously one within the other like a nest of boxes, and he had also studied experimentally the conditions which induce and regulate moulting and metamorphosis. His consequent and fatal assumption that no *new* parts are formed, and that the perfect insect is there *all the time*, led him to adopt, if not to promulgate, the Preformation Doctrine, the long and evil reign of which lies so heavily on his reputation. Swammerdam severely criticizes Harvey's uninspired views on metamorphosis, esteeming him at " little less than nothing ", and stating that his work on generation contains almost as many errors as words. Harvey's philosophy of generation may have been, as Vallisneri says, " encrusted with Aristotelian pitch and heavy with rust ", but it was the deadly obstruction of Preformation that stopped the clock.

The picture which the evidence justifies us in forming of the personality of Swammerdam is that of a man with an introspective and incandescent mind which found expression occasionally in diffidence or arrogance, but more often in the barren distractions of mysticism and spiritual exaltation. He belonged to a type which, whilst humbly but loudly proclaiming its own ignorance, arraigns without mercy the ignorance of others. Add to this a consuming energy which produced, in some ten years before the age of forty, one of the most remarkable works on comparative anatomy which has ever been written, and also an indifferent physique, wasted from time to time by fevers, and the result could only be a life which rapidly burnt itself out. His strength in research lay in observation and experiment. In reflection he lacked restraint and intuition, and apart from the fact that he was a convinced and uncompromising opponent of spontaneous generation, which was incompatible with his belief in preformation, no important generalization can be associated with his name. No portrait of him is known to exist. That reproduced in the English edition of Michelet's *L'Insecte* (1874), and later by numerous other authors, is a forgery based on one of the figures in Rembrandt's " Anatomy Lesson " of 1632 — painted five years before Swammerdam was born. Rembrandt died in 1669, and it was only then that the genius of Swammerdam was beginning to show itself. There is hence no ostensible reason why Rembrandt should have painted a portrait of the naturalist, except on commission, nor do we know that they ever met. Swammerdam, however, was acquainted with Dr. Nicolaas Tulp, the professor of the " Anatomy Lesson ", who was his senior by some forty-four years.[1]

XXV

RUYSCH

Frederik Ruysch was born at The Hague in 1638 and was thus contemporaneous with Swammerdam, who was one of his intimate friends, and by whom he was made familiar with injection technique. Ruysch soon became famous as the apostle of the injection method — in fact it was often referred to in the literature of the time as the " Ruyschian Art ". He was occupied with anatomy between 1665 and 1728, and his main object was to produce finer, more complete and more lasting injections than his predecessors. His preparations illustrated the detailed structure of the vascular system rather than its more obvious features, and he was the first to show that the injected ramifications of the vessels exhibit specific and widely diverse patterns in the various organs of the body ; for, to quote Boerhaave, " in the liver they appear like small pencil brushes, in the testicles they are wound up like a ball of thread, in the kidneys they are inflected into angles and arches, in the

---

[1] Cf. the lengthy historical novel based on the life of Swammerdam and his anatomical contemporaries by Klencke.

intestines they ramify like the branches of trees, in the uvea they form circles and radii, in the brain they are waved in and out in a serpentine course, in the omentum they are disposed something like the meshes of a net, and in almost every other part of the body they assume a different and peculiar structure ". Galen and the ancients believed that there were parts of the body, to which they gave the name " spermatic ", which had no vascular supply of any kind. This belief survived until the seventeenth century, when the wax injections of Ruysch demonstrated the occurrence of blood vessels in almost all the tissues of the body — even in the thin fibrous periosteum of the auditory ossicles (cf. p. 120). The very blood vessels themselves had their vasa vasorum. The more closely were the tissues explored with the microscope, so were finer and still finer vessels brought to his astonished and delighted notice. Ruysch therefore may claim the credit of having established the ubiquity of the blood vascular system. It is natural that he should exalt the significance of his own discoveries, and regard this most subtle and intrusive system as having an existence *per se*. He claims, for example, that by establishing the high vascularity of the cerebral cortex he has disproved its glandular nature as asserted by Malpighi, and his substitution of the vascular for the glandular interpretation of the cortex tends to obscure the important advance that was actually made. He is unusually and unfortunately emphatic on this point. Of all the discoveries which he has been making for forty years, he says in 1705, the most important is the proof that the cerebral cortex is not a gland but a mass of blood vessels.

Ruysch's discoveries, indeed, were his own undoing. His injections were so pervasive that the parenchymatous elements were overlaid and obscured by the multitude of blood vessels which sprang into view. This, and doubtless certain extravasation effects, induced him in 1696 to support the doctrine, first enunciated by Erasistratus, that the tissues were nothing but vascular networks variously arranged. Ruysch was not the first to formulate this view in modern times, although it is generally attributed to him, but he was certainly responsible for its general acceptance by his contemporaries. King (1669) had stated in the clearest language that there was no parenchyma in the tissues, but only " congeries of vessels of various sorts and their several liquors ", and other

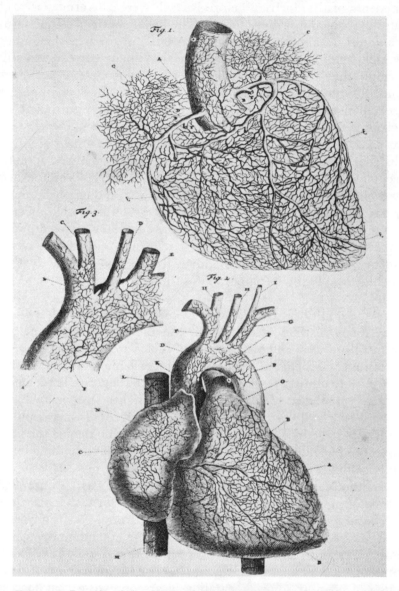

FIG. 129.—Ruysch, 1696. *Homo.* Injection of the coronary arteries and of
the nutrient vessels of the aorta

exponents of the doctrine before Ruysch were Duverney (1679),[1] Muralt (1682) and Ridley (1695). The belief survived in one form or another for almost two hundred years, and towards the end of the eighteenth century even the critical William Hunter was still teaching that the glands consisted of a concourse of vascular elements and excretory ducts without any definite or apparent parenchymatous continuum. Ruysch, however, did not hold that a gland was formed *wholly* of blood vessels, but admitted that between the extremities of the vessels there existed a neutral pulpy substance, and that in certain glands a small quantity of non-vascular substance might be present. He claims to have established this by injecting the arteries and veins with wax, and, after thorough maceration in water, finding nothing but a tangle of the injected vessels. Two exceptions to this vascular autocracy were admitted — the ovary and the testis, the fabric of which Ruysch was never able to prepare by injection, and he concluded reluctantly that red blood did not penetrate into the essential parts of these glands.

As the result of Ruysch's experiments there arose a " despotism of injections ", which attributed the functions of the tissues to the specific disposition and peculiarities of their vascular supply. This disposition was believed to be infinitely variable, and the diverse activities of the tissues were explained as the logical consequence of such variations. Malpighi, who had rightly understood the relations of glandular tissue and blood supply, was thus opposed and thrust aside by Ruysch and his followers, who denied the very existence of glandular tissue, and could see in a gland nothing but a subtle complex of blood capillaries.

Another consequence of Ruysch's injections was an improved method of preserving bodies from putrefaction, for which his name became famous in all the anatomy schools of Europe. His museum was described as a " perfect necropolis, all the inhabitants of which were asleep and ready to speak as soon as they were awakened ". Entire bodies of infants and adults were mummified by the injection of possibly some preparation of arsenic, with what success there is the abundant and convincing testimony of those anatomists

---

[1] The manuscript of the " little Treatise " dated 1679, in which Duverney attempted to establish this point, has been lost. It was not published until 1733, and then only in abstract.

and physicians who were conducted round the museum by Ruysch himself. The following paraphrase of an eloquent passage in Éloy's *Dictionnaire historique de la médecine* may be accepted as a faithful digest of their impressions : [1]

All the bodies which he injected preserved the tone, the lustre and the freshness of youth. One would have taken them for living persons in profound repose — their limbs in the natural paralysis of sleep. It might almost be said that Ruysch had discovered the secret of resuscitating the dead. His mummies were a revelation of life, compared with which those of the Egyptians provoke but the vision of death. Man seemed to continue to live in the one, and to continue to die in the other.

The methods devised by Ruysch which made these astonishing results possible are to a large extent unknown. Rieger in 1742, eleven years after the death of Ruysch, made a partial revelation of the injection methods, said to be based on a set of instructions in Ruysch's autograph, but the document evidently conceals more than it discloses. This secretiveness, or fear lest others should rival their performances, was unhappily characteristic of the early biologists generally, and was doubtless one of the reactions to the plagiarism which was rife at the time. Besides injection, Ruysch was very successful with inflations, especially of the lymphatics, the valves of which he did not indeed discover, but he was the earliest to publish a careful study of them in his first memoir issued in 1665. During his lifetime Ruysch's mummies were apparently immune from corruption, and he himself was not prepared to set a limit to their duration, but after his death the combined effects of natural dissolution and human neglect soon put a period to their survival. In 1794 Jessé Foot remarks : " I saw the preparations, belonging to Ruysch, which are deposited in the Museum at Petersburg, going apace into decay ", and Lieber-kühn, the most important of the pioneers of *histological* injections, had in 1748 expressed the opinion that Ruysch's finer injections

---

[1] Obviously, however, inspired by a passage in Fontenelle's *Éloge de Ruysch* issued in 1731, which in its turn owes something to the rhetoric of Ruysch himself. Hazen has recently expressed a belief that the biography of Ruysch printed in James' *Medicinal Dictionary* (1742) is a translation of Fontenelle's *Éloge* [without acknowledgment !] *by Dr. Johnson.* No evidence is available for such a belief, and if an appeal be made to literary style, it can only be replied that the translation adds so little to the reputation of the lexicographer that Dr. James may well be left in undisturbed possession of its merits.

did not stand microscopic analysis. In 1847 Duvernoy was examining the mercury injections of Ruysch in the anatomical collections at the University of Leyden, and found that many of them had bled so much as to be quite worthless.

The three great contemporaries, of whose anatomical work a brief description has just been given, by no means exhaust the Hollanders of the great age in which they lived, but they are unquestionably the outstanding figures. It was a period in which *anatomical* methods and objectives were only less pursued than they are now. Indeed, in some respects we appear to have gone back on the achievements of these old masters. The modern morphologist, armed with highly efficient binocular microscopes, would be severely taxed to excel, or even to repeat, Swammerdam's dissections of the body louse, Leeuwenhoek's analysis of commercial cochineal, and Lyonet's beautiful preparations of the larva of the goat moth. But without the energizing influence of evolution, advance along *philosophical* lines was bound to be slow. We see in Leeuwenhoek a pure microscopist and an undaunted and tireless opponent of the obstructive fallacies of spontaneous generation, but withal a misguided supporter of the equally dangerous myth of Preformation. The statement commonly made, however, even by some who should know better, that he was so bemused with his microscope that he examined everything that came to hand indiscriminately, without attempting any organized or sustained piece of research,[1] is one that betrays only the ignorance and prejudice of the critic. If it were so it would be a blunder of the first magnitude, and particularly in a pioneer, who should fasten on the heart of his theme, and leave its exploitation to other and lesser minds. Swammerdam was one of those delightful miniaturists of whom his country produced so many in various walks of life in the seventeenth century. The mind of such a craftsman is not at home in large-scale intuitive speculation — he sees in it only dim meaningless outlines and perverse distortions of his beloved detail. And yet *given* a promising idea, he can often contribute to it a useful setting. Thus Swammerdam has produced an admirably comprehensive and accurate system of insect anatomy and metamorphosis. As early as 1667 he had satisfied himself on the latter subject, and was claiming that the world had now no reason to

---

[1] This reproach might be levelled at Hooke, but not at Leeuwenhoek.

lament the loss of Harvey's treatise on insects. We can only regret
that his own work contains errors of speculation, which are
occasionally difficult to understand and even to forgive. Ruysch
taught us that the mere *technique* of science had its victories, and
therefore should be studied for its own sake. It is fitting that we
should hold in honour the memory of these three disciples of
Nature, for we learned much from them, and it was long before
the world saw their like again.

VII

ACADEMIES AND SOCIETIES: THE ANATOMY
LESSON

XXVI
COLLECTIVE RESEARCH

ONE of the most fascinating and difficult tasks of the historian, in addition to exhuming and studying the literature of his subject, is to enquire into the circumstances and tendencies which *initiate* research, and finally lead to publication. It is not enough to trace the slow and erratic growth of speculation and observation towards the light — we must follow the parent stem back to its source underground.

In 1632 Rembrandt painted his famous picture "The Anatomy Lesson", which has found a national home in the Royal Picture Gallery at The Hague. Many critics who are concerned only with the artistic values of the painting believe that the stark uncompromising realism of its subject originated with Rembrandt, but this is an error. He had numerous predecessors, and also successors, and his own special contribution to "The Anatomy Lesson" was to show how such a picture could illustrate the birth of a new method in science. The earliest drawings of anatomy lessons date from the fifteenth century. In most of them the professor is represented standing in a rostrum some little distance from the body with a text of Galen or Mundinus before him. On the floor of the theatre a small company has assembled to follow the course of the dissection, the object of which was not research, but to confirm and illustrate the entrenched tradition of ancient authority. The accuracy and competence of the text were not open to question — indeed, in some schools any departure from it was sternly forbidden by statute. In such an atmosphere research could not flourish, and was not even contemplated. Now although we know that medical students attended anatomy lessons — at Bologna, for example, every medical student of two years' standing had to attend an "*Anatomy*" once a year — the young medical

312

FIG. 130.—Anatomy lesson of Cornelis van 's Gravesande painted by Cornelis de Man at Delft in 1681. Behind the lecturer on his left is Leeuwenhoek. The original painting is in the Gasthuis in Delft

student has no place in the personnel of the anatomy lesson *as painted*. Instead, his place is taken by middle-aged and sometimes elderly persons of the practitioner or professorial class. The reason for this is to be sought in the needs of the artist. Each of the figures in the picture is the portrait of some local medical celebrity who was able and willing to pay his share of the artist's fee. The obscurity and poverty of the medical student sufficiently explain his exclusion. We therefore see, at all events in most of the later paintings, why the canvas is nothing more than a collection of unrelated portraits, each member of the company gazing straight at the painter, and displaying no interest whatever in the dissection, which is introduced only that there may be some suggestion of composition and harmony in the group. The genius of Rembrandt cut out all this dull parade of self-consciousness. His " Anatomy Lesson " is also a batch of portraits, as is well known, but he never loses sight of the picture *as a work of art*. Consequently *his* professor discourses, and *his* pupils are not publicly sitting for their portraits, but are actively participating in the avowed purpose of the lesson. Nevertheless, the example of this great painting was not followed, and the old type of anatomy lesson continued to be produced down to the end of the eighteenth century. It may, however, be here recorded that it was an artist who was the first to hint at the possibilities of collective research, which, later in the century, was to produce such remarkable results.[1]

After the fifteenth century the procedure at these functions undergoes important changes. The professor descends from the rostrum and himself undertakes the dissection of the body. Also, no longer is he cramped by the limitations of an obsolete text — the body itself is his text, and what it teaches must be accepted. The engraved title of the *Fabrica* of Vesalius (1543) and the *De re anatomica* of Columbus (1559) illustrate this type of anatomy lesson, but with such heterogeneous and unmanageable companies it is difficult to see how any original work could have been attempted. It was, in fact, not until the following century that the claims of *research* began to leaven the proceedings of the anatomy lesson, with the result that the traditional objective, which sought only to

---

[1] Bacon advocated the study of science by corporate action in his *New Atlantis*, issued posthumously in 1627, but whether by this he understood a small body of experimenters working together *as a team* may be doubted.

confirm and illustrate, was replaced by a desire to discover something *new*, and so to extend the boundaries of knowledge. This development re-animated the old practice to such an extent that anatomies began to enjoy the doubtful advantage of popular approval, and by *c.* 1650 at Bologna they were made the occasion of formal and state assemblies in the presence of ladies, invitations to be present being extended to the elegant and learned of all faculties. Research, however, demanded for its successful prosecution further modifications in the organization of the lesson, and accordingly we find that attendance became restricted to a few serious and competent workers, directed by one of their number, who acted as the chairman of a committee in the work of which *all* took an active part. This scheme of collective research was quickly adopted in most European countries. It was probably initiated in London as the Invisible College mentioned by Boyle in his letters of 1646–7, and other well-known centres were the French Company known as the Parisians, the German peripatetic Academia Naturae Curiosorum, the Private College of Amsterdam and the Medical College of Copenhagen. The activities of these co-operative study groups may now be considered in some detail.

XXVII

THE ROYAL SOCIETY

Boyle's " Invisible College " developed some years later into a corporate body — the Royal Society of London, having as its main object the advancement of " experimental learning " by organized research. Nor was the Society oblivious of the spirit of decorum and restraint which the pursuit of science should impose on its professors. In the preface to Volume 29 of the *Transactions* the correspondents of the Society are desired to be brief in their communications, and " to omit all Personal Reflections ; for if such should happen to be inserted, the Publisher shall take the liberty of leaving them out ; it being his Opinion, that Disputes on Philosophical Subjects, may be managed with the utmost Candour, Respect and Friendship by the Disputants, whose only Aim ought to be the Search of Truth ".

The Society at once proceeded to put its designs into operation in two ways : (1) Fellows were asked to suggest subjects for research, and to develop their theses *practically* at meetings to be convened for that purpose. Those who were adjudged experts on any special issue were " requested to bring in experiments ", and to demonstrate them with comments in public session. Emphasis was thus laid on the *experimental* approach to science, as contrasted with the purely literary efforts of the scholastics. In this respect, however, the biological activities and methods of the French Academy of Sciences at first surpassed those of the English Society, especially in the domain of comparative anatomy — a superiority said to be due to the active support of Louis XIV in supplying material for dissection, and to the existence of a professional corps of anatomists ; (2) Fellows were also encouraged to contribute observations for publication in the *Philosophical Transactions*, and memoirs for production in book form. This applied particularly to foreign members and correspondents such as Leeuwenhoek, Swammerdam and Malpighi,[1] and resulted in establishing the international reputation for which the Society has always been justly famous. But such work was not accompanied by practical exercises or proofs, and the procedure adopted did not differ essentially from that followed to-day, when the Society concerns itself only with the expediency of publication, and responsibility for the accuracy of the results devolves solely upon the author.

From its foundation the Royal Society displayed considerable interest in anatomical studies. No extensive work, however, was published, but the dead bodies of malefactors were dissected, and attention was directed particularly to specific discoveries and the authorities originally responsible for them. Judging from the papers which appear in the *Transactions*, the Society was more interested in medicine than in pure biology. Coming now to comparative anatomy, and omitting reference to many rare and remarkable species which were described in the *Transactions* from time to time, the early volumes include descriptions of the anatomy of four worms, one mollusc, one crustacean, three insects, three fish, one amphibian, four reptiles, four birds and twenty-two mammals. Anatomical observations on many other species were discussed and recorded in the Journals and Register books of the Society, but were not printed in the *Transactions*.

---

[1] Malpighi signed the manuscript of his last work in 1694 three days before he died, and directed it to be sent to the Royal Society, by whom it was published in 1697.

The above list discovers a preponderating interest in mammalian anatomy. Some of these pioneer researches included in the early volumes of the *Transactions*, and disregarding the contributions of Leeuwenhoek which have already been examined, may now be considered with the view of illustrating the range and character of the activities of the Society at the period of its foundation up to about the year 1700. It should be noted that from the beginning the Society was alive to the importance of keeping abreast with scientific developments on the Continent, and the early numbers of the *Philosophical Transactions* include numerous reviews of foreign biological memoirs and books.

King (1669) discusses the structure of the testis in connection with his belief that the tissues have a purely vascular structure and are without a parenchyma. He refers to de Graaf as having been the first to unravel the seminiferous tubules of the rat. The testis of this animal is well known to be amenable to this operation. De Graaf in 1668 had shown that when the investing membranes were removed the seminiferous tubules could be shaken out and displayed without further dissection, in virtue of the absence of any connecting substance or parenchyma. In a letter to the Royal Society dated July 25, 1669, he describes this experiment, and forwarded to the Society in confirmation of his statements a specimen preserved in spirit of wine, which was figured in the *Transactions* of that year. Leeuwenhoek, among others, repeated the dissection in 1680 and Tyson in 1683, and the preparation was duplicated in most anatomical museums of the period.

Ray's notes on the anatomy of a young porpoise (1671), which animal he "met", far from its native haunts, in West Cheshire, have been reviewed on pp. 199, 204. He saw the compound stomach, lobulated kidneys, pancreas, spleen, liver without gall-bladder, heart, brain and urogenital system — all constructed, he says, exactly on the mammalian plan, except, and here he is in error, that the brain lacks the corpora quadrigemina and that there is no external auditory meatus. His account of the intranarial epiglottis is particularly good, but he does not appear to have any suspicion of its real function. Respect for contemporary authority, combined with a timid reluctance to accept the teaching of his own discoveries, obvious though it was, sufficed to ensure a further lease of life to one of the most flagrant and persistent anomalies in the classification of animals. And so the Cetacea continued to wallow in taxonomic seas disguised as fishes.

An anonymous writer in 1675 considers the question of the role of the swim-bladder in facilitating the motions of fish in the vertical plane, and he suggests that this might be effected either by the taking-up and release of air, or by the control of the *bulk* of the bladder as the result of muscular action. Boyle, to whom this question was submitted, devised an experiment to test whether a fish rises or sinks in the water by " expanding or constricting himself ", but there is no record that this experiment was ever attempted. Neither did it occur to the anonymous author and Boyle, or to Ray (1675), who was stimulated by the correspondence to write an interesting note on the comparative anatomy of the swim-bladder in fishes, that the gaseous contents of the organ might be varied by glandular secretion and absorption.

Passing over Goddard's unenterprising notes on the structure of a female chameleon (1678), we find that the Society was devoting attention to the anatomy of the ostrich in 1682, when Edward Brown reported on the results of his examination of a young male of this species. The Society returned to the subject in 1725 and again in 1730, " by Order of Sir Hans Sloane, Bart. ", the President, and Ranby submitted the results of his dissections of an ostrich, in which there is a good description with figures of the structure of the eye, including the sclerotic plates and the muscles of the nictitating membrane. The latter, however, had been discovered in the cassowary and admirably figured by Perrault in 1676. In 1726 Warren confirmed Brown's statement that there were " beautiful spiral valves " in the rectal caeca, and found some partially digested copper

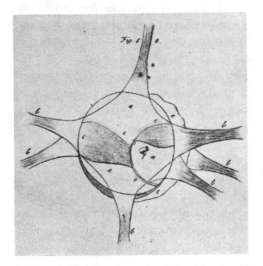

Fig. 131.—Ranby, 1725. Ostrich. Muscles of the eye-ball and nictitating membrane

and silver coins in the gizzard. The Parisians also removed coins from the gizzard of the ostrich, but they give sound reasons for believing that the abrasion of the metal is due to friction with stones in the gizzard, and not to the corrosive effect of any acid. Brown himself saw the " two most elegant " sterno-tracheal muscles, and he was the first to describe [1] the oval bare spot on the dorsum of the head, which he thinks protects the brain from the effects of noxious night dews. This highly interesting bare patch has been shown by Duerden (1920) to be a pineal or brow spot, "the ostrich being the only bird in which a permanent structure of this kind has been described ". Associated with it in the ostrich chick is a large stalked parapineal vesicle, suggesting an ancestral pedunculate eye, which, however, has an embryonic life of only a few days, and is undoubtedly a part of the pineal complex. Brown does not suspect that the organ of voice is located in the *posterior* larynx, but he distinguished the ventricles of the heart, the right having much thinner walls and " more fleshy " valves. The absence of the crop is noted, or rather he regards the pro-ventriculus as the crop, and mentions its characteristic glands. The two long rectal caeca, he says, have a " skrue or spiral Valve within them after the manner of the *Caecum* of a Rabbet ", although here he is only repeating a discovery announced by the Parisian anatomists in 1676. He adds that each valve has " about twenty turns ", as against the twelve found by the Parisians. Brown saw the testis, pancreas and spleen, but no gall-bladder, and he refers to two problematical oval glands situated *behind* the kidneys which might be the adrenal bodies but for their alleged position.

Allen Moulin's [2] work on the elephant (1682), although not published in the *Philosophical Transactions*, was nevertheless one of the publications which the Royal Society may claim to have initiated and made its own. Moulin dissected a male Indian elephant whose carcass had been damaged by fire in Dublin, and considering the crushing difficulties under which the dissection was conducted, the results are truly admirable. His memoir was a pioneer effort, and indeed was the first comprehensive account of

---

[1] Not perhaps the first to note. Perrault remarks that the head of the ostrich is " absolutely bald at the top ".

[2] In the *Dictionary of National Biography* this author appears under the name of Allan Molines or Mullen, but in the Charter Book of the Royal Society, into which he was elected in 1683, he signs himself Allen Moulin.

the soft parts of the elephant to be published. His description of the bones is good, but the figure of the complete skeleton is not too successful, and the excessive number of phalanges attributed to the toes shows that he could not have examined carefully the skeleton of those parts. He realized that the " *Proboscis* was only a nose prolong'd ", though he did not suspect the part played by the upper lip in its formation. He saw the nasal cartilages and the great ligamentum nuchae, or " taxwax ", bearing the weight of the ponderous head, which ligament, he says, is " placed not flat, but edgewise, like Planks used as Joices to bear up Floors ". The fusion of the lungs with the chest wall was correctly described, which was " contrary to what I ever observed in other Quadrupeds ", but in which " they resembled the Lungs of Fowl ". The conformation of the pharyngeal region of the elephant is peculiar. Moulin saw the perforation of the root of the tongue leading from the pharynx to the oesophagus, and from this he rashly concluded that " there was no communication between this [the pharynx] and the passage into the Lungs, contrary to what we may observe in Men, in all Quadrupeds and Fowl that ever I had an opportunity to dissect ". Hence the trachea is " destitute of an *Epiglottis* ; there being no danger of any things falling into the Lungs, from Eating or Drinking ". Moulin is wrong in this. An epiglottis is present in the elephant, and, although it may not be so obvious as in other mammals, there are no fundamental characters of the larynx which discriminate the elephant from other mammalian species. Moulin measured the gut and found it to be about seventy-five feet long, and, if his description of the parts is often defective, allowance must be made for the fire, decomposition, hasty investigation by candle-light, and the acquisitive habits of the general public, who took away portions of the body.[1] He finally disposed of the ancient belief that the elephant has an os cordis. In his account of the male generative organs his knowledge of comparative anatomy led him to look for the testis in the scrotal position, and he described one of the accessory glands (prostate or Cowper's gland) as the testis. Later he acknowledged his error, having found the true testis near the kidney somewhat damaged by

[1] The manager of the booth was compelled to procure a " File of Musqueteers " to protect the body from the attentions of these human jackals. Such troubles, says Moulin, " may satisfie the Royal Society how difficult it was to give a satisfactory Anatomical Account of the Elephant ".

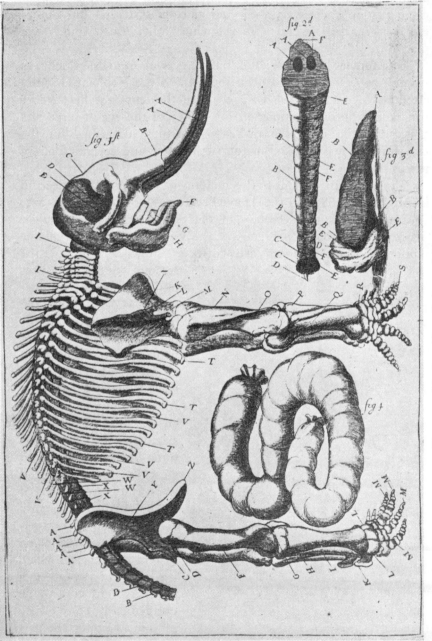

FIG. 132.—Moulin, 1682. Skeleton of the elephant

his butcher assistants, whose " forwardness to cut and slash what
came first in their way, and their unruliness withal did hinder
me from making several Remarks which otherwise I would have
made ".

In a further communication on the eyes of animals, which was
printed as an appendix to the memoir on the elephant, Moulin
attempts to establish the vascularity of the lens, but he is more
trustworthy in observations on the pecten of the eye of birds and
the campanula Halleri of fishes — structures described for the
first time by the Parisian anatomists in 1676 and seen by Peiresc
in 1634.

Tyson's contributions to the anatomy of various vertebrates and
invertebrates (1683 and 1698) have been considered on pp. 198-221.
In 1693 Waller, the secretary of the Society, gives an excellent
figure of the urogenital organs of the water-rat, in which all the
more important features are correctly shown, including the
seminal vesicles and pre-
putial glands. The pros-
tate, os penis, and the
unusual relations of the
penis itself, although not
figured, are recognized,
and the absence of a
gall-bladder is recorded.
Allen Moulin (1693) dis-
covered that the Eu-
stachian tubes of all the
birds he examined united
so as to be connected
with the mouth by a
single median aperture.
He saw also the naso-
pharyngeal or tympanic
system of air sacs, which
he investigated in a
number of birds, and he
concluded that they
played an important part
in hearing. The variable

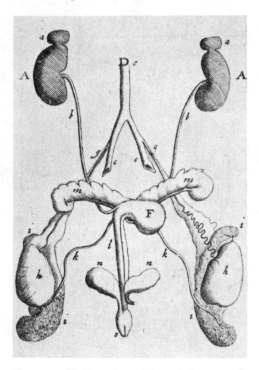

FIG. 133.   Waller, 1693.   Urogenital system of
the water-rat

strength of the trigeminal nerve in birds in response to the variable development of a tactile bill was another point which did not escape the notice of this acute observer of Nature. Waller's dissection of a parrot (1694) reveals no points of special interest, and Thomas Molyneux's description (1697) of " Scolopendra Marina " (*Aphrodite aculeata*) is interesting chiefly on account of the subject. He did not, as he imagined, discover the species, which had been previously examined and figured by Rondelet (1554), Swammerdam (*c.* 1675), Jacobaeus (1675), Redi (1684) and others.   Redi was the

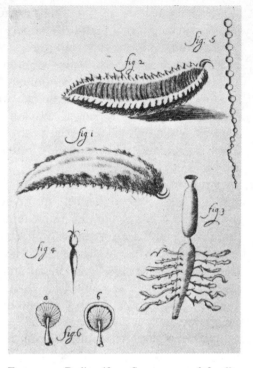

Fig. 134.—Redi, 1684. Sea mouse, *Aphrodite.* Anatomy of gut and nervous system (Redi's series of " little hearts ")

only one of Molyneux's predecessors who made any serious contribution to the anatomy of the animal.   His description and figure of the gut are admirable for the time, and he even gives details of the sieve associated with each of the intestinal caeca.   He fell, however, into one picturesque and instructive blunder.   The blood of *Aphrodite* is colourless, but the ventral nerve chain, and especially the ganglia, are in the living and freshly-killed animal a bright scarlet — due, as Lankester showed in 1872, to the presence of haemoglobin.   This led Redi to assume that the nerve cord was the main blood vessel, and that the ganglia were " little hearts ".   Molyneux's account of the gut, written in ignorance of Redi's work, is much less satisfactory.   His interpretation is equally at fault, since he compares the pharynx with the gizzard of birds, and, moreover, he entirely overlooks the intestinal caeca.   He remarks that the muscles are the most noteworthy feature of the internal anatomy, but describes only the dorsal longitudinal and the superior oblique muscles.

Preston's notes (1697) on the *Structure of the Internal Parts of Fish* fail to reflect the knowledge even of his own time. There was no justification for the statement that the respiratory current of fishes could pass forwards as well as backwards, and his observations on the heart, viscera and swim-bladder simply confirm existing beliefs without adding to them anything new. Lister's anatomy of the scallop *Pecten* (1697) has been noticed on pp. 243-5.

Poupart's researches on the dragon-fly are noteworthy as one of the few attempts to dissect an invertebrate. He describes the ejection of water from the anus of the larva, and found that it could be " shot out a great way ". He believed the object of this unusual habit to be the cleansing of the intestines prior to moulting.[1] He dissected the muscles of locomotion, and arrived at the novel result that " the same muscles which flutter the wings serve also to move the legs ", the upper tendon of the muscle entering the wing and the lower tendon passing " a good way into the legs ; yet the contrary motions of these organs are not at all hindered ; for as long as the wings play, the feet lie still . . . and when the feet are in action, the wings are quiet ". Having expounded this difficult problem to his satisfaction, he proceeds to register another *tour d'imagination*. Each eye of the dragon-fly, he says, is supplied with two tracheae, through which air is driven into the eye to give it greater convexity, and so adapt it to near vision ; but when air is discharged the eye flattens, and becomes adapted to distant vision. " This conjecture ", he adds, in anticipation of criticism, " is not altogether frivolous ", for when air is forced into the tracheae the eyes become considerably tumefied, but flatten again when the air is allowed to escape.

These observations may be taken as representative of the quality of the zoological work which was stimulated and published by the Royal Society in the seventeenth century. Considered as one of the products of the Olympian age in which it was produced, when detailed and accurate anatomical research was already an actuality, even under the most adverse conditions, the record is one with which some disappointment may be permitted and

---

[1] The ejection serves the double purpose of respiration and locomotion.

expressed.[1] Nor can it be compared with the numerous and ably conducted public demonstrations of the Parisian comparative anatomists covering the same period. After 1700, however, the English results are not only more numerous but also more impressive, as will be evident from the two examples which may now be considered.

Patrick Blair (1710), having Moulin's account of the elephant in front of him, had an easier task when he was writing his imposing memoir on the anatomy of that animal, since he could compare as well as describe. His specimen was also an Indian elephant, but a female, which had died near Dundee on April 27, 1706. Blair set out to study and prepare for exhibition the skin and skeleton, to dispose of certain errors which had arisen regarding its habits, and to compare the parts of its skeleton with " large Bones, supposed to be those of *Elephants*, found many Feet deep in the Ground ". He also, like Moulin, had his difficulties, and of very much the same character and magnitude. The Dundee butchers appear to have been as reckless and insubordinate as their Irish brethren, and their crimes are the occasion of the following lament : " I caus'd the *Abdomen* to be open'd, and then the *Thorax*, and that by the unweildy Hands of unruly Butchers, who at opening the first, would have wholly cut through the *Ossa Innominata*, had I not hinder'd them ; and at last, whether I would or not, did so slash the *Sternum*, and mangle several of the *Cartilages*, as to render them useless, cutting and tearing wheresoever they came ". Blair was not able to start operations until late on a Saturday, and, the Scottish Sabbath intervening, work could not be resumed until the following Monday, by which time many of the parts had dried up, and the remains were in an unpleasant state of putrefaction. The weather was " mighty hot " for the season, and the labours of the anatomist were compromised by a turbulent rabble intent on carrying off the carcass in pieces. They even succeeded in stealing an entire fore-foot, " which, after much Pains and the earnest Care of Provost Yeaman, we recover'd about 6 Weeks afterwards. . . . This, I hope, will be a sufficient excuse for the Lameness of the following Account."

Blair gives a close description, accompanied by good drawings,

---

[1] It should, however, be added that in the seventeenth century, and for long after, only a minority of the fellows of the Society were scientific men.

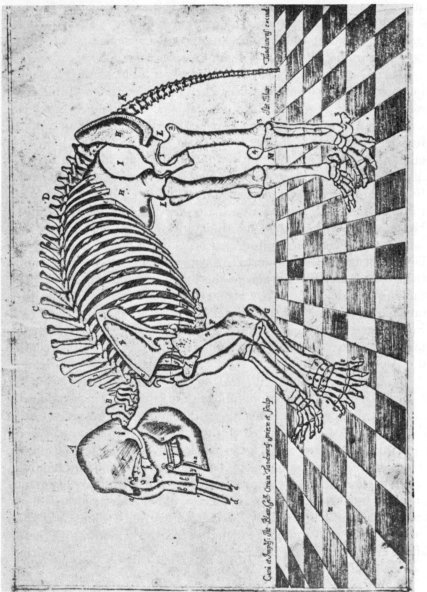

FIG. 135.—Blair, 1710. Skeleton of the elephant. Note that the scapula figured is the *right* and that the first digit is wanting in the hind limbs

of the skeleton, which constitutes the main part of his work. He was the first to see the prepollex (though he gave it joints) and the auditory ossicles. The latter are figured natural size. His method of articulating the skeleton is explained at considerable length. He looked for the semicircular canals of the ear, but found only the cochlea. The cavities of the skull, he says, serve to lighten its weight, in which he agrees with Moulin, but they cannot be filled with water from the proboscis, nor are they associated with smell. The brain exhibits human characteristics, and the female genitalia, including the broad ligament, are figured. He gives a good account of the proboscis with its muscles, blood vessels and nerves, and notes the absence of a gall-bladder. Blair very properly has grave doubts of the accuracy of Moulin's interpretation of the pharyngeal region, but the head of his own elephant had been " so mangled at the cutting off, that I was neither able to receive, nor to give you any satisfaction about it ".

In a later contribution to the *Philosophical Transactions* dated 1718, but published in the following year, Blair returns to the organ of hearing in the elephant, which he has in the meantime further investigated, using for the purpose the os petrosum of the right side, which came away from the skull " after the Head was taken out of the Caldron " in which it had been boiled. He now describes and figures, for the first time, the *bony* labyrinth, and re-describes and figures in greater detail the auditory ossicles. The cochlea, he says, has a spiral of three turns, and is perforated

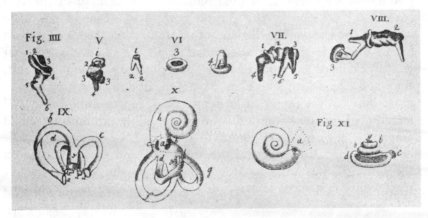

FIG. 136.—Blair, 1718. Elephant. Auditory ossicles, cochlea and bony labyrinth

at the apex to receive a branch of the auditory nerve, " which accompanies and passes along all its *Gyres* ".[1] He admits that, owing to the hardness and solidity of the periotic bone, he " could not so exactly trace the three Canals or Ducts of the *Labyrinth*, so as to give a true Idea of the manner of their several Turnings ", and he mentions being " at some Pains to file down a great part of the *Os Petrosum* " in order to expose the membranous labyrinth. He claims to have traced and measured the canals by inserting bristles into the bony tubes and noting their length. Blair's figure of the osseous labyrinth must be viewed with some dubiety. He admits that in his description of these parts he was " directed exactly " by Valsalva's figures of the human ear, and it is difficult to believe that the labyrinth of the elephant can so resemble its human counterpart that figures of them would be almost indistinguishable. We can only conclude that Blair did not see the complete bony labyrinth of the elephant, nor indeed does he specifically claim that his figures of it are based on that animal.

Gynandromorphism occurs most frequently in insects, and less so in birds, and it is therefore remarkable that the first recorded instance of this phenomenon should be in a group in which it has not since been recorded. In the *Philosophical Transactions* of 1730 Nicholls describes and figures a lobster, which, he says, " if split from Head to Tail, is Female on the right Side, and Male on the left Side ". Nicholls examines first of all the external characters of the normal male and female, and directs attention to the following characters, which make it possible easily to distinguish the two sexes : (1) position of the genital apertures — that of the male being on the last walking leg and of the female on the last but two ; (2) in the male the first abdominal appendage is styliform, that is, " thick, hard and void of Hair ", but in the female it is " soft, thin, and edged with long Hair " ; (3) the abdomen is broader and more hairy in the female in order to provide accommodation and attachment for the spawn. In the specimen described *both* sets of diagnostic features were found to exist side by side, the female characters occurring only on the right side and the male only on the left. It was unfortunate that Nicholls did not completely dissect the genitalia of this unique specimen, urging as his reason a

---

[1] This is an error. No branch of the auditory nerve enters the cochlea at this point, nor is there any vessel which might be mistaken for a nerve.

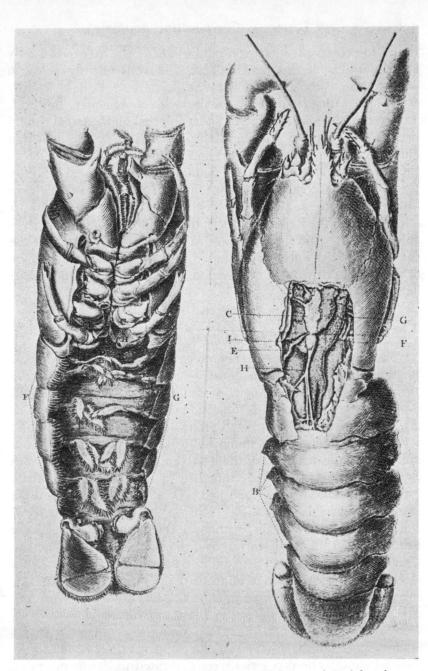

Fig. 137.—Nicholls, 1730.   Gynandromorph lobster, male on left and
female on right

desire to preserve it intact for exhibition. Thus he exposed only short portions of the genital ducts, and consequently failed to observe the complexity of the vas deferens, which is sharply contrasted with the simplicity of the oviduct. Enough, however, was seen to demonstrate the fact that internal and external genital organs wholly agreed in being exclusively female on the right side and male on the left. The heart and some of the arteries, such as the dorsal abdominal, were also figured, but not referred to in the text. Nicholls concludes : The animal was " steeped in three different Spirits, and carefully disposed in a Glass . . . that it may remain in the Repository, as an undeniable Proof of so remarkable a Fact ". And there it did remain until 1781, when, with the rest of the Royal Society's collection, it was interred in the vaults of the British Museum, where we lose sight of Nicholls' ambiguous lobster.

XXVIII

PRIVATE COLLEGE OF AMSTERDAM

We have seen that the seventeenth century witnessed two changes in the outlook of anatomists — animal anatomy or zootomy having at last acquired a vogue of its own, and collective research supplementing, and even to some extent replacing, the labours of the independent worker. Small groups of medical men now met at frequent intervals to study by dissection and microscopic examination the structure of the lower animals. The results were checked by all present, and embodied in illustrated descriptions which were subsequently published. One of the earliest of these groups to be established was the " Private College " of Amsterdam, which produced two little anatomical tracts accessible in Great Britain only at the Bodleian Library, the British Museum and the University of Glasgow. There can be no doubt that these works are the rarest and least known of all the early literature on comparative anatomy. Indeed if it were not for the fact that the " Amsterdammers " were freely quoted in Blasius' *Anatome Animalium* of 1681,[1] their work would be entirely unknown, and, as it is, no modern writer on the history of biology appears to be familiar with their labours. The two series were dedicated to the members of the " College ", and were published at Amsterdam in 1667 and

---

[1] He uses the Latin form—*Amstelodamenses*.

1673.[1] No others appeared, but a premonitory quotation from Julian, which introduces the second tract, already contemplates, in spite of a busy and successful final year, the approaching dissolution of the Collegium Privatum :

> *Etsi alterum pedem in sepulchro haberem,*
> *adhuc addiscere quaedam vellem.*

The Private College of Amsterdam was a society without a home, and is thus comparable with the " Invisible College " of London mentioned by Boyle in his letters of 1646–7. It was founded, probably early in 1664, by Gerard L. Blasius or Blaes (?–1692), professor at Amsterdam, and himself the author of three original works on comparative anatomy which have already been noticed. The most distinguished member of the College was Swammerdam, who joined it in 1665, and whose influence is very conspicuous in their published works.[2] After Blasius and Swammerdam the only member who distinguished himself in any way was the Englishman Matthew Slade. Under the anagram of Theodorus Aldes he wrote a polemic against Harvey's work on generation in 1667,[3] and also produced a treatise on embryology in 1673.

The date of each meeting or session held by the College is duly recorded, except in the case of eight observations which are not *separately* dated. The others range from February 22, 1664 to May 23, 1672, but are not printed in strict chronological sequence. Apart from four meetings held in May in 1671 and 1672, the summer months were observed as vacation, nor are there any entries for 1667, 1669 and 1670. The activities of the College reached their peak in 1665 when there were twenty-one sessions, and even in their last year, 1672, they met eleven times. The records do not suggest that they followed any definite plan in their work, but that the material was examined as it came to hand whatever its nature. Their final interest, however, was certainly fishes, and especially the biliary apparatus and digestive glands of those animals, for which investigation they must have collected the appropriate types. The species examined throughout their

---

[1] Reprinted in 1938 by the University of Reading.
[2] It is probable that Blasius and Swammerdam were largely, if not entirely, responsible for the memoirs of the Amsterdam College. In their later writings both *claim* the authorship of considerable sections of them.   [3] Some copies dated 1666.

studies comprise three flat-worms, sixteen fishes, one amphibian, one reptile, four birds and seven mammals.

Of the comparative anatomists of the seventeenth century the Amsterdammers were almost the first in the field. Their only notable predecessors were Severino (1645) and Steno (1664–7). Swammerdam was one of their own body, and Walter Needham was a contemporary of 1667. De Graaf and Charleton started publishing on comparative anatomy in 1668, the first important work of the Parisians and of Malpighi appeared in 1669, and Willis in 1672 was one of the earliest to dissect invertebrates. Leeuwenhoek's first letter is dated 1673, and therefore did not come under the notice of his fellow-countrymen in Amsterdam. T. Bartholin and the Copenhagen Society date from 1673 (for 1671–2), but Hoboken (1672), Duverney and Martin Lister (1678), Perrault and Tyson (1680), Grew (1681), Muralt (1682), Redi (1684) and Samuel Collins (1685) are all clearly later. It is significant that the comparative studies of all these workers should have embraced a period of no more than forty years. We may say therefore that the Amsterdammers produced the first *extended* work on comparative anatomy after Severino, but that their researches were unsystematic and confined to the vertebrates, as would be natural with men who were making their first contact with zootomy through medicine — a limitation, however, from which Swammerdam must certainly be excluded.

The ground covered by the Amsterdammers was, it is true, restricted, but it must not be forgotten that they were pioneers, and therefore confined by the barriers of imperfect knowledge. What they say of their methods makes us regret that they do not say more. The pancreas and its duct in *Anguilla* were found to be so involved that they spent two hours in cleaning it up. They were familiar with the advantages of Power's frozen sections of such organs as the brain and eye, since in this way only could the relations of the humours of the latter organ be studied. Their most important contribution to biological methods, however, relates to anatomical injections, and it is in their work that we find almost the earliest description of injections with mercury. Malpighi (1661) was actually the first to exploit this medium, and Duncan (1678), who also practised mercury injections, states that Swammerdam had previously attempted the method, which would suggest that it was he who was responsible for the experiments of the Amsterdammers dated February 8, 1672. Their first effort was to inject mercury into the bronchial artery of a calf. The result, they say, was an amazing spectacle, and the arteries *appeared to consist of silver*. Not only the larger but the smaller vessels were filled, and the mercury passed through all the vessels until it reached the pleura. A second experiment with the pulmonary artery of the sheep was not so successful, owing to extensive ruptures. They also investigated " ope siphonis " the relation of the fourth ventricle of the brain to the vessels of the choroid plexus, and they found that the renal artery and vein of the sheep became very conspicuous if they were injected with a coloured liquid, but the statement that the fibrillae of the kidney [uriniferous tubules] became tinted if coloured water were thrown into the emulgent vessels admits of more than one interpretation.

The Amsterdammers attempted, not altogether unsuccessfully, some rough histology, especially of the mammalian kidney, but their physiological experiments were much more important. These were obviously inspired and doubtless carried out by Swammerdam, in which case they enable us to assign a date to perhaps his earliest experiments, when he was only twenty-eight years of age, on nerve-muscle physiology. On October 20, 1665, the spinal cord of a frog was irritated by a rod being thrust into it from the head, in consequence of which the limbs were drawn together. If the nerve passing to one of the limbs were ligatured, a stimulus applied *above* the ligature produced no response in the part to which the nerve was distributed, but if the stimulus were applied *below* the ligature the muscles supplied by the nerve were very strongly contracted. This result held good after the body of the frog had been cut through transversely [that is, when the brain was excluded]. Two simple but pregnant experiments in the history of functional neurology.

The liver fluke is mentioned several times in the literature of the sixteenth century, and was first recognizably described by Anthony (or John) Fitzherbert in 1523. It was, however, *known* long before then. In 1698 [1] Bidloo describes and figures in this

---

[1] This is the date of the reputed first edition — in Dutch, but at least some copies of the first Latin edition are dated 1697. There are also references to an even earlier edition, the existence of which, however, may be questioned.

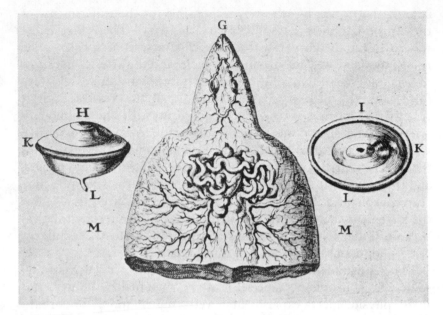

Fɪɢ. 138.—Bidloo, 1697.   Liver fluke, *Fasciola*, showing two large " eyes "

parasite two very well-defined and prominent eyes, like, as he says, the eyes of a flatfish.[1]   The Amsterdammers also concede eyes to the liver fluke, but they admit only " two black spots which have the appearance of eyes ".   It is difficult to say what it was that Bidloo actually saw, unless it was an abnormal paired ventral sucker, especially as he omits to mention whether he saw these eyes more than once.   And even if we accused him of inventing them, we should be more considerate than the writer of his life in the *Biographie Médicale*, who says : " The animalcules which Bidloo describes in this letter addressed to Leeuwenhoek are evidently the products of his imagination ".

Fishes appear to have been the main preoccupation of the Amsterdammers, and they devote particular attention to the abdominal viscera such as the pyloric caeca of the bony fishes, or " pancreas " as they call them, of which they give an excellent account in a number of species.[2]   In the sturgeon they describe and figure the gut and " pancreas ", which latter is peculiar in this fish

[1] The Dutch name for the liver fluke (*Botjes*) means a " little flounder ", which seems to have suggested the comparison to Bidloo.

[2] This part of the work is claimed by Swammerdam in the *Biblia Naturae*.

since the numerous pyloric caeca are aggregated into a single compact glandular mass by connective tissue and an external investment of peritoneum.   The whole of the caeca unite to form a single wide duct " into the opening of which in the gut a finger or thumb could easily be inserted ".   All this was discovered by the Amsterdammers,[1] who saw also and figured the " beautiful " ligamentous spiral valve in the terminal section of the intestine, and understood its delaying action on the passage of the food.   The figure of the gut, swim-bladder and pneumatic duct of the allis shad, *Clupea alosa*, is correct, but the interpretation is faulty. They criticise Needham (1667) for making the pneumatic duct open into what they call a " continuation of the oesophagus ",

whereas Needham rightly regards this part of the gut as the stomach, and is moreover sounder in his location of the pylorus.   The *Panharinck* of the Amsterdammers is presumably the pilchard or sprat, which is recognized as a Clupeoid especially as regards the structure of its viscera and the air bladder, which latter in these fishes, they say, is silvery and leaves this colour on the hand.   In the carp they properly failed to discover any trace of a true stomach, and therefore they oppose de Graaf who had stated that the bile duct opened into the stomach in this

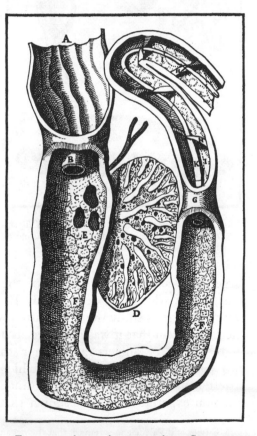

[1] This section on the sturgeon was also probably the work of Swammerdam.

FIG. 139.—Amsterdammers, 1673.   Sturgeon. Gut, showing pyloric caeca and spiral intestine

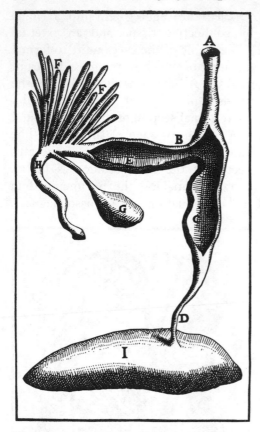

FIG. 140.—Amsterdammers, 1673. Shad, *Clupea alosa.* Gut, swim-bladder with pneumatic duct, gall-bladder and pyloric caeca

fish, which, however, they themselves had asserted in the first part of their work. The description of the texture of the single auricle and ventricle of the heart is brief but accurate, but they were most interested in a feature which was quite new to them. "That which is commonly named the tongue", they say, "is not the true tongue, since it belongs to the upper palate and the food passes *underneath* it. This so-called tongue consists of a kind of glandular substance, is white, soft and tumid, and when pricked or injured in any way it exhibits rapid movements of an extraordinary character." There is no doubt that the Amsterdammers are here describing the large and peculiar *palatal organ* of the Cyprinoids, the gustatory function of which was deduced by E. H. Weber in 1827. They were the first to study the organ after Aristotle and Rondelet, and their statement that it was regarded as a tongue refers perhaps only to the belief of anglers and other casual ichthyologists.[1] The two parts of the swim-bladder, which are separated by a constriction in the carp, and the pneumatic duct are investigated by inflation experiments. It is shown that air blown into one of the

[1] Cuvier and Valenciennes remark that the organ has often been called the "tongue of the carp". According to Monconys (1665), the palatal organ of the carp was discussed at a meeting of the Royal Society held on May 13, 1663.

vesicles passes via the constriction into the other, and hence in one operation the whole swim-bladder can be inflated from the opening of the duct into the gut.

The tench, *Tinca*, according to the Amsterdammers also has no stomach, the bile duct opening into the gut anteriorly in the abdominal cavity, and hence there is no pylorus. The section on the pike is claimed by Blasius, who dissected this fish in 1666 and 1672, and gives a good account of its general anatomy. He describes an experiment which is interesting as one of the earliest attempts to follow the circulation of the blood in fishes. Air was blown into the [cardiac] aorta, and, having passed first through the " heart " [ventricle], it reached the auricle, from thence entering the veins of the liver and afterwards penetrating into all the veins of the body. The inflated ventricle, bulbus arteriosus and afferent branchial vessels are figured. The presence of many " Vermes lati " [probably the cestode *Triaenophorus nodulosus*] in the intestine not far from the pylorus is recorded, and the " obvious " opening of the pneumatic duct of the swim-bladder into the oesophagus was noted. Mention is made of an examination of the ovary *under water*, and of the difficulty of dissecting the pancreatic duct, which was attempted seven times " with a sharp knife ", but without success. The pyloric caeca

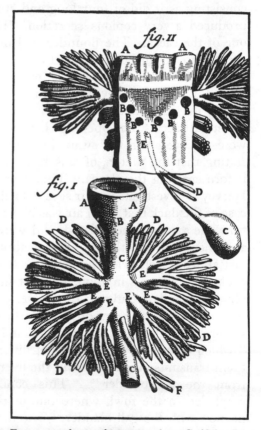

Fig. 141.—Amsterdammers, 1673. Codfish. Pyloric caeca and gall-bladder with their openings into the gut

of the teleost were very carefully examined and figured in the codfish. They were enclosed, we are told, in a thin membrane [peritoneum] which kept them together, and they united to form six " pancreatic ducts " which opened separately into the duodenum.[1] The number of the caeca was about 299, and they were supplied with arteries, veins and nerves which accompanied the bile duct, the latter discharging into the gut in the neighbourhood of the six apertures of the caeca. In the perch the Amsterdammers were correct in finding only three caeca, each opening separately into the duodenum. Of the Pleuronectidae three species were examined — the halibut, flounder and turbot. All present regarded with admiration the large " pancreas " of the halibut, consisting as it did of four long and ample glandular caeca which produced a very copious secretion. The bile duct opened into one of the caeca. The flounder had only two very small pyloric caeca, and the turbot two short ones.

On birds there are only a few casual observations. In the bittern (*Botaurus*) they found that one valve of the heart was clearly muscular, and they noted the absence of a gall-bladder and intestinal caeca. The openings between the air sacs and the lungs were discovered in the swan, but our anatomists were wrong in stating that the ureters of this bird united to form a single duct before entering the cloaca. They were the first after Coiter and Harvey to investigate the air sacs of birds, and they observed in the case of the duck that when air was blown into the trachea through a tube the sacs became inflated, and were found to extend into the abdomen. On the other hand when air was drawn *out* of the trachea just above the clavicles the natural sound [quack] of the duck was evoked. Fabricius in 1600, Perrault in 1680 and Muralt in 1683 obtained similar results in the goose, duck and kite. They figure what appears to be a double sterno-tracheal muscle — a rare feature in birds, and they describe and figure also two bile ducts " which convey to the gut bile of different qualities, since one of them transmits green bile from the liver and the other blacker bile from the gall-bladder ". This condition is to some extent paralleled in the fowl, where one of the bile ducts has no connection with the gall-bladder. The more detailed section on the pigeon is the work of Blasius, who includes an account of the great

[1] The openings are usually about half this number in *Gadus*.

pectoral and the bipinnate subclavius wing muscles, and describes the difference in the colour of their muscle fibres. The air sacs of the pigeon, he says, resemble those of other birds and are two or three in number. He points out how they receive their air via the lungs, and therefore can be inflated from the trachea. When a sac is ruptured the outrush of air will extinguish a candle. The glandular proventriculus is mentioned, the lining of which recalls in appearance the seed of the poppy, and the rectal caeca are no larger than a grain of wheat. Embedded in the " cornea tunica oculi " [sclerotic] there is an osseous circle, and there are two uropygial glands from which an oleaginous material can be expressed.

After the fishes, mammals receive most attention. In the respiratory organs of the sheep the origin (from the aorta) and distribution of the bronchial artery are examined, and the variable structure and mutual relations of the tracheal cartilages, and also of the bronchial cartilages internal and external to the lung, are carefully investigated. Five dissections of the calf and one of the ox are recorded. In the stomach of a newly-born calf a pyloric valve was observed, and they discovered the oesophageal groove and its extension which bring the oesophagus into relation with the reticulum and omasum. All the four chambers of this compound stomach and their characteristic linings are described and figured. They note that the fourth chamber or abomasum is not directly connected with the oesophagus, and that food can reach it only through the third chamber. In an interesting dissection of the heart of the calf a vein was found which did not open into the vena cava, but passing forwards close to the great artery discharged separately into the right auricle. This vessel is obviously the azygos vein, which in the ox opens usually into the great coronary vein, but may pass directly into the right auricle. The Amsterdammers evidently hit upon the variation. In the same specimen they discovered the urachus hidden away among the umbilical vessels, and state that it was connected with the bladder by an aperture large enough to admit a medium-sized style. The lumen of the urachus of the adult ox, however, is generally closed. In the heart of the calf and ox they confirm Steno in finding a hard and hollow os cordis, the interior of which was occupied by a red marrow.[1]

---

[1] There may be *two* ossa cordis in the ox. The Amsterdammers probably found the right member of the pair.

Their " qualia stamina ", found in both ventricles, which did not end in the valves or even reach them, were doubtless the papillary muscles.

The dissection of the rabbit was carried out by Blasius. He gives a good description and figure of the gut, which he says is eleven times the length of the body, and he saw the spiral valve of the very large caecum. He does not distinguish the appendix from the caecum, and even states that the spiral valve extends into it, nor did he observe the glandular (Peyer's) patches first noted by Severino in the dog and cat in 1645. Grew's figure of the gut of the rabbit is later (1681), and somewhat better, but he seems to have been unaware of the work of his Dutch predecessor. The civet-cat, *Viverra*, was a favourite subject for dissection with the early anatomists, and the Amsterdammers examine male and female specimens of this species. They produce a useful account of the general anatomy, directing attention to the unusual development of the temporal muscle, which " almost surrounded the entire head ", and the " bony " penis, but they are specially interested in the paired perineal scent glands, which they found in both sexes, and their cavities were believed to contain amygdalin. The structure of these glands is described and the nature of the secretion tested. In the Molossian dog the lymphatic vessels on the left side of the neck were observed to unite and discharge into the venous system at the junction of the jugular and axillary veins. The thoracic duct also discharged at the same place, and at this point there was an obvious valve. If the lymphatic vessels were inflated and dried they were seen to be provided with well-defined valves, the arrangement of which is described. The oesophagus of the dog exhibited the spiral muscles so characteristic of the ruminants. The fibres of these muscles were collected into two spirals coursing in opposite directions, so that the bundles of one spiral decussated with those of the other. Hence there were two antagonistic tracts — one ascending from the stomach to the throat and the other following the opposite course. The Amsterdammers were aware that such a system had already been described by Steno in 1664.

From this digest it is obvious that the work of the Amsterdam Private College is a collection of somewhat unrelated observations on animal anatomy, physiology, histology and biochemistry viewed from the traditional angle of the medical biologist. Such an attitude was inescapable at that early stage, with no unifying principle such as evolution to guide and illuminate the labours of the pioneer anatomist. None the less this modest little book has no small historical importance, and deserves to be rescued from the oblivion which has unjustly overtaken it.

XXIX
ACADEMIA NATURAE CURIOSORUM

The Collegium Academiae Naturae Curiosorum compares with no other body of its period. It was neither a College nor an Academy, but resembled rather the modern research board in that it promoted research among its members and the well-affected, but did not itself operate as a team. Its headquarters, if it can be said to have had any, revolved about the residence of the member who happened to be president, and hence the Academy, as its ambit spread from town to town, became of necessity a peripatetic body. In these respects therefore it differed from the Royal Society, the French Academy and the Amsterdam Private College, but had more in common with the Copenhagen School. That it exercised a discriminating influence on biological and medical research is certain. Thus in 1683 Muralt says : " What I have so far written on various animals has not been in vain, since I have been specially invited to continue this work by the officers of the Academy, who have judged these matters to be weighty and far reaching ".

The Academy was instituted in 1652 at Schweinfurt by J. L. Bausch, who became its first president. In its initial year ten members only were enrolled, and even after seventy-five years of existence the complete list of members, living and deceased, comprised but 397 names. The publication of serial *Transactions* began in 1670 and, subject to some intermissions and the customary modifications of title, has continued to the present day. An examination of the early volumes up to 1700 provides some astonishing reading. Medicine and surgery predominate, and what little there is of pure science is usually the work of younger men whose main preoccupation was medicine. There are almost innumerable observations on curious abnormalities, such as a goose with three hearts, on startling monsters, and on fortuitous and meaningless resemblances of plant roots to sacred persons and animal types. We are invited to believe in flying Crustacea, based on perversions of obviously typical insects, and in the discharge of worms, scorpions, serpents, small fish and many other animals and strange objects, such as goose feathers, from the apertures of the human body. Even the resemblance of a nut to the face of a monkey is solemnly recorded. Such observations have no scientific interest or value, except in so far as they testify to deceptions artfully practised by the public on guileless physicians, unaccustomed to assess the value of questionable evidence, or even to suspect its existence. They also provoke the suspicion that the early *scientific* work of the Academy must be judged with this failing constantly in mind.

Other activities of these ingenuous students of Nature are less open to criticism, if we except the belief, universal at the time, that substances of mysterious therapeutic value could be distilled from animal tissues and products. They recognized the importance of the diffusion of scientific learning, and hence published extracts, reviews, reprints and translations into Latin of important English and French memoirs and books. Their *Anatomiae Practicae*, or dissections of cadavers, from which results having a direct bearing on pathological science might be expected, are combined with their dissections of beasts, the practical value of which, if less obvious, could nevertheless be predicted. It remains to add that the identification of the animal types examined by them raised many problems, and perchance has produced some errors, but, it is hoped, none of serious consequence to the interpretation of their results.

The researches of the Academy during the seventeenth century cover a sufficiently wide field, and this is their chief and indeed almost their sole merit. Of the 104 animal types examined the internal organs of a moderate number only were dissected, and in most of these cases the author contents himself with listing the organs found, which are more or less accurately named, but no figures or adequate descriptions are included. Still less are there many attempts to compare one animal with another. The contribution of the Academy to the fabric of *comparative* anatomy, therefore, is unimportant, especially if we contrast it with the almost contemporary record of Perrault and his colleagues at Paris, who brought to the task not only greater skill but a more prophetic vision of the speculative aspects of anatomy. Apart from one mollusc, three crustacea, two arachnids and thirty-six insects, the German Academicians were more interested in the vertebrates, and their list includes the lamprey and eleven teleosts, three reptiles, sixteen birds and thirty-one mammals. Most of the classic orders of insects and mammals are represented. Muralt was incomparably the leading zootomist, contributing almost half of the total observations published, Seger coming next with eight, then Peyer with six, Wepfer, Schelhamer and König with four each, and Volckamer and Harder with three each. Since comparative values are ignored in almost all these memoirs, the interest of which is purely zootomical, it will be more convenient, in our brief survey, to take them in zoological sequence.

Muralt practised medicine and surgery during the greater part of his life, although in his earlier years he was attracted rather to zootomy. In his medical and scientific work he was a close follower of Severino, but his dissections are less detailed, and he makes scarcely any use of illustrations, with the result that the interpretation of his descriptions is often difficult and occasionally impossible. The papers by him on the anatomy of insects which were published by the Academy were reissued in book form in 1718. These observations, however, are just scraps of anatomy which conveyed nothing to a mind essentially objective and averse from philosophical issues. His account of the anatomy of *Helix* (1682) includes the statement that the brain is so small that it is scarcely visible — an error explained by his peculiar views on the position of the invertebrate heart, for he identifies the supra- and sub-oesophageal ganglionic masses as the auricle and ventricle of the heart. He evidently saw the salivary glands, and describes the liver and the course of the gut. Having found a large number of eggs in the hermaphrodite gland, he concluded, not unreasonably, that it was the ovary. Compared with the dissections of Gastropods by Redi and Lister of almost identical date Muralt's observations seem negligible. However, the anatomy of the wood-louse, *Oniscus*, is somewhat better. The external characters are recognizably described, and he perceives that the brood pouch of the female is formed by four of the anterior feet. He examined the gut and saw the digestive glands, but his identification of the heart as a small fleshy point at the posterior end of the abdomen is wrong.

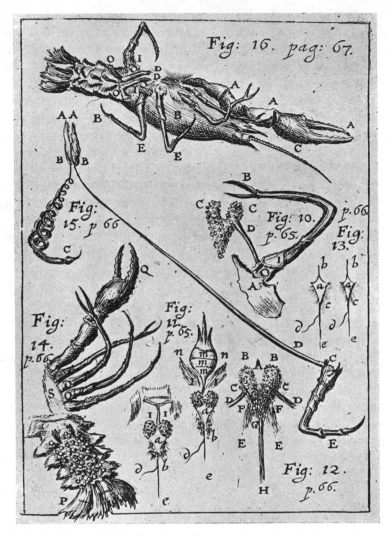

FIG. 142.—Portius, 1687.   Crayfish.   Structure of gut, heart and genital system

The noteworthy dissections by L. A. Portius (Porzio) of the male and female genitalia of the crayfish, *Astacus* (1687), are accompanied by four somewhat crude but useful plates.   The author is diffuse, irrelevant and even garrulous, yet withal a sound observer.[1]   He compares usefully the crayfish with "*Pagurus*", probably the

[1] This paper was Porzio's sole contribution to comparative anatomy, and it is creditable that a man of his alien interests and occupations should have written it.

crab *Portunus*, which he has also dissected carefully, and he distinguishes correctly between the male and female as regards external characters, recognizing the importance of all the significant points. There is a detailed consideration of the abdominal appendages and of their ovigerous function, and a comparison with the modified appendages of the crab. He saw the heart, watched it beating, and points out that the vessels arising from it are difficult to see owing to their white colour, but that they become visible when they palpitate. In this way he discovered the ophthalmic, dorsal abdominal and sternal arteries, and in the heart he found two dorsal ostia which were missed by Willis in his description of the heart of the lobster of 1672. Portius does not mention Willis. He noted the alae cordis, and a good account is given of the ovary with its short straight oviduct, the testis with its long convoluted vas deferens, the gastric mill, liver and proctodaeum. A solid and reliable piece of work, but how different from the beautiful memoir on the anatomy of the crayfish published by Rösel in 1755.

A lycosid spider was dissected by Muralt in 1683. He describes the eyes, but overlooked the anterior series of small eyes which are less conspicuous, and, although he evidently paid some attention to the chelicerae and pedipalps, he states unaccountably that there are only six feet. The account of *Locusta viridissima* is almost confined to the external characters. He recognizes that the wing covers and wings belong to the second and third segments of the thorax, and that the wings have a distinctly characteristic type of venation. The description of the gut is very brief. The structure of the field cricket, *Gryllus campestris* (1682), is treated in more detail, especially as regards the external parts. He is again unfortunate in his identification of the heart, which he professes to have found pulsating in the head. The motion of the elongated dorsal vessel of the cricket, the cardiac region of which extends into the thorax, has been observed to occur throughout its entire length, which may perhaps explain Muralt's error. The gut is described, and the male and female genitalia are distinguished. He is not far from the truth when he claims that stridulation in the cricket is the result of the right wing cover, which is the uppermost of the two, vibrating against the left, and the stridulation of the cricket is compared with that of the cicada. He gives Malpighi the credit of being the first to observe and describe the tracheae of insects, and he points out that such a respiratory system has no connection with the mouth, as in man and other animals. He doubts therefore whether it takes any part in sound production. This paper is one of Muralt's more detailed and thoughtful contributions to the Ephemerides of the Academy.

That quaint and convergent insect known as the mole cricket, *Gryllotalpa*, was first figured in modern times by Ferrante Imperato in 1599.[1] Afterwards it attracted the attention of many naturalists, and was dissected several times before Muralt examined it in 1682. He says that the animal has only four feet, regarding the very specialized fossorial first pair as " arms ". He does not appear to have understood the remarkable shearing action of the tibia and tarsus — an adaptation for severing roots when burrowing. He describes the small wing covers and the elongated folded wings projecting beyond the extremity of the abdomen. His account of the gut is not good, and his views on the nerve cord and heart are incomprehensible. It should be noted that Jacobaeus, in 1677, had already correctly figured the heart, and had given a better description of the gut, accompanied by a passable figure. In 1683 Muralt returns to the mole cricket, and now gives a fairly good figure of the entire insect and two figures of the gut, showing the crop, the armature of the gizzard and the two

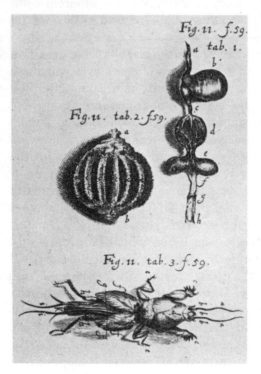

FIG. 143.—Muralt, 1683. Gut of mole cricket, *Gryllotalpa*

[1] According to Boehmer this is the second edition, the first appearing in 1593, but no copy of the earlier edition has been traced.

enteric caeca, which he interprets as the first, second, third and fourth stomachs, discussing the possibility of comparing these four stomachs with the compound stomach of ruminants. His observations on the earwig call for no comment beyond that he failed to discover the wings, and, after referring to the long list of animals which "have so far been brought under my scalpel", he remarks: "I may now, at the request of our Celsus [1] [L. Schröck, jun.], proceed to the examination of the larger Zurich aquatic fly ", which was apparently the larva of the stone-fly, *Perla marginata*. The external characters of the adult fly, as well as those of the larva, are well described and figured, and although he dissected the genitalia, he failed to notice the unusual union of the gonads characteristic of these plecopterous insects. He traced the emergence of the larva from the oval black eggs laid in the water, and witnessed the metamorphosis of the larva into the fly. A few brief notes on internal anatomy are added, and his statement that a small heart occurs at the origin of the thorax is but another instance of his failure to recognize this organ in the Insecta. He did not discover the thoracic tracheal gills, but figures the gills attached to the bases of the tail filaments, which, however, are not mentioned in the text. The gills persist in the adult, but are not conspicuous in that stage, and it is not surprising that they were overlooked.

Muralt (1682) gives a satisfactory account of the external parts of the louse, *Pediculus*, but the internal anatomy was beyond him, although it yielded to the superior and astonishing skill of Swammerdam and Leeuwenhoek. The heart of this parasite is situated at the posterior end of the abdomen, and is not, as Muralt says, concealed in the thorax. What he saw was probably the mycetome, known to Power, which is visible through the ventral skin. He suggests that Hooke's " liver " is not the liver but blood in the gut seen through the transparent skin. Hooke's " liver " is obviously the mycetome. The figure of the may-fly, *Ephemera*, given by Menzel and Ihle (1682), is a caricature, but their account of the life-history, though brief, is correct. No anatomy is attempted, nor were they acquainted with Swammerdam's monograph on the *Ephemera* of 1675, which would indeed have made their own paper superfluous. Muralt (1683) wrote two notes on the dragon-flies

[1] The members of the Academy, on election, were accorded a complimentary pseudonym by which they were known among the brethren.

Anax formosus and *Libellula cancellata*. In both species the external characters receive the most attention, and he is particularly struck by the " very large glassy eyes consisting of a countless number of mirrors ". Of the internal organs the gut and genitalia are dismissed in a few words. The aquatic larva of *Anax*, he says, has its stomach and intestine full of air. It is true that the larval gut acts as a respiratory organ, but the medium is water and not air. Mentzel (1684) also examined the external parts and life-history of *Libellula*, and he is responsible for a well-illustrated but diffuse account of the aquatic larva and its metamorphosis. He regards the eye as its most remarkable organ, and considers that its pronounced curvature and very large size give this insect a wide field of vision. His prolix description, however, gives no indication that he was familiar with Hooke's observations on the compound eye. Muralt (1683) enumerates the diagnostic features of the heteropterous insects *Scutellera lineata* and the aquatic *Naucoris cimicoides*, and in the case of the latter species he examined the three-jointed rostrum, the natatorial hairs of the posterior feet, the gut and the ovary. He also investigated in 1682 an homopterous insect which was probably *Cicadella (Tettigonia) viridis*. The external characters are described sufficiently closely to indicate the identity of the species. He saw and understood the function of the haustellate rostrum, and the faceted eyes are compared with the skin which the French call *chagrin*. The two somewhat long round and green " tails " which project from the end of the abdomen are doubtless the two stout spines which occur on each side in this position. The intestine he says is visible [? through the skin] and is disposed as in locusts.

Of the lepidopterous larvae described by Muralt in 1683 only one is identifiable — the larva of the lackey moth, *Clisiocampa neustria*, but the number of segments and the position of the pro-legs are not accurately stated. He distinguishes, however, between the thoracic appendages and the pro-legs. In the cabbage white butterfly, *Pieris brassicae*, he saw the gut and the male and female genitalia, but they are not described. His dissection of the swallow-tail, *Papilio machaon*, revealed the pedunculated crop and the enlarged mid-gut, which reminded him of the ruminant stomach, and he found also the ovary with " ova in catena ". The beautiful plume moth, *Pterophorus (Alucita) pentadactylus*, was described and figured by Mentzel in 1683, Hooke having previously illustrated the species in 1665. The latter saw only two divisions in the posterior wing, and Mentzel also *figures* two, but in the text he states correctly that there are three, and that the third is concealed under the second. He appears to have seen the species first in 1661, and was unacquainted with its life-history. The scales of the death's-head moth, a species figured by Moufet in 1634, were examined microscopically by L. Schröck, jun. in 1688, who described also the external characters of this insect. Already, in 1664, lepidopterous scales had deeply impressed Power.

Muralt (1683) mentions the clustered eyes of the larva of the great water-beetle, *Hydrophilus piceus*, which are not conspicuous and do not appear to have been seen before, and in the previous year he had described the apterous female of the glow-worm, *Lampyris*. He was probably the first (1683) to examine the leaping mechanism of the click beetle, *Agriotes*, and to suggest how it operated. He had previously been attracted by the similar habits of the flea beetle, *Phyllotreta*, and had noted that the last leg was larger and specially adapted for springing. He must have been one of the earliest, if not the first, to discover that the cerambycid beetle, *Clytus*, stridulates by means of a *thoracic* mechanism, which is unusual among the Coleoptera. The external features and mouth parts of the male and female stag beetle, *Lucanus cervus*, are well described by Muralt, but he fails to recognize that the " horns " are nothing but enlarged mandibles, although he correctly identifies the much smaller mandibles of the female. Of the internal organs he admits that he was unable to find the heart, and the ovary is mentioned without any details being added. His observations on the rose chafer, *Cetonia aurata*, are confined to the external characters and a brief reference to the gut, and the same applies to his notes on the chafer, *Trichius nobilis*, except that here he adds a brief account of the female genitalia. In his anatomy of the cockchafer, *Melolontha vulgaris* (1682), he describes the characteristic lamelliform antenna, and appears to have seen the prothoracic tracheal sacs, which he is not prepared to accept as lungs. He again fails to discover the heart — a failure due to his preconception that the insect heart must be a small compact body, and he never got the length of suspecting that it might be an elongated tube. He found the ovary and the

" bicornuate uterus ", to which the genital duct of the female cock-chafer has a strong adventitious resemblance. The simple coiled gut is mentioned, but a stomach such as he describes does not exist.

Polisius (1685) gives an account, illustrated by two plates, of the external parts of the wood-wasp, *Sirex*. He saw the three simple eyes, and the serrated borer and sheath of the ovipositor. The " sting " of this insect, he says, is known to be fatal to men and animals. An anonymous contributor to the Academy (1678–9) also believes that its bite is fatal to men and beasts, and the species is still dreaded by the German people residing in its habitat. It is, however, a harmless insect, and the so-called " sting " is nothing more formidable than a boring apparatus for depositing its eggs in timber. Muralt (1682) describes the three " splendid " simple eyes of the hornet, *Vespa crabro*, but does not apparently recognize them as eyes. He deals satisfactorily with the external characters, including the compound eyes and the structure of the sting, and of the internal organs he examined the gut and the muscles of the wings and legs. Instead of lungs, he says, " snow-white " tracheae are present.

The small size of a cecidomyid fly compelled Muralt (1683) to confine himself to the external features. He notes the characteristic structure of the antennae in this family, and also the halteres and claspers of the male. In the same year Mentzel describes and figures the pupa and imago of that much-investigated insect, *Drosophila*. He saw two out of the three simple eyes, but does not admit that they are concerned with vision. He had already briefly described this species in the preceding volume of the Academy. Muralt (1682) attempted the difficult task of dissecting the fly *Calliphora*. He found the simple eyes arranged as a constellation of three, and not only failed to accept them as eyes but even criticizes de La Hire (1678) who had claimed that they were. He made very little of the internal anatomy apart from a few observations on the coiling of the gut and the genitalia, which latter left him in doubt whether the animal was hermaphrodite, since he thought he had found ovaries in the male. The penis is mentioned, and he again discovers an impossible heart in a " single conical ventricle enclosed in a pericardium " situated in the abdomen. His attempt to dissect a flea resulted only in the exposure of the gut.

Waldschmid's anatomy of *Lampetra fluviatilis* is dated June 3, 1698. He refers to his predecessor Bronzerio (1626), who described at length from a functional point of view the liver of this species, and also some other parts. Waldschmid notes that the heart consists of a single auricle and ventricle, which continue to beat after removal from the body, and that a gall-bladder is wanting. He found the respiratory diverticulum of the oesophagus which he interpreted as the stomach, but does not mention that the gills are associated with it. He remarks that the common people believe that the branchial apertures are eyes, and hence the lamprey is known amongst them as Neunauge.[1] In 1671 an anonymous author in discussing the problems of respiration in vertebrates points out that mammals use the diaphragm in respiration, but that birds respire and yet have no diaphragm. On the other hand fishes possess a diaphragm [pericardio-peritoneal septum] although it is not used in respiration. It is surprising how much knowledge of the mechanics and manner of respiration in fishes and birds is recorded in the early work of Malachy Thruston published in 1670.

Do fishes hear ? asks Seger in 1673–4. He concludes that they do not, and quotes Kircher to the effect that a definite auditory organ is certainly wanting in these animals. Hence the discovery by Casserius in 1600 of a typical vertebrate ear in fishes had not even then reached the German Academy. The presence of an ear in the aquatic or fish-like mammals, however, is conceded. Seger assumes that the difficulty of hearing under water explains the absence of an auditory organ in fishes. Peyer (1682) dissected a salmon in Wepfer's house in August 1675. He deals with the structure of the gills, on which he says " Willis [1672] deserves to be read ". The heart was found to be densely muscular, to possess a single auricle and ventricle, and in addition a " marvellous structure " [bulbus arteriosus], which he thinks reinforces the propulsive action of the heart. As usual at the time he calls the pericardio-peritoneal septum the diaphragm. He describes very carefully more than sixty pyloric caeca [they vary from fifty to eighty], which, he says, progressively decrease in size like the pipes of an organ, being more numerous and longer near the stomach. The air bladder and pneumatic duct are duly examined, and he

---

[1] The nine is arrived at by counting the true eye and the nasal opening.

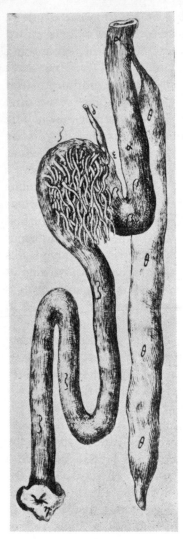

Fig. 144.—Peyer, 1682. Salmon. Gut, bile duct, pyloric caeca, swim-bladder and pneumatic duct

refutes Gesner who concluded that the swim-bladder was a part of a compound stomach and that *Salmo* was therefore a ruminating fish. The pancreas, which is apparently wanting in the salmon, was not found, but it is difficult to explain how he could have overlooked the mesentery, and imagined that the arteries and veins of the gut coursed freely in the body cavity. The comparative anatomy of the caecum in fishes, birds and mammals is considered, in which discussion he holds that *all* blind diverticula of the gut are homologous and comparable. The gut, pyloric caeca and swim-bladder are figured.

The notes on the anatomy of the lake trout, *Salmo ferox*, by Muralt (1682), on the stomach of *Silurus glanis* by Hartmann (1688) and on the head of *Esox lucius* by Hain (1671) treat of simple anatomical facts of no special interest or novelty, but König (1686) on the stomach of "*Lupus piscis*" [? *Labrax lupus*] and *Mugil* directs attention in the latter fish to a very singular type of stomach which resembles in structure and appearance the gizzard of a graminivorous bird. This discovery, however, had already been recorded in 1551 by Belon, who remarks that no other fish known to him possesses a stomach like it. Muralt's dissections of the burbot, *Lota vulgaris* (1682), cover most of its organs. He describes a segmentally valved lateral lymphatic vessel which was probably the lateral line canal. A lymphatic trunk is actually present in the position indicated, but it is not valved, and the cranial course of the vessel as described by Muralt would suggest the sensory canal system. He records observations on the renal organs, ureters and urocyst, genitalia, gut and pyloric caeca, swim-bladder, heart with its auricle, ventricle and bulbus arteriosus, and the ventral aorta breaking up in the gills. " The branchiae of Fishes ", he says, " are nothing but arteries dissolved into minute branches." He saw also the six eye muscles, but the brain is dismissed in a few words. The swordfish, *Xiphias gladius*, was investigated and figured by Hannaeus (1689) and later by Faber (1693). This fish was known to Aristotle, and had attracted the attention of all the great sixteenth- and seventeenth-century naturalists before Hannaeus. His specimen must have been an old fish, since it differs from others described in lacking the posterior remnants of the dorsal and anal fins. The species undergoes remarkable changes in its jaws, teeth and fins during the adolescent period, which takes the sting out of many of Hannaeus' criticisms of his predecessors. His strictures of Jonston (1649), however, are more just, since that assiduous compiler misplaced the pectoral fin and produced two pelvic fins from the rich stores of his imagination. Another member of the Academy, Schelhamer, published an anatomy of *Xiphias* in 1707, which the Academy reprinted in its Ephemerides for 1712. Hannaeus confined himself to an examination of the external characters, and noted the absence of teeth. Faber's paper is much more to the point. He discusses and severely criticizes Bartholin's work on the anatomy of the species (1654), directing attention to his error of interpreting the swim-bladder as a second stomach, which recalls a similar lapse on the part of Gesner. Faber, whose limitations are indicated by his adoption of the expression " Anatome Animalium aut Zootomia ", describes the structure of the skin and states that there are no scales concealed or otherwise, which is correct so far as macroscopic observation is concerned. He found no lateral line pores in the skin, or teeth in the rostrum, nor was there the least perforation of the swim-bladder and hence no pneumatic duct. The kidneys are described and figured as two large club-shaped " processes of the uterus " opening by a single pore. They were seen by Bartholin, who assumed that they were the testes. Faber's " renal " organs must have been the genital glands, which otherwise he failed to notice. There is no *obvious* pancreas in *Xiphias*, which explains its absence from his account. The remaining organ systems are systematically examined and well described, and in his anatomy of the heart he distinguishes the bulbus arteriosus as the " tubulus cartilagineus ". He concludes that *Xiphias* is *not* a cetacean but a true fish, which would indicate that the fundamental distinction between these two aquatic types was clearly understood by him. König's anatomy of a female *Lophius piscatorius* (1694) contains nothing new, and was more than anticipated by Ent's work of 1668, which König quotes. His " auditory ossicle " is evidently one of the otoliths. He agrees with Ent that there is no animal in which more wonderful things are to be found.

Seger's paper on the structure of a snake (1670), probably the grass snake, *Tropidonotus*, includes notes on all the more obvious organs without disclosing anything novel or important. Lachmund, however (1673–4), raises and settles an interesting point. He refers to the popular delusion that a tortoise can emerge from its box like a snail from its shell, especially if glowing charcoal be heaped upon it, and he demonstrates the indissoluble relations of skeleton and box by making and figuring a preparation of the entire skeleton.

Muralt's paper on the heron, *Ardea* (1686), includes a large figure of the entire skeleton. He does not distinguish the coracoid as a separate bone, and is evidently unacquainted with Belon's work on the avian skeleton, or his own interpretation of the wing and foot would have been sounder and more detailed. On the bittern, *Botaurus stellaris* (1683), he gives poor figures of the hyoid and anterior larynx under the impression that the voice is produced there, but he is better informed on the structure of the heart and great vessels, which he perceives conform rather to the mammalian type. Peyer (1683) gives close attention to the " stomach " of the stork, *Ciconia*, and discusses its resemblance to the ruminant stomach. A good figure of the proventriculus (showing its glands) and the gizzard is included, and he quotes Aelian as asserting that this species regurgitates its food for the benefit of its young, which Peyer regards as a form of rumination. The stomach of the stork, he says, differs from that of the goose and fowl, which do not regurgitate on account of the greater condensation and drawing-together of the muscles of their stomachs, whereas in the stork

these tissues are looser and slacker in order to
facilitate the reception of food into a *larger* cavity.
During regurgitation the food is prevented from
being thrust into the intestine by a valve at the
pylorus.   He admits that the structure of the
stomach of this species had already been carefully investigated by his brother Academician
Wepfer in 1671.   Schelhamer (1687) maintains
that the stomach of the stork resembles closely
neither the graminivorous nor the carnivorous
types of other birds, but partakes of the nature of
both, being much more slender than the one and
more solid than the other.   It is, he says, muscular, and its lining resembles that of the ruminant omasum.   Details are also given of the bile

Fig. 145.—Schel-
hamer, 1687.
Marginal minia-
ture of termina-
tion of gut of
stork showing
rectal caeca

and pancreatic ducts, and the two small rectal caeca are illustrated
in a quaint marginal figure.   He describes the cancellous or
" honeycomb " structure of avian bone, which he properly interprets as an adaptation to flight.

The " admirable structure of the windpipe ", which in the
swan exhibits a striking preclavicular loop, housed in a gradually
lengthening excavation of the sternum, is figured in a young
swan by Wedel (1671), but in this he had been anticipated by the
Emperor Frederick, c. 1248, and Aldrovandus (1603).   Lachmund
(1673–4) correctly figures *in situ* and dissected out the hyobranchial
apparatus of the swan.   Peyer (1682) gives Harvey the credit of
having announced the discovery of the rectal caeca of birds in his
" golden " work on the generation of Animals, " quem ruminare
poteris ".   They were, however, known to Aristotle.   Peyer
describes and figures the bursa Fabricii of the goose, which was
discovered by Fabricius and published in his posthumous work on
the chick of 1621.   The urogenital organs and gut of the goose are
admirably figured by Peyer, but Muralt's anatomy of the eagle,
Aquila (1682), contains nothing of importance not previously
described by Borrichius and Steno in 1673.   On the kite, *Milvus*
(1683), he is more successful.   The rectal caeca, he says, are very
small but represent the larger caeca of other birds.   When air is
injected into the trachea near the lungs a sound is produced which
exactly resembles the call of the living bird.   This experiment

brings to mind similar results obtained by Fabricius on the goose (1600) and by the Amsterdammers on the duck (1667). Muralt saw the tympanum of the ear " with the ossicles showing through it ", and also the pineal gland, on which his views have a strongly Cartesian savour. He says it has four obvious anterior and posterior roots by which relations are established with the various parts of the brain, concluding that the seat of the soul is in the brain and especially in its pivotal organ — the pineal gland. There is, he argues, surgical and experimental evidence that the brain may be damaged without death to the organism, but a slight touch of the pineal gland is fatal. Zambeccari (1697) studied the reactions of the fowl and pigeon to the removal of the rectal caeca. In the latter bird, he says, they are so small as almost to escape observation, being no larger than a small grain of wheat. He refers to the variable number of pyloric caeca in fishes, and claims that similar caeca are to be found in Cephalopods.

Muralt (1682) dissected two owls — " *Noctua vulgaris* ", which resists identification, and a species of barn owl, *Flammea guttata*. He dealt with the biliary apparatus, rectal caeca, the glandular proventriculus which " encircles the stomach ", the heart, cloaca, sterno-tracheal muscles and the eye. He did not succeed in finding the ventricles of the brain. He remarks correctly that although the figure of the eye is not round but flattened at the base, the lens is quite spherical. The skeletal support of the specialized and prominent cornea is noted, but the statement quoted in the Scholium that the usual eye muscles are wanting in owls is incorrect, these muscles, though small, owing to the progressive immobility of the eyeball, being present. König (1685) also dissected an unspecified species of owl, and briefly describes the gut, its glands and their ducts, the urogenital organs and the eye. The latter organ, he says, in some respects differs from that of any quadruped. It agrees in having a membrana nictatoria, in structure and size like that of the goat, and it differs in its situation which is not outside the cranium but within it near the brain. These two statements are sufficiently remarkable, and in the case of the second he appears to have been deceived by the peculiar disposition of the sclerotic osseous ring, and by the shape, very large size and immovable character of the eyeball. Nebel (1695), when writing on the Harderian lachrymal gland of mammals, mentions that his attention

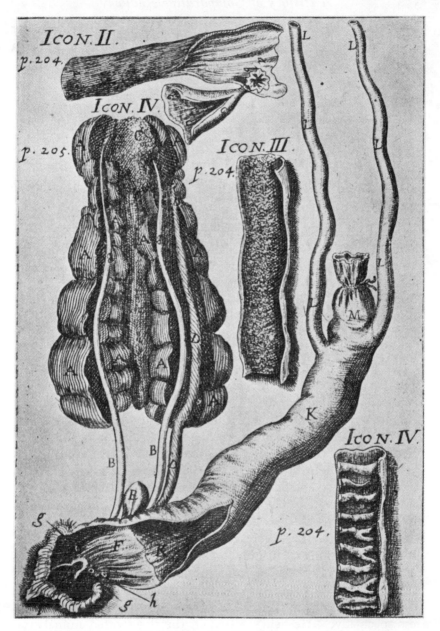

Fig. 146.—Peyer, 1682. II. Salmon. Opening of the pneumatic duct into the gut.
IV. Goose. Urogenital system, bursa Fabricii, rectum and rectal caeca

has been drawn to a similar gland in the raven, and he does not doubt but that such glands will be found in other birds, and also in reptiles. Caldesi (1687) had already found the gland in the tortoise.

Major (1672) dissected a female porpoise, *Phocaena*, which had been captured in the Baltic in the preceding year. He examined the structure of the naso-pharyngeal passage, through which, he says, *water* is ejected from the mouth. He returns to this question in 1677, but still fails to understand that the function of the blow-hole is to transmit *air* direct to the lungs via the intranarial epiglottis. He even compares the passage with the pallial siphon of carnivorous Gastropods, under the mistaken impression that this structure is exhalent in function. Like Bartholin, Ray and Tyson, he describes the three obvious chambers of the compound stomach, but excusably fails to distinguish the fourth. He found the pancreatic duct, as also did Ray, who, however, made it open into the third chamber of the stomach. The compound nature of the kidneys and the absence of a gall-bladder are confirmed. He denied the existence of a neck, and affirms that there are no cervical vertebrae apart from the atlas — a mistake which no comparative anatomist could have made. The muscles, vascular and nervous systems were not examined, nor is the systematic position of the species discussed — again an opening which the speculative anatomist must have seized. Peyer (1682) enters upon a considera-tion of rumination as a function in herbivorous animals, fishes and insects, but he is as yet only preparing the ground for his important monograph on the subject which was published in 1685. In his paper on the Rete Mirabile (1686) Peyer attributes to Galen the first investigation and detailed account of this " miracle of nature ", but says it is certain that the rete does not occur in man, in spite of Riolan's attempt to establish the contrary in his misguided vindication of Galen. Nor, he adds, is it present in the greater number of mammals, but only in those which are the most stupid, and hence it can make no contribution to the attributes of the mind. For example, he could not find it in the horse, cat or dog. He is scornful of the theories of others, but his own is none the less speculative. He believes that in the slow-witted animals the rete retards the flow of blood entering the brain, so as to allow time for that overworked organ to secrete the spirits appropriate to their placid and less ardent minds.

The musk deer, *Moschus moschiferus*, was the subject of a paper by Schröck in 1677, which was later expanded into an elaborate scholastic treatise published in 1682. The paper contains very little anatomy even of the musk glands, the *secretion* of which is the subject of the work. He examined microscopically, however, the structure of the hairs in various parts of the body, and mentions making sections of them with a small very sharp knife, in which achievement he was anticipated by Leeuwenhoek. In 1687 Wepfer completed two papers on *Cervus*, the first of which was on the male genitalia which had been unskilfully removed by the huntsman. He investigated chiefly the contents of the vas deferens and seminal vesicle, and he describes a valvular mechanism which prevents the

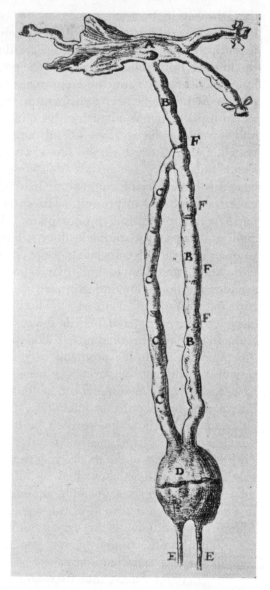

Fig. 147.—Volckamer, 1687. *Cervus*. Receptaculum chyli, thoracic duct (as two vessels) and its opening into the " axillary " vein

former from discharging into the latter. His object is to show that the contents of these two vessels are different, and that the seminal vesicle generates its own liquor. In the vas deferens he professes to have found prominent transverse valvules in both the stag and bull, which were doubtless the plications of the mucous lining characteristic of the posterior region of the duct. The second paper is on the lachrymal organs, by which he means not the definitive lachrymal apparatus, but the suborbital or lachrymal fossa, which is lined by a glandular fold of the integument, and secretes, as he says, a *solid* substance of the consistency of the wax of the auditory meatus. He says this fossa is present in the sheep but not in the goat. Volckamer (1687), unlike Wepfer, concerns himself with the general anatomy of *Cervus*, in which he covers a good deal of ground. His specimen was a male, and his paper is illustrated by five poor plates. He describes the cisterna chyli and the valved thoracic duct, which on inflation was found to open into the left subclavian vein. He saw also the mesenteric lymphatics, their valves and glands or pancreas Asellii, but his account of the compound stomach is defective. He noted only three out of its four chambers. The first, he says, which adheres closely to the oesophagus, is the omasum, and it differs in no wise from that of other ruminants. The second is the abomasum — " comparable with the stomach of man ", and the third is the reticulum, which *arises* from the oesophagus. He does not state how and where the duodenum is linked up with the stomach. There are, however, four chambers in the stomach of *Cervus*. Volckamer's third is the paunch, and he missed the reticulum, which is not as conspicuous in the deer as it is in most ruminants. No bile or pancreatic ducts are mentioned, and he did not find an os cordis as had been described by others. Some of the tissues were examined microscopically, including the lungs and tongue, in which he saw the alveoli of the one and the taste papillae of the other. Observations on the larynx and hyoid, salivary glands and their ducts, brain, cranial nerves, pineal and pituitary glands are included in this conscientious study.

Reisel (1679) claims that there are parietal valves in the postcaval vein of the sheep, of which it need only be said that if they are there they are very difficult to find. Volckamer (1686) gives an account of his dissection of the head of a calf, which is accompanied by a good figure. It is an admirable and detailed piece of zootomy — the work of an accomplished human anatomist, and

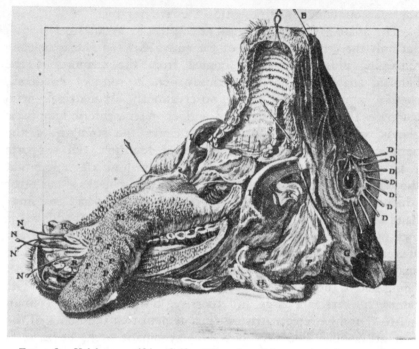

FIG. 148.—Volckamer, 1686.   Calf.   Dissection of the mouth : *A*, naso-palatine canals ; *E*, *N*, openings of salivary ducts ; *F*, *H*, parotid gland ; *Q*, tonsil

one of the best published by the Academy.   He investigates carefully the muscles of the jaws, eyelids, nasal openings, pinna of the ear, and lips, as well as other cephalic muscles, and also the lachrymal and salivary glands and their ducts, which latter he injected with a blue liquid.   The six normal eye muscles and their nerves are described in some detail, and he does not overlook the seventh or choanoid or retractor oculi muscle, which is best developed in ruminants, and is *not* present in man, as he points out.   He then turned his attention to the macroscopic structure of the eye, nose, external and middle ears, auditory ossicles and Eustachian tube, but made no attempt on the labyrinth.   He also describes and figures the paired *separate* openings into the mouth of the submaxillary and sublingual glands, which is the normal condition in the ox, although the two ducts may fuse and have a single aperture only on each side.   He concludes with some detailed observations on the hyoid and larynx, tongue, naso-palatine canals, brain and its membranes, and the cranial nerves, but unaccountably fails to mention the rete mirabile.

The anatomy of the lion was attempted by Wolfstrigel (1671), but only the very poor figure of the entire skeleton is original, the remaining illustrations being copied from the memoirs of the Parisian anatomists, nor does this paper, in spite of its detail, contain any important new observations. Wolfstrigel, who compares the lion with the dog, describes the mesenteric lymphatic glands or pancreas Asellii, and correlates the strength of the temporal musculature with the powerful dentition. He supports J. C. Scaliger's criticism of Aristotle's astonishing statement that the lion and wolf have only one bone in their necks — an error which is inconsistent with the belief that Aristotle had any first-hand knowledge of the anatomy of the lion. Wolfstrigel adds to his own memoir a translation into Latin of the first description of the anatomy of the species by Perrault, dated June 28, 1667. The anatomy of the " tiger " was Wolfstrigel's next essay, his subject being one of the larger Felidae but not necessarily the tiger. The bones, he says, are large and massive, but nevertheless contain marrow, and he rejects Aristotle's statement that the bones of the lion are solid. The very powerful zygomatic arch, he says, is bound up with the excessive development of the jaw muscles. Various comparisons with the lion are suggested, but the paper does not represent any advance in our knowledge of Felid anatomy. Gahrliep's attempt (1690) to dissect the auditory organ of the wolf was largely frustrated by the strength and unyielding coherence of the parts, which required to separate them a " sharp iron wedge ". He saw, however, and perhaps discovered, the laminae which radiate from the annulus tympanicus, and form the mastoid cells which are specially developed in the Carnivora, and he was interested in their function, but makes no further contribution to the structure of the ear. His figures are indifferent and badly lettered.

Wedel (1671) refers briefly in passing to the curiously perfect articulation of the lower jaw in the badger, *Meles taxus*, and Muralt (1686) is responsible for an account of the general anatomy of this species. He examined the muscles, thoracic and abdominal viscera, thoracic duct, male and female genitalia, os penis, recurrent nerve of the neck, and the brain, which latter, he says, agreed with that of the dog, except that he was unable to find the pineal gland.

In the gut he noted the absence of a caecum and sacculated colon. He adds negligible figures of the gut, and of a skeleton which had been articulated by a surgical colleague. He quotes also from a letter by E. König, who thinks that the thymus is a kind of pancreas and should therefore have a duct, which he professes to have found, but admits that as it was too small for inflation it *might* have been a blood vessel or nerve. T. Bartholin (1670) finds it difficult to banish from his mind the image of that substantial fairy, the *Homo marinus* — a being whose existence rests insecurely on picturesque distortions of the smaller marine Carnivora.

For many reasons the common seal, *Phoca vitulina*, attracted the attention of the early anatomists. Severino published a treatise on its structure in 1655 (preface dated 1645), the Parisians dissected it in 1676, and two members of the German Academy followed their example. They were Seger (1678–9) and Schelhamer (1699–1700). The former examined a female specimen which was in a bad state of preservation, having been captured in November 1675. His paper includes very brief notes on the thoracic viscera, gut and pancreatic duct, compound kidney, genitalia and the skeleton. He notes the absence of an external ear, but saw the meatus. The muscles, nervous and vascular systems (except the heart) are ignored. Schelhamer, whose memoir on a male seal covers much of its anatomy, and is accompanied by one unlettered plate, describes the structure of the compound kidney, which, he says, compels the greatest admiration and is a most wonderful spectacle. The plate illustrates the anastomosing vascular plexus on the surface of the kidney, which has been inflated to render it more conspicuous. The dissected kidney was injected so as to demonstrate the formation of the ureter by the union of factors from the component renules, and the uriniferous tubules of each renule are shown converging to constitute their own factor of the ureter. There is hence no common pelvis in such a kidney. Schelhamer asserts that both colon and caecum are absent in the seal, the gut having a uniform diameter throughout. This is to ignore the two divisions of the stomach, and a caecum, though small, is present and by no means difficult to find. His statement that a compact or definite spleen is wanting is again an error, since a spleen of the usual type occurs in its normal position. The heart is correctly recognized as having two auricles and ventricles, and the fact that the foramen ovale and ductus arteriosus are closed, explains, he says, why the animal cannot live a long time under water. This inaccurate conclusion must not be ascribed to Schelhamer, who was only adopting a belief accepted by most anatomists of the time. The genitalia are adequately dealt with, and he rightly points out that a prostate is present but that there are no seminal vesicles. An os penis similar to that of the dog was found, and the structure of the penis itself was closely examined. Although Schelhamer realized that the limbs of the seal conformed to the *terrestrial* type, and that on this ground alone its aquatic status was ambiguous, he made no attempt to arrive at a solution of this very simple problem. It must, however, be borne in mind that the early anatomists, who not only had no conception of evolution, but had thrust upon them views on species which were a direct and formidable negation of evolution, found any progress towards a settlement of such questions barred at the outset. It is interesting to note that Schelhamer refers to the meeting of Hippocrates and Democritus, and to the dissections of various animals which that philosopher is presumed to have made.

The paper by Nebel (1695) on the Harderian lachrymal gland in rodents is illustrated by figures of the rabbit, hare and squirrel. He injected its *duct* with wax, and found that the injection passed into the vesicles of the gland, and also into all the blood vessels, from which he concluded that there existed a *direct* connection between the blood vascular system and the glands. This was the view of Frederik Ruysch, who maintained that blood was poured directly into the factors of the efferent ducts of the glands, since his wax injection when thrown into an artery emerged by those channels without entering any intermediate non-vascular tissue or spaces on the way. Velschius (1670) describes the pancreatic duct of the alpine rodent, *Arctomys marmotta*, which duct had been first seen in an *animal*, the turkey, by M. Hofmann in 1641.[1] Wepfer (1671) gives an admirable account, but without figures, of his dissections of two male beavers, *Castor*.[2] The first specimen had been obtained in 1667 swimming in the Rhine, and its head had been

---

[1] Hofmann was then a student at Padua under Vesling. He did not claim the discovery until 1648, when the duct had already been figured in man by Wirsung in 1642.

[2] The French translation of this paper by J. Gautier (1756) is in places grossly careless and inaccurate.

seriously battered by its captor. Wepfer dissected it with the assistance of his colleague, E. Hurter, and an English physician, H. Sampson. Nothing, he says, is more admirable than the structure of the stomach. He describes the subdivision of this viscus into two parts by a constriction, and he made a very careful examination of the gastric gland, which he thought at first was an inflammation. He found the highly vascular follicles of the gland and their numerous openings into the stomach, through which they could be inflated. They were clustered together like the grains of maize in the cob. The second specimen was undamaged, and was dissected in more detail in the presence of three physicians. The foramen ovale was closed, and there were pyloric (sphincter), ileo-caecal and colic valves. He saw the ano-preputial sac, and he gives an excellent description of the castoreal glands. They are enclosed, he says, in a " common, fibrous and almost muscular membrane " [it *is* muscular], which surrounds the parts and compresses them. He examines very carefully the large anterior pair of glands, inflates them, and shows that they are *sacs* with a corrugated lining, and not a parenchymatous gland as he had at first thought. He failed to unravel the posterior pair of castoreal sacs, which he says resemble the anterior pair, and he did not discover that each member of this pair consisted of three separate glands. The paper includes a fairly comprehensive description of the urogenital system, including the os penis and all the more important features, and he examined also the nature of the contents of the parts. Wepfer's version of the anatomy of the species is thorough as far as it goes, but there is hardly anything on the skeleton and muscles, and nothing on the vascular and nervous systems. He appears, moreover, to have been ignorant of the memoir on the anatomy of the beaver published by the Parisian anatomists in 1669, but it must be allowed that he was close on the heels of his French predecessors, and it is possible that his dissections may have antedated their own.

In the opening sentence of his note on the hamster, *Cricetus*, Clauder (1686) remarks that " the muses love change " — evidently thinking that the reader would welcome a respite from the depressing medicine favoured by the Academy. He reports the discovery by J. Hilscher [1] of the very large cheek pouches of this

---

[1] No publications by this observer have been traced.

rodent, which are used for storing and transporting food. Although they fill him with astonishment, Clauder contents himself with the bare announcement of their existence. There is, however, a good figure. Peyer (1682), in his memoir on the anatomy of the leveret, expresses admiration of the comparative studies of Muralt. He himself carried out experiments on the reactions of the gut in the living animal. He gives a figure of the male genitalia in which spermatic arteries and veins fuse before reaching the testis, which seems to be the intention of the author ; and the vas deferens completely fuses with the ureter, which may perhaps mercifully be laid at the door of the engraver. Moreover, the prostate is called the vesicula seminalis, and the perineal glands are identified as the prostate. The dissections of the salmon and goose, already referred to, are also discussed in this paper, which, in spite of its errors, is one of the best of the early essays in comparative anatomy published by the Academy, and it surpasses the random notes by Muralt in having at least some philosophical bias. Schelhamer (1697–8) describes the rectal and perineal glands of the hare, and rejects as a fable the statement that this animal is hermaphrodite.

One of Muralt's best contributions to the Academy, but still a medley of anatomical jottings with no attempt to deduce anything from them, is his anatomy of a female hedgehog, *Erinaceus* (1682). Harder's account of the male (1687), apart from his description of the urogenital organs and associated glands, adds nothing of importance. Muralt works through the thoracic and abdominal viscera, touching, as will be seen, not always accurately, on such features as the pancreas Asellii, absence of caecum, heart without a pericardium, valves of the veins, lacteal vessels, pancreas and its duct, salivary and ductless glands, hyoid and its muscles, and the auditory ossicles. He has not written much on the skeleton, but the figure is good, and the genitalia are almost ignored. He discusses the unique orbicular muscle, which, however, had been known since the time of Coiter.

Seger's notes on the anatomy of a male mole, *Talpa* (1671), reveal little that is new. He thinks the eye, despite its degenerate condition, must serve some purpose, if only to be sensitive to light, in which he agrees with the " acute Scaliger ". He directs attention to the absence of an obvious external ear, but found the auditory meatus, which led to a cavity in the petrosal bone. Schelhamer's

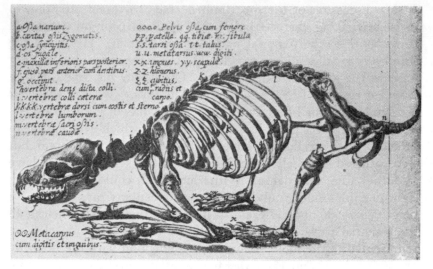

FIG. 149.—Muralt, 1682.   Skeleton of the hedgehog

paper on the same species (1682) is much more detailed.   He starts
off with an elaborate justification of his interest, as a medical man,
in comparative anatomy.   He holds that it is not merely expedient
but imperative to study the lower animals.   Therefore, he concludes, " I shall proceed briefly, and in the simple language of
sincerity, to expound to you the anatomy of *Talpa*, and at the same
time I shall refer to the points in which it differs from other
animals ".   His observations include the integument and panniculus
carnosus muscle, abdominal muscles, gut, biliary apparatus,
pancreas Asellii and urogenital organs.   He does not mention the
pancreas, which is a dense compact gland in the mole, and his
account of the male genitalia is dwarfed by the earlier and excellent
figure published in 1681 by Blasius, who attributes it to Swammerdam.   Schelhamer saw the seminal vesicles and the large Cowper's
glands of this species, but failed to identify them.   He did not
dissect the more interesting genitalia of the female, having examined
only a single male specimen.   He found no mediastinum in the
thorax, the heart being confined to the left side and " the lung "
below it on the right.   As his operations had been confined to one
specimen he questions whether such a disposition of the thoracic
organs might not be abnormal, but his statement is correct in one
respect — the heart of the mole is situated entirely on the left side.

In his discussion of the fore limb he fails to direct attention to its most striking characteristic — the forward position developed in response to the fossorial habits of the animal. He found the three very small auditory ossicles and the tympanum, but did not observe the vestigial external ear. On vision his voice has an eloquent ring : " I hear you asking ", he says, " what of the eye ? Verily nothing, for I found no eyes.   Hints only of eyes did I see— if you can call eyes those hard black globules which even appear to be solid.   Neither did they have any connection with the brain, but were situated externally, concealed under the skin, and, as I have said, they were very dense.   With such eyes it could not see, nor is sight possible underground.   Whether indeed any rays of light are perceived we do not know, and it matters little."   Nevertheless, all the essential parts of the vertebrate eye and the eye muscles are represented in the mole, including a connection with the brain by an optic nerve, and even the fusion of the eyelids is not quite complete, a very small gap being left by which the eyeball main-

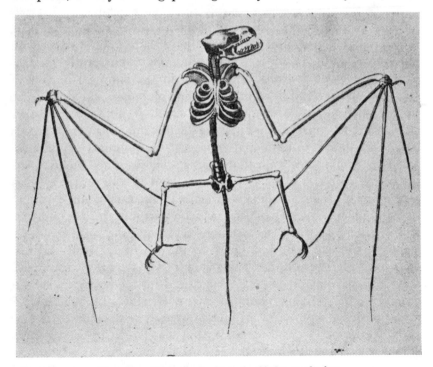

FIG. 150.—Lachmund, 1673-4.   Skeleton of a bat

tains direct but precarious relations with the outside world.

In order to demonstrate that bats have tails, which, he says, some have denied, Lachmund (1673–4) made and figured a preparation of the entire skeleton, and thus amply established the point. Muralt's dissection of a young female bat, *Vespertilio* (1682), confirms the existence of a tail, and shows how it has been absorbed into the mechanism of flight. He describes the muscles of flight, and the modifications undergone by the limbs to ensure an adequate framework for the extensive wing membrane, but the internal parts are disregarded. In the same year he examined the baboon, *Papio*, and gives an account of the thoracic and abdominal viscera, in which he has notes on the heart and great vessels, the pancreas (but no duct is mentioned) and the female genitalia. His account of the circulation is Harveian. Observations on the brain, muscles of the neck and limbs are also included, and he claims to have found in the brain a rete mirabile which could not have been there.

XXX
ACTA MEDICA HAFNIENSIA

The Copenhagen biologists, under the quickening influence of Thomas Bartholin, produced five volumes of transactions known as the *Acta Medica et Philosophica Hafniensia*, which are now very rare and almost entirely forgotten. The first volume was printed in 1673 and the last in 1680. There is little difference in the quality of the work published by the German, Dutch and Danish groups, except that in its early days the German Society could not boast of a Steno or a Swammerdam. All agreed in selecting medicine as the main objective, but pure biology, anatomy and zootomy, if not actively encouraged, were not lightly regarded. In one respect the methods of the Danes tended to approach those of the Parisians, in that their dissections were not always the work of *one* person but were sometimes undertaken by small groups. The genius of Bartholin was not profound, and displayed little searching of the mind, and it is clear that he relied for inspiration on his famous colleague Steno, to whom he frequently refers in terms of unusual praise as a " great Prosector, subtle of hand and modest in speech ". Indeed, in some papers under his name he is only reporting on dissections and drawings made by Steno, who is

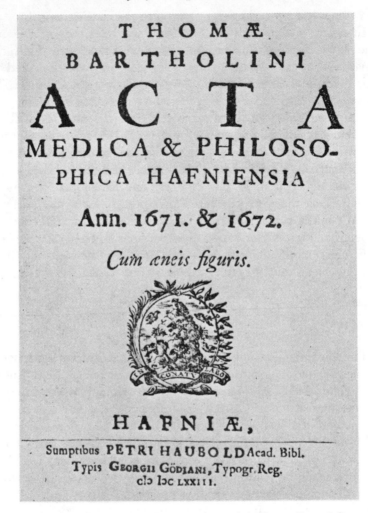

THOMÆ
BARTHOLINI
A C T A
MEDICA & PHILOSO-
PHICA HAFNIENSIA
Ann. 1671. & 1672.
Cum æneis figuris.

H A F N I Æ,

Sumptibus PETRI HAUBOLD Acad. Bibl.
Typis GEORGII GÖDIANI, Typogr. Reg.
cɔ Iɔc LXXIII.

FIG. 151.—Title-page of the first volume of the Proceedings of the
Copenhagen Medical Society

occasionally mentioned in the *Acta* as the source of much work to
which his own name was not attached.

The types examined in Copenhagen included one worm, one
crustacean, two insects, one arachnid, the lamprey, ten fishes, two
Amphibia, four reptiles, twelve birds and twenty mammals. Here
again the emphasis is on the vertebrates, particularly the mammals,
and to a lesser extent fishes and birds. The leading authors were

Jacobaeus (Jacobsen), seventeen papers, Steno, nine, T. Bartholin, eight, Borrichius (Borch), seven, and C. Bartholin, six. Thomas Bartholin, as a human anatomist, does not venture outside the Mammals, but his son Caspar is disposed to transcend the parental limitations. Steno is the most venturesome, and recognizes the necessity of extending his researches to *all* the vertebrate groups, whilst Jacobsen and Borch are the pure zootomists, whose only concern is to record observations and to shun speculation.

C. Bartholin (1680) is the author of an early discourse on methodology, and he dwells on the importance of preparations in the teaching of anatomy, and on the need of adequate and relevant instructions for each exercise in the dissecting room. Jacobsen (1675) publishes one plate of the polychaete *Aphrodite aculeata*, but there is no text beyond a few lines descriptive of the figures. He saw the respiratory spaces in the dorsal integument, and mentions some of the various foreign bodies, such as molluscan shells, which become embedded in the felt. His " intestine " is a vascular plexus on one of the intestinal caeca, rendered conspicuous by a deposit of developing sex cells. Swammerdam, writing after Jacobsen, also figures one of these plexuses, but he refers to it somewhat ambiguously as the " fabric of the gut ". The isopod *Aega psora* is described by Borch (1677–9) apparently for the first time, and he gives one figure of the dorsum of the head which assists in the identification of his species. " If we may believe Ovid," he remarks, " the head of Argus is beset with 100 eyes, but if the head of our animalcule be examined with a smicroscopium even more than 100 eyes will be seen." He claims that there are at least two hundred elements in each eye, but owing to its curvature, and the difficulty of focussing more than a small portion of the surface at a time, it was not possible to count them accurately. " Long and often ", he tells us, " have I thought on these eyes with the assistance of my smicroscopium, and not without admiration for the complexity of so abject an animalcule." He compares its eye with the honeycomb of bees, and is of opinion that each element is a tube which opens internally into a large air space, the mesial wall of which therefore receives a visual contribution from each tube. He remarks that, unlike those Crustacea in which the eye is situated on a movable stalk, the eye of *Aega* is fixed, and there are no traces of muscles — a handicap offset by the large size of the eye and the prodigality of its elements.

Jacobsen (1676) gives us an exceptionally interesting account of the mole cricket, *Gryllotalpa*. The wings, wing covers and the highly modified fore-legs are accurately described, and he mentions that he has not noticed saltation or flight in the species, nor does he refer to the ear-slit situated on the first pair of legs. He saw the respiratory movements of the abdomen,

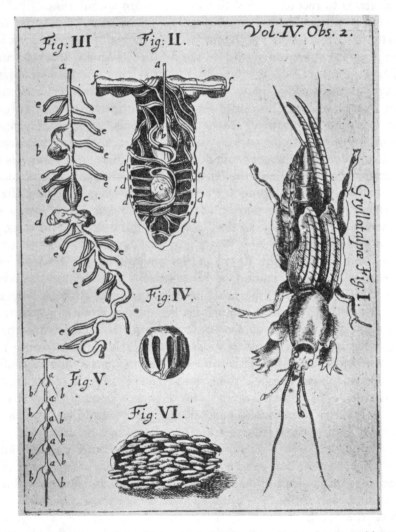

FIG. 152.—Jacobsen, 1676.   Mole cricket, *Gryllotalpa*.   Gut and segmental heart

and figures the stigmata. When the body was opened, he was surprised to observe myriads of silvery tracheae, which he compared with lungs. The stigmata having been blocked with oil, general paralysis of the body was produced and all motion ceased. He considers this was due to suffocation by preventing access of air to the internal parts, as Malpighi had found before him. His account of the stomach is good but not complete. He saw the unpaired lateral pouch of the oesophagus (food reservoir or crop), the armoured gizzard, which is also separately figured, and the very large paired enteric caeca. He calls these structures the first, second and third stomachs, and he found them again in the locust, but he is not prepared to say whether such insects should be placed among the ruminating animals, as Swammerdam was disposed to recommend. Jacobsen's paper is important as being one of the first in which the elongated segmental heart of insects is described and figured, and he saw five enlargements or " separate hearts " in this structure. The fat body or " omentum ", which is compared with that of frogs and tritons, is also described, and he found 164 eggs in one female. Roots encountered when burrowing are cut by the shearing action of the first pair of legs, but this was not known to Jacobsen, who attributes this function to the " iron-like teeth " of the mandibles. It is evident that this memoir on the anatomy of the mole cricket is a commendable piece of zootomical research, and it is all the more outstanding because the subject of it was an invertebrate. There are, however, striking omissions to be recorded, such as the author's failure to discover the prominent salivary glands and the still more prominent Malpighian tubes. In a later paper on *Anguilla* Jacobsen mentions that the severed head of a mole cricket continued to exhibit movements for two days.

The external characters of the scorpion are correctly described by Jacobsen (1677–9), whose material was obtained from Redi in a good state of preservation. He was struck by the pectines and the stigmata, and unhappily by the superficial resemblance of the segments of the metasoma to vertebrae or " Spondyli ". The last of these " vertebrae ", he says, lodges the poison gland and poison sting. He, and Redi, found it difficult to locate the poison apertures even with the microscope, and these apertures are in fact so minute as almost to justify Aelian's statement that they elude the sharpest

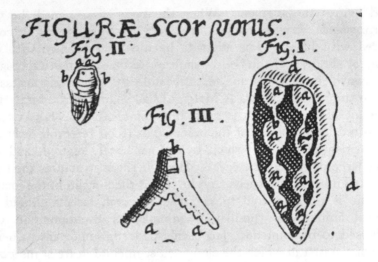

FIG. 153.—Jacobsen, 1677–9.   Scorpion.   Viviparous development and pectines

eye. Jacobsen discourses on the nature of the poison, and notes also that the scorpion is viviparous and that development is direct, the young resembling the parent.[1]   Two of his figures, though crudely and inaccurately drawn, have some historical interest.   One of them represents an ovarian tube opened up so as to expose the contained follicles which lodge the developing embryos, and the other is a small sketch of a young embryo showing the " eyes ", pedipalps and developing appendages, the metasoma being bent forwards under the body.   A third is a drawing of the pectines and genital aperture.

So much for the invertebrates, and the results obtained make us wish that the Danish anatomists had paid more attention to them.   Their vertebrate record is much more comprehensive and imposing.   In the lamprey Jacobsen (1677–9) describes the suctorial buccal funnel and teeth, the single median nostril, which is compared with the cetacean blow-hole, and the gills and gill openings, from which water is said to be discharged.   He tried blocking the branchial openings with oil in imitation of Malpighi's experiment on the spiracles of the silkworm, but it failed owing to the large size of the apertures.   With wax he was more successful and the motion of the gills was slowed down, but was actively resumed on

[1] This was known to Aristotle and Redi (1668).

removal of the wax. The gut,
he says, passes straight, and with
no convolutions, from mouth to
anus, the liver has but a single
lobe and the heart is enclosed in
a hard cartilaginous case.

Steno (1673), stimulated by
his important researches on the
female genital ducts and ova of
the terrestrial vivipara, turned
his attention to fishes, and the
types selected were *Mustelus laevis, Acanthias vulgaris, Torpedo*
and *Argentina sphyraena*. He
gives an account in the shark
Mustelus of a functional placenta
which is connected with the gut
of the foetus by a vitelline duct,
as in the embryo of birds, forming
by convergence a structure comparable with a cotyledon of the

FIG. 154.—Steno, 1673. *Acanthias.*
Spiral intestine

ruminant placenta. He gives us also, in *Mustelus* and *Acanthias*,
one of the earliest descriptions and figures of the elasmobranch
spiral intestine, which he names "intestinum cochleatum". He
found the abdominal pores and discusses their function, and dissected out the auditory organ, noting that the semicircular canals
were disposed in the three planes of space. He recognized that
this structure was a sense organ consistent with the ear enclosed
in spongy bone in birds, and in solid bone in man and quadrupeds.
He discovered the sacculus, and correctly compared its soft white
otolith with the stony otoliths of other fishes. He had already
described in the selachian fishes the lateral line canals and the
cartilaginous optic peduncle, and now informs us that the oviducts
of *Acanthias* resembled those of the placental shark in that one of
them contained a foetus, but that in this species and in the torpedo *there was no placenta*. In *Torpedo* he describes the electric
organ, its characteristic vertical prismatic columns, and nerve
supply. He does not, however, clearly associate the shock produced by the animal with the electric organ.

That peculiar elasmobranch *Centrina Salviani*, which was first figured by Belon in 1551 and by Rondelet and Salviani in 1554,[1] but was known to the ancients, was described by Jacobsen in the 1677-9 volume. The external characters and teeth are carefully examined, and he points out that the large spiracle is considered by some to be the ear—a morphological anticipation. Oppian is quoted as believing that the spines of the dorsal fins produce a septic wound, and Jacobsen was familiar with the fact that spines also occur in the same fins of *Acanthias*. The gut is described, and its rectal gland compared with the caecum of the terrestrial animal. There is one serious error — the statement that the oviduct is thrown into innumerable convolutions as in the frog, salamander and *Torpedo*, his specimen being obviously a male, and the duct in question the vas deferens. Jacobsen is guilty of the same mistake in his account of the torpedo, although the vas deferens and oviduct of this species had already been figured and accurately identified by Lorenzini. In the same volume Jacobsen describes his dissections of the torpedo, in which he refers to Lorenzini's work of 1678, without, it must be confessed, having carefully studied it. He discusses the nature of the shock produced by this fish, and mentions the lateral line sensory canals. Some of the ampullary organs are figured, but he is aware that the canals of this system were known to Lorenzini and others before him. Borch is quoted as having seen four eyes in one torpedo, the second pair being doubtless the spiracles. Other anatomical points noted by Jacobsen are the cartilaginous optic peduncle, already known to Steno, the spiral intestine, which he says resembles that of the skate, *Mustelus*, and dog-fishes, and a small rectal gland. His figure of the bile ducts is copied without acknowledgment from Lorenzini. Besides the internal and external openings of the gill slits he describes and figures five small intermediate oval apertures, which he compares with the stigmata of insects. These " stigmata " are simply parts of the normal gill slits, as he would have discovered had he continued his incision further into the branchial pouches. He also figures ten hemibranchs instead of nine, and distinguishes two

[1] Salviani's work was published in parts, the engraved title being dated 1554 and the colophon 1558 or 1559. In the writer's copy the latter date is 1557. The *Centrina* figures appear somewhat late in the work, and it is therefore improbable that they were issued in 1554. Salviani's figures of the species are greatly superior to those of his two contemporaries.

AQVATILIVM ANIMALIVM HISTORIÆ,
LIBER PRIMVS, CVM EORVMDEM
FORMIS, ÆRE EXCVSIS.

HIPPOLYTO SALVIANO TYPHERNATE,
ROMÆ MEDICINAM PROFITENTE
AVCTORE.
ROMÆ · M · D · LIIII ·

Fig. 155.—Salviani, 1554. Engraved title of the first edition of his work on
Fishes, with inscription to Cardinal Nicolaus Oliva

auricles and one ventricle in the heart, but admits that Redi had found only one auricle, and Lorenzini might have been quoted to the same effect. No attempt is made to justify the existence of two auricles, which shows that he could not have examined the interior of the heart. He confirms Redi that the heart continues to beat for some hours after removal from the body. There are three plates to this memoir.

Steno (1673), after Severino (1659), noticed the retia mirabilia in the swim-bladder of a fish, the species being the Salmonid *Argentina sphyraena* and the eel. He did not unfortunately follow up this observation, which possibly did not appear sufficiently promising to arouse his curiosity. In the 1671–2 volume there are two figures by Steno, without text, the first of which is a section through the abdomen of the carp showing that the swim-bladder, ovaries and kidneys lie outside the cavity which lodges the gut and liver. This figure would suggest that Steno, like Galen, had some conception of the relation of the abdominal viscera to the coelom. The second figure, also of the carp, represents the swim-bladder subdivided by a constriction, and the pneumatic duct connecting its cavity with that of the gut. Jacobsen (1677–9) severed the head from an eel and was surprised that it should continue to exhibit movements for over an hour. He examined the parts of the brain, the two optic lobes of which, he says, are analogous with the corpora quadrigemina of higher animals, and he describes the third and fourth ventricles, the infundibulum and pituitary. The gills are poorly figured. Borch (1673) gives an account of his dissections of male and female specimens of the swordfish, *Belone*.

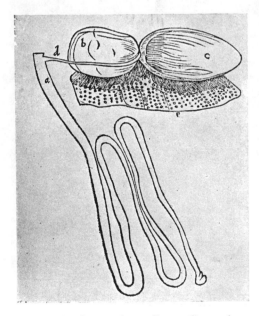

Fig. 156.—Steno, 1672. Carp. Gut, swim-bladder, pneumatic duct and ovary

The intestine was a simple tube without convolutions, pyloric caeca or special parts. Amongst some anatomical notes of no special interest he describes the elongated kidneys, liver and gall-bladder, and the heart with a bulbus arteriosus at the root of the ventral aorta. The swim-bladder, he says, can be variously charged with air to facilitate movements in the vertical plane when swimming. Borch was apparently the first to observe the very singular green colour of the bones in this genus, afterwards recorded by Willughby (1686) and many other ichthyologists. He mentions that the colour can be dissolved out with spirit of wine. Wille (1674) discusses the reactions of the poison secreted by the weever, *Trachinus draco*, and correctly locates the poison apparatus in the spines of the dorsal fin, but he was not aware that the opercular spines were equally venomous.

The salamander of the early anatomists was often the related genus *Triton*, and this is clearly the case in the memoir on " Salamandra " by Jacobsen (1676), a revised edition of which appeared as part of a separate work in 1686. A comparison of his figures with *Triton* dissections makes it certain that he had not dissected a salamander, and if any doubts remain they vanish before his experiment of keeping a " Salamander " alive without food for almost a year *in pure water* — an ordeal which no true salamander could have survived. Jacobsen's paper was angrily criticized by Swammerdam, whose own work on the frog, it is true, was on a far higher plane. Jacobsen very briefly describes the spleen, liver (which he states wrongly has four lobes), gall-bladder, bile duct and the lungs, the extent of the lungs being exaggerated in the figure. He found that the heart continued to beat for many hours after removal from the body, just as in frogs, toads and the torpedo. He failed to see the inter-auricular septum of the heart, and in the kidneys he figures what might be scattered supra-renal nodules, but from their disposition were possibly the Malpighian bodies. Other organs described are the paired ovaries and oviducts, the latter being completely separated from each other at both ends, the two-lobed testes with attached fat bodies as in the frog, the bifid frog-like urinary bladder, and the regions and geography of the gut. Most of these structures are figured.

Jacobsen (1677–9) dissected three snakes — *Tropidonotus*, the viper and an unspecified " male snake ". In the grass snake he

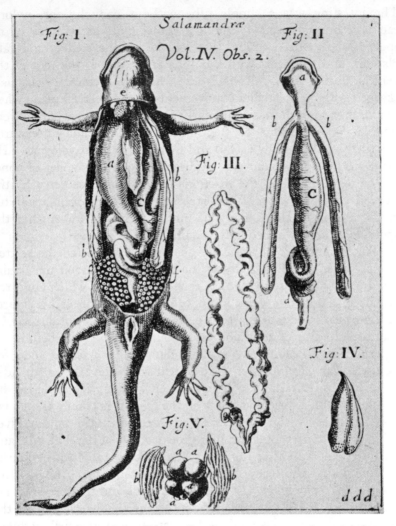

FIG. 157.—Jacobsen, 1676. *Triton.* Gut, heart, respiratory and urogenital organs

draws attention to the asymmetrical lay-out of the kidneys and testes, to which Steno had already referred in 1673, but neither author associates such an arrangement with an elongated and attenuated body form. The organ described as the thymus was doubtless the thyroid, and in the crude figure of the biliary apparatus and pancreas the gut is omitted, to the mystification of the reader. The cloaca was examined and the opening into it of the ureters and the vasa deferentia noted. In the viper Jacobsen discusses the nature and effects of the poison, and in this connection gives some attention to the teeth and their relation to the bones of the skull. He also briefly describes the external features of the brain and fourth ventricle, the eye, trachea, elongated lungs, and some of the muscles, including the external and internal inter-costals. He again failed to recognize two auricles in the heart, which may be contrasted with his generosity in bestowing an extra one on the fish, but his account of the ovaries, oviducts and their ostia is better. In one female he found that the ostium of the right oviduct was expanded and embraced the extremity of the ovary, and he saw also the two anal sacs. In his nameless snake Jacobsen gives a surprisingly accurate description of the biliary and pancreatic apparatus and spleen, which are singularly complex in some Ophidia.

Borch's paper on the tongue, pharynx and hyoid of birds (1673) is illustrated by two plates, and covers species of *Anser, Anas, Aquila, Numida, Gallus* and *Psittacus*. He says that his previous study of the eagle attracted him to this subject, and the above material happened to be available. Nevertheless, no description or discussion is appended, since he is of opinion that his figures are sufficiently expressive to enable the reader to do that for himself. As his examples represent four different tribes of birds, we can only regret that he declined so favourable an opportunity of writing a minor essay in comparative anatomy, nor is his con-cluding apology a sufficient justification of his failure to do this.

The papers by Jacobsen and C. Bartholin on the anatomy of the heron,[1] stork, peacock and screech owl (1673-9) call for no comment since they are hardly more than lists of the features dissected, and disclose no new points of major interest. Borch (1671-2) bases his memoir on the eagle, *Aquila*, on figures by Steno, who made a detailed study of the structure of this bird. The mouth, tongue, pharyngeal region and glottis were carefully examined, but the nictitating membrane and " its muscle " are almost ignored. The capacious crop is not mentioned, but the bile and pancreatic ducts, and the very small rectal caeca character-istic of the Falconiformes, are described. Most attention, however, is devoted to the stomach, of which there is a good figure. The

---

[1] The figure is by Steno (1673).

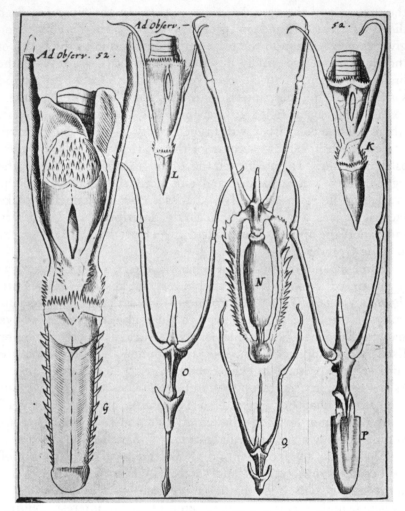

Fig. 158.—Borch, 1673.   Goose, *G, N* ; fowl, *K, L, Q* ; guinea fowl, *O* ; duck, *P.*
Tongue and hyoid

stomach of this carnivorous bird, he says, consists of a broad zone
of glandular tissue which is separated from the oesophagus by an
obvious sphincter-like band, and a muscular region from which
the duodenum arises *laterally*, so that a posterior blind or caecal
portion is formed.   Steno's dissections of the muscles of *Aquila*
were completed in April 1673, but Aldrovandus in his memoir on
the eagle published in 1599 had already examined the muscles of

this bird with some care.   Steno's work is one of the most remarkable essays in zootomy published up to his time, and it is perhaps
more detailed and reliable than almost any other, but the unfortunate absence of illustrations handicaps the labours of his
interpreters.   Among the muscles of the head he describes those
of the nictitating membrane with considerable precision, and shows
how the tendon of the pyramidalis muscle is threaded through that
of the quadratus, in order that the optic nerve may not be stretched

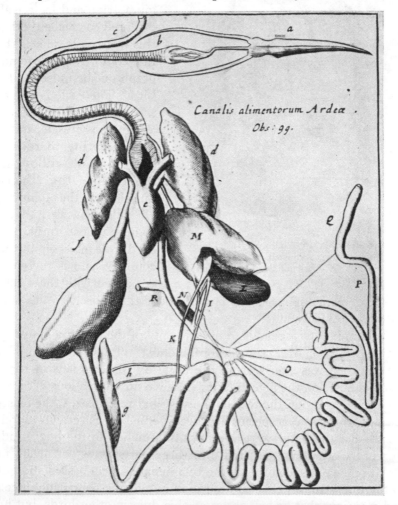

FIG. 159.—Jacobsen, 1673.   Figure by Steno.   Heron.   Hyoid, respiratory organs,
   heart and abdominal viscera.   The single, short wide caecum not shown

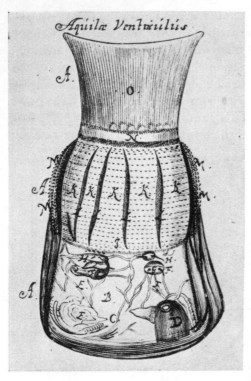

FIG. 160.—Borch, 1671 (pub. 1673). Figure by Steno, 1673. Stomach of eagle. *O*, oesophagus; *K, M*, glandular region of stomach; *D*, pyloric opening

or abraded when the nictitans is drawn across the eye, in all of which he was anticipating the Parisian Academicians. The muscles of the tongue, hyoid, crop, trachea, neck, thorax, abdomen, uropygium, wing and leg are all explored so efficiently that this work must rank as the first comprehensive monograph on the muscles of a bird. The eagle is one of those species which possess an ambiens muscle, and it is interesting to record that Steno describes this muscle as the " 15th muscle of the femur ". He fails, however, to recognize that it forms part of the flexores digitorum complex, and is therefore associated with the mechanism of perching. He did not discover this muscle, which was known to Aldrovandus.

Jacobsen (1673–5) adds to his anatomy of the parrot a figure by Steno (1674) of the " sinus rhomboidalis " of the spinal cord of *Anas*, which was later seen in the eagle by the Parisians in 1676. Borch's notes on the pigeon (1671–2) have no illustrations except a very small figure in the text of the vestigial rectal caeca, the existence of which in this species had been denied by Severino. Borch found that the lungs were very firmly attached to the body wall, that each kidney was divided into three parts, and that owing to the absence of a gall-bladder the two hepatic ducts passed straight to the duodenum. He also described one of the pancreatic ducts, and mentions that he had observed this duct in other birds, but he was mistaken in claiming that the septum ventriculorum of the

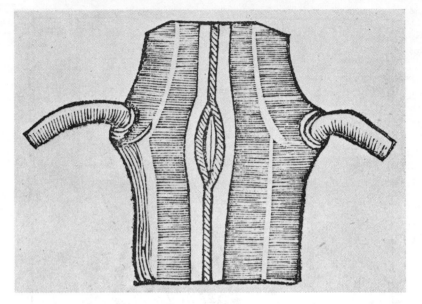

Fig. 161.—Steno, 1674. Dissection by Jacobsen. Duck, *Anas.* Sinus
rhomboidalis of spinal cord

heart was perforated. The paper on the parrot, *Psittacus*, by
Jacobsen is based on dissections and figures by Steno (1673). It
covers only a few special points, however, and leaves the general
anatomy of the animal untouched. These points are : the
abdominal air sacs ; complete absence of rectal caeca ; sterno-
and broncho-tracheal muscles ; freedom of the upper mandible ;
tongue and structural details of the tympanum and columella auris.
He claims that the upper jaw is capable of independent movement,
and in this respect compares the parrot with the crocodile. It is
true that the fronto-nasal joint, which permits a vertical motion
of the upper mandible, is well developed in certain parrots, but
there is nothing comparable with it in the crocodile. He attempts
to explain Psittacine speech by asserting that the tongue of the
parrot is human in form, and hence appropriate to this purpose.

The tongue of the black woodpecker, *Picus martius*, was dissected by Jacobsen (1677-9), and he was the first after Coiter to
investigate this fascinating and unique anatomical puzzle, but he
was apparently unaware that he had been anticipated. There are
two poor figures, without, however, an adequate explanatory text.

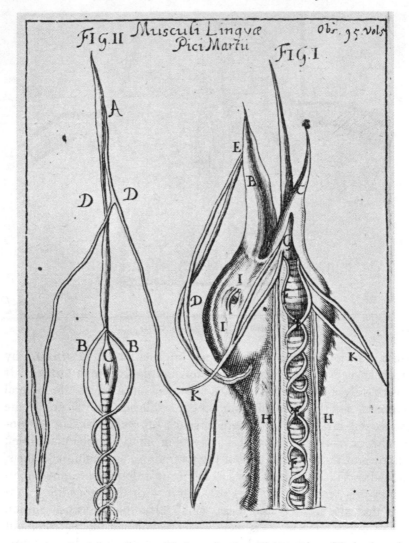

FIG. 162.—Jacobsen, 1677-9. Black woodpecker, *Picus martius*. Mechanism of tongue

He describes the elongated tongue, and appreciates its use in extracting insects and their larvae from places difficult to reach, although this was already well known. He followed correctly the prolongation of the very long and muscular ceratobranchial cornua up and round the occipital region of the skull, and then forwards in grooves to their insertion in the *right* nostril, and he understood that by this mechanism the tongue was capable of considerable protrusion. The figure is inaccurate in showing both the cornua *arising* on the right side, and is also confusing in representing the genio-hyoideus muscles detached and severed from the anterior portion of the apparatus. The tracheo-hyoideus muscles are figured correctly as forming two spirals round the trachea, although the spiral portions are exaggerated, and he has not grasped their function. The tracheo-laryngeus muscles are also figured.

A few contributions by the Bartholins, Paulli and Jacobsen (1671–9) on musk and fallow deer, the horse, ass and dog are not of sufficient importance to call for special notice, and the mammals may be introduced by two papers of a somewhat general nature by T. Bartholin and Steno. The former (1677–9) investigated the nasal chamber especially in the sheep and dog, but also laid under contribution the deer, ox, bear, hare, mouse and hedgehog. He remarks that the olfactory organ must be studied rather in *animals*, in whom this sense is better developed than in man. He is of opinion that it can be estimated by studying the *complexity* of the turbinal bones, the air current serving the double purpose of respiration and olfaction. In the hunting dogs, which have an exquisite sense of smell, the leaves of the turbinals are unusually numerous and dendritic, and indeed almost beyond computation, but he is not aware that the dorsal and more extensive portion of the nasal chamber in the dog is olfactory and the ventral portion respiratory. The assumption that the *whole* of it is olfactory detracts materially from the force of the argument. Nevertheless, the essay is both welcome and instructive, inasmuch as the author makes use of the facts of comparative anatomy to attack a physiological problem, and therefore turns aside from the traditions of the purely descriptive anatomist.

Steno's memoir on the homologies of the genitalia of the vivipara (1673) is more searching in its methods and on a larger scale. In 1667 he had suggested that the so-called " testes " of women were comparable with the ovaries of the egg-laying animals. He now proceeds to enlarge on this theme, but his later observations are a part only of a work he had in contemplation which was never completed. In the meantime, in 1672, de Graaf and Swammerdam had published important contributions to the same problem. Steno's types are the salamander, tortoise, sheep, cow, hare, rabbit, dog, bear, hedgehog and man. He supports the Harveian doctrine of the origin of animals from eggs, which he acclaims as a " divine truth ". His work, however, is embryological, and only incidentally touches on questions of comparative anatomy. In the salamander he describes the two ovaries and fat bodies and also the two oviducts, which latter, he says, contained ova and were not joined at either end, but in the tortoise, he adds incorrectly, they coalesced posteriorly to have a common external opening. The orifice of the testicular [Fallopian] tube in the cow is considered to be identical with the ostium of the oviduct in birds, and the Graafian follicles of the hare are compared with the intra-ovarian eggs of the fowl. In the bear the testicle [ovary] consisted of many white granules, and thus resembled the ovary of fishes. He appends a good account of the female genitalia of the hedgehog, in which he describes the peritoneal sac of the ovary, designed, he thinks, to prevent the escape of the eggs into the abdominal cavity when they break from the ovary.

T. Bartholin (1673) describes and figures the tough thickened sclerotic of the eye characteristic of the Cetacea, which greatly puzzled him. He says it is not horny, fleshy or fatty, but a solid substance *sui generis*. He thinks he has " positive and certain evidence " that spermaceti is excreted and discharged by the *brain* in the larger whales, but how it escapes from that imprisoned organ he cannot explain. The anatomy of a male and female reindeer, *Rangifer*, was undertaken by T. Bartholin (1671–2), who added three plates to his description. Steno dissected this animal and publicly demonstrated its anatomy in 1672. Bartholin describes the geography of the gut, and the ascending and descending spiral muscular bundles of the oesophagus, which attracted the attention of Steno, Grew and Willis at about this time. Bartholin compares the reindeer with the bear and calf, but his paper is nothing more pretentious than a cursory and partial review of the general structure of the species. He adds some details and figures of the brain of the bear by Steno, and his plates include figures of the gut of the squirrel and fallow deer, and the brain of the squirrel, drawn in 1672 also by Steno. C. Bartholin (1675) discusses the statement by Blasius that the commonly accepted belief in the absence of a gall-bladder in the horse is false, since a vesicle of the size of a man's fist is to be found *embedded* in the substance of the

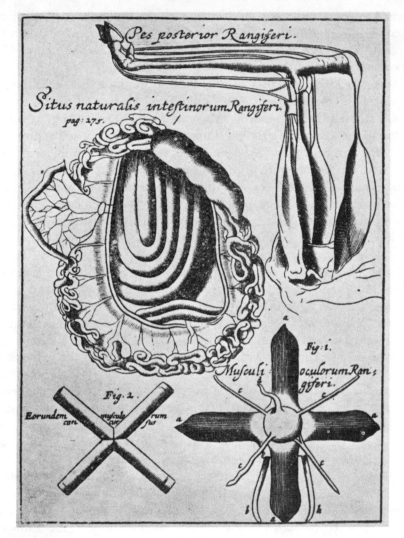

FIG. 163.—T. Bartholin, 1671–2. Reindeer. Abdominal viscera, myology of hind limb and muscles of the eye, including the choanoid or retractor oculi (Fig. 1, *c, c, c, c*)

liver. This, if true, would explain why it had been overlooked. Bartholin refers to the dissections of Paulli, who after a most careful search had failed to discover any signs of a gall-bladder in the horse, and he suggests that it may be present in some individuals and not in others. A few years later Jacobsen confirmed the findings of Paulli, and added that the ass also was without a gall-bladder. The negative result is the correct one, and Blasius' vesicle may well have been a cyst of *Echinococcus*, although this parasite is rare in the horse.

A " homicidal " lion having been brought into the anatomy theatre to be submitted to the anatomical knife, and finally to be skeletonized for the Royal Museum, was dissected by Paulli, Borch and Caspar Bartholin, Thomas Bartholin himself reporting on the proceedings (1671–2). He had already published an account of the anatomy of the lion in 1657, and the Parisians' first dissection of the species was in 1667. Bartholin now states that the inter-auricular septum of the heart is *not* perforated, and he could find only one valve at the origin of the descending aorta. He adds notes on the abdominal and thoracic viscera, without, however, disclosing any noteworthy feature. He is struck by the large dorsal conical papillae of the tongue, forming recurved spinules, but these had been mentioned formerly by himself, and also described and figured by the Parisians, whose " splendid observations " he quotes. Steno (1673) publishes a series of nine drawings, one of them by Swammerdam, of the thoracic duct and associated lymphatics of the dog, with the object of illustrating the variations in the lymphatic system which may occur in a single species. There is no text beyond a list of the parts figured, and this paper in fact is a fragment only of an extensive work on the system which was planned but never carried out. C. Bartholin in his anatomy of the hare (1672) discusses the statement in Leviticus (xi. 5) that the " coney ", by which he understands the rabbit, chews the cud. The coney of Leviticus, however, is not the rabbit or the jerboa, but *Procavia* (*Hyrax*) *syriaca*. In spite of the fact that the stomach of the rabbit consists of a single chamber only, and that there are incisor teeth in *both* jaws, he is disposed to accept the Biblical pronouncement, because, he says, the stomach contains two kinds of excrement — dry on the right and moist on the left.[1] He points out that small animals could not accommodate a large compound stomach, but he neutralizes this assertion by admitting that, in compensation, the rabbit has an exceptionally large and capacious caecum, which is small in other animals. He saw the

---

[1] This may be interpreted as the first reference to the peculiar coprophagous habits of the species.

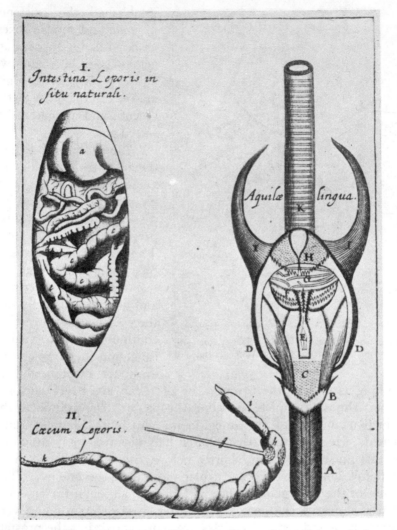

Fig. 164.—C. Bartholin, 1671–3. Figures by Steno, 1672. Abdominal viscera of hare. This plate includes a figure by Steno, 1673, of the tongue, mouth cavity and hyoid of the eagle in illustration of a paper by Borch on the anatomy of the eagle in the same volume

diffuse pancreas and the bile and pancreatic ducts, together with their openings into the duodenum. The helicine caecum, appendix and sacculus rotundus are described, and he noted and figured the honeycomb pattern of the two latter. Reference is made to the belief that the hare is hermaphrodite, but he claims to have

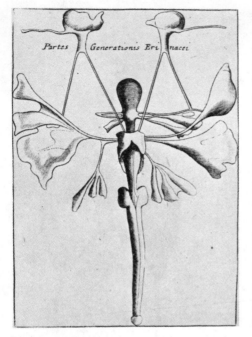

FIG. 165.—Borch, 1671–2. Dissection and figure by Steno, 1672. Male genitalia of hedgehog, *Erinaceus*

established the existence of male and female sexes, each with its appropriate genitalia. There are two figures — one of the gut *in situ*, and the other of the caecum and appendix dissected out.

The paper by Borch on the anatomy of a male and young female hedgehog, *Erinaceus*, is based on dissections and a figure prepared by Steno in 1672. The abdominal viscera, biliary apparatus and lacteal vessels are described, and he states correctly that there is no caecum. The continuous utero-vaginal canal and the large cornua uteri are recognized, but nothing is said of the oviducts, or of the ovary being enclosed in an almost complete peritoneal capsule. He examined the orbicular muscle of the integument, and noted how it functioned. He was aware that Coiter had devoted some attention to this muscle, and that, shortly before, Steno had displayed its anatomy and provided an explanation of its economy. The figure of the male genitalia has no lettering, although the explanation assumes its existence. The testes, vasa deferentia and very large compound seminal vesicles are represented, but the prostate and Cowper's glands, though included, are ñot mentioned in the text. T. Bartholin's notes on the male mandrill, *Papio maimon* (1671–2), are accompanied by three anatomical plates and a figure of the entire animal, which had died of disease in the Royal Menagerie. Bartholin was surprised that he could find no illustration of the species in the literature. There is no occasion to discuss the text of this paper, except the statement that the interventricular septum of the heart was perforated, which Bartholin admits was not what he expected. Of the two figures of the male genitalia one has clearly been badly engraved, since the spermatic arteries and veins and the ureter fuse to constitute a vessel unique in the annals of anatomy, and the renal artery arises from the vena cava. The figure of the penis is better, and shows the seminal vesicles, prostate, Cowper's glands and the bulb of the corpus spongiosum.

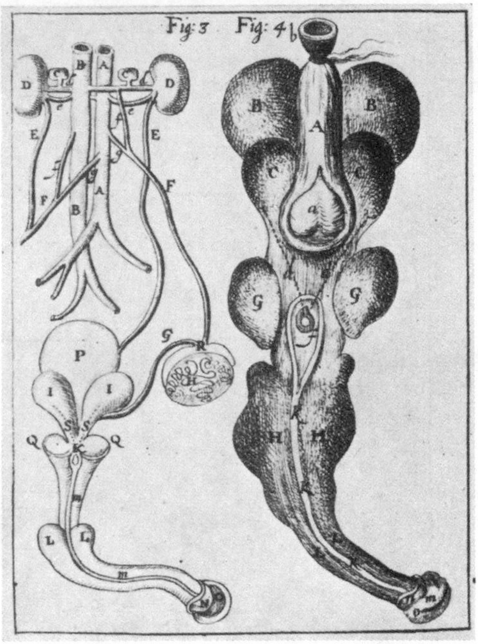

FIG. 166.—T. Bartholin, 1671–2. Mandrill, *Papio maimon.* Urogenital system of male. In Fig. 4 *C* is the seminal vesicle, *G*, prostate, and *H*, bulb of corpus spongiosum. Cowper's glands are shown between *G* and *H*, but not lettered

XXXI
L'ACADEMIE ROYALE DES SCIENCES

The constitution of the French Academy of Science in 1666 established a school of morphology to which the modern development of comparative anatomy may be directly traced. The Academy divided its forces into mathematicians, who met on Wednesdays, and Physicists, as biologists were then called, who met on Saturdays. As we gather from contemporary engravings, and from the reports of their proceedings, the Academy was in no sense comparable with the scientific society of to-day, but was rather a *laboratory* for the practical examination and discussion of natural phenomena. In the subsequent decline of the Academy up to its reconstitution in 1699, the biological section alone retained its vitality, and the small but vigorous group of comparative anatomists was never disposed to dissipate its energies in calculating the odds of a game of chance, or to exercise its genius on the details of ornamental gardens. Descartes, however, found

FIG. 167.—King Louis XIV and his Chancellor Colbert visiting the French Academy of Science in 1671. The figure between the King and Colbert is Perrault

a philosophical use even for the gardens. The engineers had employed water-power to operate highly ingenious mechanical devices, often to the discomfiture of the visitor, which Descartes compared with the functioning of the vascular and nervous systems in the *human* machine.

The longevity of the early Parisian anatomists was remarkable, only one of them dying before the age of seventy-five years.

Their leader was the veteran Claude Perrault, a member of a versatile family, who abandoned the profession of arms for the diversions of art, and became one of the leading architects of his age. But he is no less distinguished as an anatomist and a physician, and it was due mainly to his influence that a number of the early members of the French Academy, who are often conveniently referred to in the literature of the period as the " Parisians ", laid the foundations of our modern knowledge of comparative anatomy.[1] The

FIG. 168

[1] Perrault himself shared the common discipline of the French scientists in having to submit to the satire of the literary lions. He was contemptuously attacked by the poet Nicolas Boileau, who at first stigmatized him as a bad physician, but admitted that he was a competent

FIG. 169

Perrault, 1671. Historiated capitals from the first folio edition of the Parisian Memoirs introducing the Temptation and the Animals entering the Ark

principal members of the " Company " were the " acute and lucky Pecquet ", as Robert Boyle used to call him, Louis Gayant, the great but leisurely Duverney, Moyse Charas, Philippe de La Hire, the Jesuit Father Thomas Gouye and the argumentative Jean Mery. It is a commonplace both in literature and science that a great book seldom fails to attract an adequate illustrator, and the Parisians were fortunate in enlisting the services of Sébastien Leclerc, again a well-known architect, and an engraver on copper of outstanding merit. Some of the plates are unsigned, but there are no peculiarities of execution sufficient to justify the belief that other engravers were employed. Of the numerous subsequent editions none approach the first folio in the excellence of the illustrations, and it is therefore all the more unfortunate that so few copies were printed, since the work is now practically unobtainable. We learn from Alexander Pitfeild that it had become very scarce even in the seventeenth century, and in recent times only a single complete copy has come into the market for many years.[1]

A brief survey of the publications of the Parisians, the details of which are somewhat elusive, has a relevance of its own, but it will serve also to illustrate the bibliographic refinements with which the historian of science must be familiar. It will be noted that, contrary to contemporary practice, the works of the Parisians were published in their own language. Their first venture was the anonymous issue, without title-page, of a small tract of twenty-seven pages and two plates, published at Paris in 1667. There is good reason to believe that this pamphlet was written by Perrault. The Company had dissected a " large fish " [*Alopias vulpes*] on June 24, 1667, and a lion on June 28 of the same year, and the tract contains a description of their results. Two years later they published a larger anonymous work of 120 pages and five plates, dealing with the anatomy of the chameleon, beaver, dromedary, bear and gazelle.[2] They were now definitely committed to a more

architect. Later he withdrew his approval even of his art, and saw no good in him whatever. When reproached for ingratitude on the ground that Perrault had cured him of two serious illnesses, he replied denying the fact, and added : " The proof that he was never my physician is that I am still alive ".

[1] All copies of this edition were exhausted by the King and the Academy as presents " to Persons of the greatest Quality, and were hereby rendered unattainable by the ordinary Methods for other Books ".

[2] Both these publications are very rare — the former particularly so. The writer possesses one of the few copies which have survived.

ambitious enterprise, and, encouraged by the interest which these papers had aroused, they projected an extensive work on comparative anatomy on a scale not hitherto attempted. This was published anonymously at the charges of the King as two magnificent elephant folios in 1671 and 1676, but in the latter year both parts were re-issued as one volume with Perrault's name on the new title-page.[1] The preparation of the work was begun by Perrault, Pecquet and Gayant, and the species they dissected had died of sickness in the Royal Menagerie mostly during the winter months. Gayant died in 1673 and Pecquet in the following year, but the work being still un-

DESCRIPTION
ANATOMIQVE
D'VN CAMELEON,
D'VN CASTOR,
D'VN DROMADAIRE,
D'VN OVRS,
ET D'VNE GAZELLE.

A PARIS,
Chez FREDERIC LEONARD, Impr. ordin. du
Roy, ruë S. Iacques, à l'Eſcu de Veniſe.

M. DC. LXIX.
AVEC PERMISSION.

FIG. 170.—Perrault, 1669.   Title-page of the first quarto edition of the Parisian Memoirs

completed, Duverney was happily invited to assist in the final stages, and his services are specially commended by the leader.

Perrault died on October 9, 1688, aged seventy-five, of an infectious disease which he is said to have contracted at the dissection of a camel. All the members of the Company who were present at this dissection were similarly infected. His death left Duverney in charge of the work, but habits of procrastination did not favour a vigorous application to the discharge of this trust. In later years he failed again to publish the manuscript of Swammerdam's *Biblia Naturae*, which he had acquired by purchase. He discovered among Perrault's papers descriptions of sixteen new

---

[1] Not quite the same book.   There are differences in three plates, and some new matter is added to the chapter on the " Coati Mondi " [*Nasua rufa*].

MEMOIRES

POUR SERVIR

A L'HISTOIRE NATURELLE

DES ANIMAUX.

A PARIS,
DE L'IMPRIMERIE ROYALE.

M. DC. LXXI.

Fig. 171.—Perrault, 1671.   Title-page of Vol. I of the first folio edition of the
Parisian Memoirs

animals, but he neglected to edit them, and his attempt to bring
out a new edition of the series resulted only in the publication of
the first section in the year 1700.   Urged by the Academy to greater
efforts, he undertook the preparation of a revised and extended
edition in three volumes quarto, but he died in 1730 before this
was accomplished.   The Academy itself, being stimulated by the
appearance of an unauthorized and unrevised Dutch edition at

The Hague in 1729, now entrusted the task to Winslow, Petit and Morand, who examined the papers bequeathed to the Academy by Duverney, compared them with previous editions, and completed their task by December 1730. They included the sixteen unpublished descriptions of Perrault, and added a chapter on the viper by Charas, which had appeared separately in 1669. A fourth volume based on material left by Duverney, Mery and La Hire was not completed, and according to Vicq-D'Azyr (1773) the observations of Duverney and de La Hire on the anatomy of fishes " are not yet printed ".

In 1686 and 1689 Father Tachard published descriptions of the two missions dispatched by the Jesuit Fathers to Siam. The fathers interpreted their mission in a liberal spirit, and, at the suggestion of the King, they supplemented their ecclesiastical duties by observations on the natural history of the country. Many animals were dissected on the spot, and others were forwarded to Paris, where they aroused the lively interest of the members of the Academy of Science. In this way the Academy received in 1687 a crocodile, a " Toc-Kaie " [*Gecko verticillatus*], a camel and a leopard, and the anatomy of these animals by Father Gouye is included in the edition of the Memoirs we are now considering, which was at length published in three volumes dated 1732-4.[1] This, therefore, is the most complete edition of the monographs of the Parisians, and, though rare, can still be obtained, but it does not compare in interest or typography with the first extended edition of 1671-6.

The booksellers of the seventeenth and eighteenth centuries provoke the resentment and tax the labour of the bibliographer by their casual methods of publication. In the first instance the sheets were issued under the name of the original printer, but he retained the right of farming them out to all who chose to apply for them, the purchaser being permitted to revise the title, and to publish the work from his own town under his own imprint and date.[2] These are the so-called title-page editions, an early example of which is the *Introductorius Anatomiae* of Massa, which first

[1] So in most copies, but some are dated 1731-4, others 1733-4, whilst others again are undated. All the re-issues are title-page editions.

[2] For example the 1688, 1701 and 1702 English editions of the Parisian Memoirs are identical except for the title-pages and dedication. The paper has the same watermark in all three.

appeared in 1536, and again as a title-page edition in 1559. This practice was naturally exploited for the profit of the unscrupulous, and the case of Dr. William Cowper, who produced an edition of Bidloo's *Atlas of Anatomy* under his own name *as author*, is familiar to students of the history of anatomy.[1] Thus it happens that the same work may be published a number of times, from many centres and under a variety of dates. Copies even of the same issue may bear varying dates, an altered date or no date at all. The conscientious bibliographer, who dares not assume the identity of editions which he has not personally examined, can only arraign the practice and submit to his fate. And it does occasionally happen that he is rewarded by the discovery of some obscure but precious detail which has escaped the vigilance of his predecessors. The monographs of the Parisians are a tedious example of the prevalent vice, and it is doubtful whether, even now, *all* the ramifications of this sprawling publication have been traced. After the issue of the completed first editions in 1676, another edition was published at Paris in 1682. The first English version appeared in 1687, the text translated by Alexander Pitfeild and the plates etched by Richard Waller. The engraved title was plagiarized in Valentini's *Amphitheatrum Zootomicum* of 1720. Unhappily the English plates are little better than caricatures of the masterly originals of Sébastien Leclerc, but one must not forget that they represent the maiden efforts of the engraver. An imperfect French edition, together with Father Gouye's independent observations, appeared in 1688. Then follow numerous issues, some of them incomplete, in English, Dutch, French and German, down to 1758, when the work ceased to be printed after a memorable life of almost a century. In 1778, however, a travelling showman, who had acquired a living specimen of the cassowary, reprinted at Bury the chapter on the anatomy of that bird, in which he not only forgets to acknowledge the source of his information, but prints the following mendacious statement at the foot of his title-page : " Books to be had of Pidcock the Proprietor, and of no other Person in England ". That a showman should consider it his duty to initiate his patrons into the secrets of comparative anatomy is a unique event in the annals of the circus.

[1] Cf. Cole, *Oxford Bibliographical Soc.* 5 (1938–9), *Dict. Nat. Biog.* art. " Cowper ", and F. Beekman, *Ann. Med. Hist.* N.S. 7 (1935).

Fig. 172.—Perrault, 1671. Dissection of an otter at a session of the French Academy of Science

In the just applause of their own procedure the Parisians happily disclose their methods of work, and they are the only contemporary research group to do so. The dissections were carried out, not by any individual, but in session of the whole Company, and nothing was committed to paper which failed to command the ready assent of all present. They say : " That which is most considerable in our *Memoires* is that unblemishable evidence of a certain and acknowledged Verity. For they are not the Work of one private Person, who may suffer himself to be prevail'd upon by his own Opinion. . . . This so precise exactness in relating all the particulars which we observe, is qualified with a like care to draw well the Figures, as well of the intire Animals, as of their external Parts, and of all those which are inwardly concealed. These Parts having been considered, and examined with Eyes assisted with *Microscopes*, when need required, were instantly designed by one of those upon whom the Company had imposed the charge of making the Descriptions ; and they were not graved, till all those which were present at the Dissections found that they were wholly conformable to what they had seen. It was thought that it was a thing very advantagious for the perfection of these Figures to be done by a Hand which was guided by other sciences than those of Painting, which are not alone sufficient, because that in this the Importance is not so much to represent well what is seen, as to see well what should be represented." Characters presenting no feature of special interest are hardly more than catalogued, but they explore with patience and curiosity any departure from the commonplaces of anatomy, such as the compound stomach of the gazelle and the claws of the lion. Their limitations are well defined, and not always consistent. They expect too close an agreement with the human type, an attitude which constrains the imagination without preventing error, for they deny, after only a casual inspection, that the chameleon has an ear. Their lack of familiarity with microscopical methods introduces other difficulties, and they hesitate to distinguish between the kidney of the chameleon and its testis. Repeated efforts are made to link up structure and function. Thus they endeavour to associate the production of voice with a vertical glottis, and its absence with a transverse one — an essay in philosophic anatomy after the manner of Aristotle. The meetings of the Parisians became famous.[1] They obtained most of their material from the Royal Menagerie, and at least on one occasion, in 1681, when the anatomy of the elephant was being investigated, the King himself looked in. Duverney was in charge of the dissection, the leader Perrault was writing the description, and de La Hire was at work on the drawings. Never, perhaps, had an anatomical dissection made so great a stir. The King enquired for the anatomist, whom he could not see, whereupon Duverney arose from the carcass, in which he was, so to speak, entombed, and showed himself.

We are now in a position to consider the range of the investigations undertaken by the Parisians and the results which they achieved. In doing so the complete three-volume edition of 1732–4 will be used, the date of which, however, is misleading, since the whole of the work was completed within the seventeenth century. The animals were dissected and described in the order in which they were obtained, and no attempt is made to classify them. The Parisians devote some attention to describing the species under examination, but it is not always sufficient, and a few of the modern equivalents which follow are somewhat speculative. Our naturalists are committed above everything to anatomical research, and only occasionally comment on matters of classification, although when they do so their views are acute and defensible. Forty-nine species are described, and the groups represented are those commonly drawn upon in stocking a menagerie. They are : PISCES : *Alopias vulpes*. AMPHIBIA : *Salamandra maculosa*. REPTILIA : *Gecko verticillatus, Chamaeleon vulgaris, Vipera aspis, Crocodilus vulgaris* s. *niloticus, Testudo indica* s. *Perraulti*. AVES : *Struthio camelus, Casuarius galeatus, Phalacrocorax carbo, Pelecanus onocrotalus, Ciconia ciconia, Ibis candida, Platalea leucorodia, Phoenicopterus ruber, Gyps fulvus, Aquila chrysaetus, Haliaetus albicilla, Crax alector, Numida meleagris, Porphyrio caeruleus, Grus virgo, Balearica pavonina, Otis tarda*. MAMMALIA : *Manis pentadactyla, Loxodonta africana, Camelus dromedarius, Cervus canadensis, C. axis, Alces alces, Bubalis boselaphus, Antilope cervicapra, Rupicapra rupicapra, Felis leo, F. tigris, F. pardus, F. serval, F. cervaria, Viverra civetta,*

---

[1] " The great Bossuet left the brilliant court of Louis XIV, to shut himself up in the anatomical theatre of Duverney, that he might master the secrets of organization before writing his treatise *De la connaissance de Dieu* " (Lewes).

Ursus arctos, Nasua rufa, Lutra vulgaris, Phoca vitulina, Arctomys marmotta, Castor canadensis, Myoxus glis, Hystrix cristata, Erinaceus europaeus, Macacus cynomolgus. In this list the mammals (twenty-five species) and the birds (seventeen species) are offset only by five reptiles, one amphibian and one Fish — a selection which largely rules out the most fruitful possibilities of comparison in the vertebrate series. Hence the publications of the Parisians are zootomical rather than comparative, although, as will be seen, they canvass the philosophical aspects of their results whenever a clear opening presents itself.

The figures of *Alopias vulpes* (1667–71) are poor,[1] but the description by Perrault is better. The boundaries of the different regions of the gut are correctly defined, the distal limb of the stomach and the extent of the duodenum and rectum being determined on morphological criteria. The Parisians usually refer to the work of their predecessors, but they appear to have been unaware that the spiral valve, which is described and figured, was known to Severino, and had even apparently been seen by Fabricius in 1600. They note the backward extension of the nerve of the lateral line, and in the heart they observed the single very large auricle, and, " as in most animals which do not breathe [air] ", the single ventricle with its muscular conus arteriosus. The interior of the heart was exposed, and they found some of the valves, such as the proximal row of the conus and the two auriculo-ventricular valves. The account of the division of the ventral aorta, assuming that the efferent branchial system was not dissected, is mysteriously inaccurate, unless for " cerveau " we read " oüyes ". The eye was carefully investigated, without, however, its very curious orientation being observed.

Perrault must have dissected the salamander before 1680, but his account of it was not published until 1734. He examined male and female specimens, and noted that the black parts of the skin when viewed under the microscope showed a large number of yellow spots which were almost invisible to the unassisted eye. There was no external auditory opening as in lizards, but the animal could hear nevertheless. The tongue, liver, teeth, gall-

---

[1] The figure of the entire animal endows the pelvic fin with a pair of conspicuous pointed claspers, the existence of which lacks confirmation. Also the anal fin is shown as a *paired* fin, which is certainly an error, although it is repeated by Pennant.

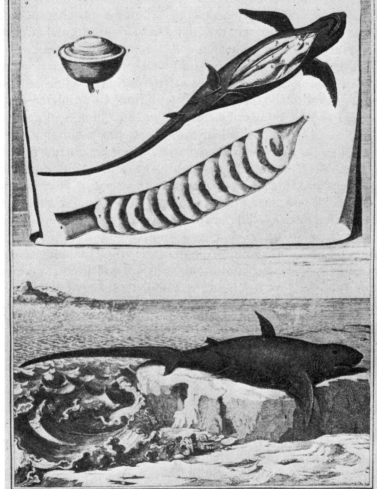

FIG. 173.—Perrault, 1671. Thresher shark, *Alopias vulpes*. The figure of the spiral intestine of this species appeared first in the folio edition, but not in the first original issue of 1667

bladder and gut are well described, but the pancreas is assigned a wrong position, and his hepatic duct seems to have been a blood vessel. Various mistakes and omissions occur in the account of the bladder and urogenital system, and his version of the female genital ducts is simply incomprehensible. It is perhaps remotely possible that his specimen may have been an hermaphrodite. He asserts that, contrary to accepted belief, the salamander is sexual and has distinct male and female organs, however peculiar they may be. He failed to notice that the young larvae had external gills, and he described the heart as having only a single (left) auricle.

The anatomy of the chameleon (1669–71) is treated at length. The Parisians note the structure of the curious eyelid, and they draw attention to the old error of attributing co-ordinated movements of the eyes to the optic chiasma, or the " joining of the optic nerves " as it was then called, for they found a chiasma in the chameleon — an animal with remarkable powers of independent movement of the eyes. The stiffness of the neck is held responsible for this free and antagonistic behaviour of the eyes. They describe the unusual character and extent of the lungs, and inflated them through the trachea. The anatomy of the tongue claims a large share of their attention, and they observed how it was used in feeding. Their discussion of the mechanism of the tongue, however, is highly ingenious but unsound. It is nevertheless interesting to note that the protrusion and erection of an organ, such as the tentacle of the snail, in response to *vascular* pressure rather than to muscular effort, was familiar to these seventeenth-century anatomists. But the tongue of the chameleon is *not* an example of this process.

Charas' account of the asp or red viper (1669, 1732) is not distinguished, although his plates are better. He describes correctly the poison gland with its muscular fascia, the poison duct, and the capsule of the fangs, but the secretion of this gland is said to consist of humours collected from the brain, eyes and neighbouring parts, and to be a " pure and quite innocent saliva ". He directs attention to the asymmetrical arrangement of the urogenital system, and he saw the paired evaginable spinose penes, their muscles and associated anal glands. The absence of one lung is noted, and his " thymus " was probably the thyroid or perhaps the vestigial left lung. The account of the heart is wholly wrong. It consists, he says, of two " ventricles ", the venous blood entering the right one and then passing to the left, which it leaves by the aorta for the body generally. The pulmonary circulation is ignored. The abdominal viscera are briefly described and also the external features of the brain, including parts of the ventricular system.

FIG. 174.—Perrault, 1671.   Chamaeleon.   Skeleton, tongue and abdominal viscera

Two lengthy chapters are devoted to the anatomy of the crocodile.   The first is by Gouye (1688, 1732), with interpolated notes by Duverney, and the second is by Duverney only (1734). An important point not considered by either author is the presence at the back of the mouth of a palato-lingual velum which enables the animal, with only the tip of its snout above the surface, to fill its mouth with water, without interrupting a continuous flow of air into the lungs. In the Cetacea the intranarial epiglottis achieves the same purpose, but more efficiently. Gouye's statement that the upper jaw of the crocodile is capable of considerable movement which can easily be observed, whilst the lower jaw is immobile, becomes even more astonishing when attempts are made to establish the existence of hinges and muscles by which this preposterous movement is supposed to be effected. Duverney, however, was not deceived, and insisted that the upper jaw was more rigidly articulated to the skull than in most other animals, and was therefore incapable of independent motion. He is indeed astonished that this should ever have been questioned, since the facts are readily demonstrable. The paired musk gland of the crocodile, which lies immediately under the skin of the lower jaw, was first described by Recchus in 1628, and is now re-discovered by Duverney. It secretes a fluid having a very powerful odour of musk. According to Gouye the nictitating membrane moves horizontally over the eye, and is so transparent that it does not completely interrupt vision. Duverney adds that it has one muscle instead of the two found in birds, the quadratus being absent, but he does not mention the choanoid muscle, which diverts the tendon of the nictitans from the optic nerve and in that respect takes the place of the quadratus. Gouye describes the large auditory meatus and the tympanic membrane, and he figures the " very delicate " columella auris, which he says consists of two parts corresponding with the malleus and incus of higher animals.[1] In his account of the gut he fails to recognize the unusual structure of the crocodilian stomach, one of the most complex to be found in the Reptilia, nor was he struck by its resemblance to the gizzard of a bird. In the second paper, however, this point is made by the more learned Duverney. The structure of the heart is examined by Duverney in two memoirs. In the first (*c.* 1681)[2] he distinguishes two auricles separated by an "imperfect" septum. Hence their bloods mix before entering the *single* ventricle, which has one cavity only, although it is subdivided into a thousand crevices by strands of fleshy tissue. The main arteries arising from the heart are briefly and accurately described, except as regards the exact origin of

---

[1] The extra-columella and stapes of the modern comparative anatomist.
[2] First published in full 1734.

subclavians and carotids, and he compares them with the corresponding arteries of the tortoise. One of Duverney's unhappy obsessions was the supposed existence in reptiles of a double posterior vena cava, the larger member of the pair [the true postcaval] discharging into the right auricle and the smaller one into the left auricle. The latter was probably an anterior abdominal vein, which does not, however, open into the heart.[1] It will be noticed that Duverney does not recognize a sinus venosus. His second account of the heart of the crocodile (1688), according to Mery, is not based on the crocodile but on the marine tortoise, and it certainly reads like a description of the heart of that animal as Duverney himself was wont to interpret it. The two auricles, he says, are related to the venous system as in the higher vertebrates, but the ventricles are *three* in number *and in free communication with each other*. Hence the circulation cannot conform to a type in which two *independent* ventricles are involved. He concludes that one-third only of the mixed ventricular blood passes to the lungs in every cycle, in which event it cannot be said that animal life is dependent on the mixture of blood and air in the lungs. In the crocodile no blood reaches the lungs beyond what is necessary for their own nourishment, and their function is rather that of hydrostatic organs, adapted to sustain the animal in the water. It is clear that at this stage Duverney was far from understanding the structure of the heart of the crocodile, and it is surprising that an anatomist capable of working out correctly the structure of the truncus arteriosus of the frog should have failed to discover the imperforate interventricular septum of the crocodile. A few years later he published the results of his classic researches on the structure and function of the circulatory apparatus in fishes, amphibia and reptiles, in which his earlier views were modified, and, whilst still failing to reach a complete understanding of the problem, he drew nevertheless some very acute conclusions. The fact which should have helped him much more than it did — that in the fish circulation the *whole* of the cardiac blood passes through the respiratory organs before reaching the system generally, he explains as being necessary owing to the small amount of air contained in water.

[1] It is not unlikely that one of the hepatic veins was mistaken for the continuation of the anterior abdominal, but neither would *it* open directly into the heart.

The memoir on the Indian tortoise supervised by Duverney is not only the first competent account of the anatomy of a chelonian to be published, but is an instructive display of the strength and weakness of the Parisians, nor do they appear to have seriously respected the wishes of the King to return the specimen sufficiently intact for exhibition in the museum. The Parisians were also the

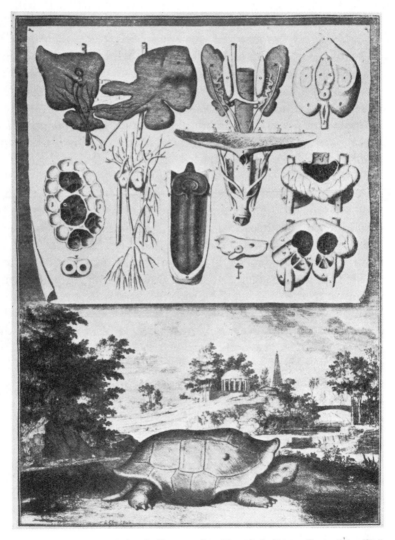

FIG. 175.—Perrault, 1676.   Indian tortoise, *Testudo indica*, s. *Perraulti*.   Brain,
columella auris, heart, vascular and male urogenital systems

first to describe this interesting species, which was a large terrestrial Indian tortoise from the Coromandel coast. It is now extinct, and has apparently not been described since the type specimen was dissected in Paris in 1676. No further specimens have been traced, and even its name is a subject of dispute among systematists. The origin and course of the cystic and hepatic ducts are worked out, the epididymides are unravelled, and the vasa efferentia disclosed by the injection of a coloured fluid. The bladder is astutely recognized as comparable with the allantois of higher animals, and the urogenital organs receive masterly treatment, accompanied by an excellent figure. Even the comparative anatomy of the lung is only partially baffling, as we gather from their happy comparison of the chambered lung of the tortoise with the almost parenchymatous lung of the mammal. Vivisection itself is resorted to in matters of difficulty, and they were among the first to investigate the physiology of the lungs in a living animal (the dog) in which respiration was maintained with a pair of bellows (1671). Similar experiments are described by Vesalius in 1543 and by Robert Hooke in 1664, but the Parisians go further than Hooke, and find that in the inflated lung an injection thrown into the pulmonary artery passes more readily through the capillaries into the pulmonary vein than in the deflated organ. They discuss also the nictitating membrane of the eye of the tortoise and its muscles, the relations of the tympanic cavity and the columella, and they realize that the extrusion of the head and neck is just as much a question of muscular *contraction* as its withdrawal — a simple deduction, but one which later biologists have not always visualized. On the other hand in spite of several ingenious, but misleading, experiments they fail to grasp the broader facts of the reptilian circulation. The auricles are correctly described, and they define three ventricles, which, however, have only incomplete walls and whose contents therefore mix. The pulmonary veins, they say, do not unite, but *open into the precavals*, so that the blood *entering* the ventricles is mixed. The aortic arches are regarded as branches of a *single* vessel, and it is claimed that the blood dispatched to the lungs is very small in quantity, sufficient only for their own nourishment. Hence the lung of the tortoise has no relation to the *general* circulation, in this respect resembling the lung of the human foetus. To establish this point the pulmonary artery of a tortoise was ligatured, but the motion of the heart was unaffected, and the general circulation proceeded as before. What then, they ask, *is* the function of the lung of the tortoise ? They admit they have no answer to this question, but consider that they are entitled to throw out a suggestion, since final solutions to the problems raised by their researches are not possible. They think the lungs might be used for the compression of the internal parts, and so assist in the concoction and distribution of nourishment, but a more important function perhaps would be to act as hydrostatic organs, like the air-bladder of fishes. Some experiments bearing on this point are described.

It must be admitted that Duverney is not at his best in this, his first, attack on the heart of the tortoise. Many serious mistakes are made, and even the right and left sides are confused, so that the text does not always agree with the figures. The sinus venosus is not described, although its existence seems to be implied. The double postcaval is not easy to explain, and the account of the relations of the great vessels to the chambers of the heart is seriously inaccurate, but he saw that there were two auricles, and that the ventricular cavity consisted of three sections. At this stage he had no clear ideas on the circulation of the blood in reptiles, and his failure even dimly to understand the function of the lungs inhibited the conception of a theory which could be developed later by others. His next step could only be to correct his own errors, and in this, as we shall see, he certainly succeeded.

The structure of birds early engaged the close attention of the old anatomists, and it is therefore in accordance with tradition that the Parisians should regard birds as of only less importance than mammals. It must be noted that the figures occasionally fail to do justice to the accuracy of the text. For example, the statement that the vas deferens and ureter may be very closely attached is contradicted by the figure which represents these ducts as being completely fused. The points in avian anatomy most attractive to the Parisians are the biliary and pancreatic ducts, stomach and urogenital system. Other matters are briefly dismissed. The parts are usually correctly identified, and when they fail to recognize an organ in one bird they are often more successful in another, as in the case of the adrenal body of the Numidian crane and the pelican. In the bustard, *Otis*, they describe and figure opening into the cloaca a third " caecum " which they remark is commonly called " la Bourse de Fabricius ". No comment is attempted respecting its function and anatomical status. This familiar and much-debated organ was first made known in 1621 in a posthumous work by Fabricius, who, in ignorance of the fact that it occurred in both sexes, considered its function to be that of the female receptaculum seminis. In 1674 and 1681 Blasius suggested that voice production in birds was associated with the *lower* larynx, and not with the anterior or larynx proper as in other vertebrates. In 1680 and 1686 (published 1733) Perrault and Duverney succeeded in demonstrating this important inference in a living duck and cock.

The Parisians detect the connection between the calibre and length of the gut and the character of the food, and they note also that the absence and relative development of the rectal caeca are determined by the same factor. They are the earliest after Steno (1674) [1] to describe *and figure* the sinus rhomboidalis of the spinal cord, first in the eagle (1676), and later extending the discovery to other birds. Its contents are found to be a " white and glutinous humour ", the removal of which by a duct is considered to be possible. The pecten of the eye is investigated and described in several birds, and its pigmented and vascular character is clearly perceived. The pecten is supposed to be wanting only in *Apteryx*, but the Parisians were unable to find it in the Numidian crane. They undoubtedly anticipate a modern view that the function of the pecten is to facilitate the " nourishment of the humours of the eye ". In the cormorant they note the absence of the caeca formerly believed to be present in all birds,[2] and their description of the curious stomach of this species is excellent. Although it differs markedly from the type commonly present in birds, they distinguish both glandular and muscular portions, corresponding with the proventriculus and gizzard of the familiar avian stomach. Its unusual form they attribute to the piscivorous diet of the species.

But the bird most grateful to the curiosity of the Parisians, of which they dissected eight examples, is the ostrich, and they give a lengthy and excellent description of its anatomy. It was not until 1712 that this bird was examined again with such care —

---

[1] Published by Jacobsen, *Acta Med. Hafn.* 2, 317, 1675.
[2] Very small caeca are, however, present in the cormorant.

this time by Vallisneri. The Parisians pursue in detail the structure of the feathers, and contrast them with the quill feathers of a flying bird. In this they appear to have been ignorant of the work of Robert Hooke on the morphology of feathers, first published in the *Micrographia* of 1665. They understood the function of the barbs and barbules, and they recognize the double advantage of a concavo-convex feather — its greater rigidity and grip of the air on the downward beat, and its diminished resistance on the return stroke. They devote ample space to an account of the air sacs, which is on the whole comprehensive and accurate, and they ascertain by inflation experiments the connection between the air sacs and the lungs. They regard the partitions separating the air sacs as a series of diaphragms, and we note with regret a tendency to go back on their acceptance of the doctrine of the circulation in the description given of the passage of the blood through the lungs. The air sacs were not discovered by the Parisians, although they were the first to investigate them in detail. Thus Mery inflated and dissected the sacs in the pelican in 1686 (published 1693). The rectal caeca of the ostrich are specially remarkable in exhibiting a striking spiral structure, which is described and figured by the Parisians and also later by Vallisneri. In the section on the brain an old tale is repeated which illustrates how the confidence of mankind was abused in a credulous and superstitious age. To demonstrate the virtues of a healing balsam its compounder would plunge a knife into the head of a bird, whose life he then professed to restore by the application of his powerful ointment ; and provided the knife had been thrust into the fissure between the cerebrum and cerebellum, the position of which the operator had prudently explored by private dissection, a miracle was duly proclaimed by an amazed and unsuspecting public. Willis (1664) mentions a " familiar experiment among boys " of thrusting a needle through the head of a hen in such a way that the bird is not killed. His explanation is that the cortex only of the brain is transfixed, which, he says, is not a fatal wound.

In the cassowary (1676), the Parisians describe and figure the long aftershaft of the feathers, and they note the peculiar nature of the stomach and its valve, although the dilated duodenum is misinterpreted as a second chamber of the gizzard. They compare, not without reason, the air sacs of the bird with the branching lung

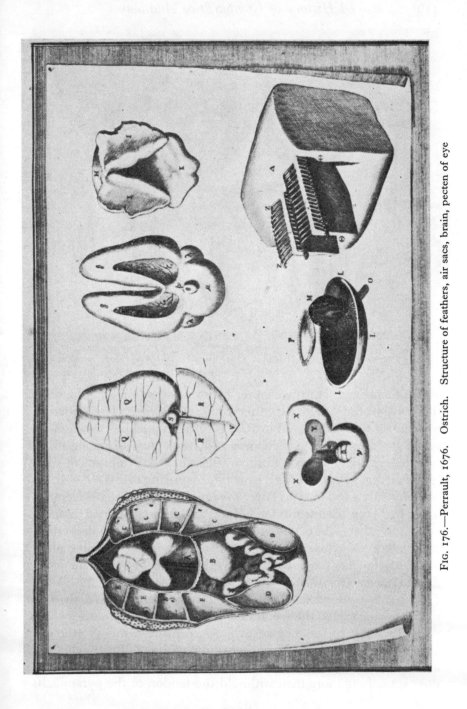

Fig. 176.—Perrault, 1676. Ostrich. Structure of feathers, air sacs, brain, pecten of eye

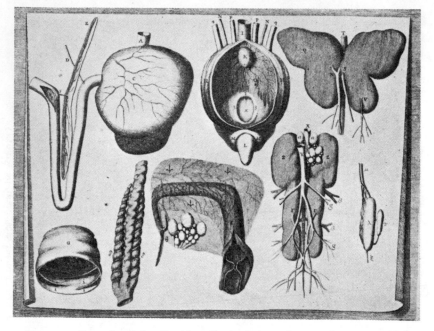

FIG. 177.—Perrault, 1676.   Ostrich.   Gut, spiral rectal caeca, cloaca, urogenital
system of female

of the chameleon, but the chapter fixes our attention on account of
the passage on the nictitating membrane of the eye.   This structure
was known to Aristotle and in 1651 it is mentioned by Harvey, who
called it the membrana nictatoria and who refers to its cleansing
functions.   Steno discovered the muscles of the membrane in 1673
(published 1675), and it is perhaps not surprising that the Parisians,
working only a few months later, were not familiar with his memoir.
They had seen the membrane before, and it is mentioned casually
here and there, but it was in the cassowary that they were
stimulated to disclose the complete facts.   It is a remarkable piece
of research, illustrated by clear workmanlike figures, and more
detailed and trustworthy than some modern versions.   They
describe, in addition to the six normal muscles of the eye, the two
muscles which draw the membrane over the cornea, and they show
that the object of the movement is to keep the surface of the eye
clean.   The mechanics of the origin and doubling of the pyramidalis
are understood and explained, and they realize that the quadratus
does something more than withhold the tendon of the pyramidalis

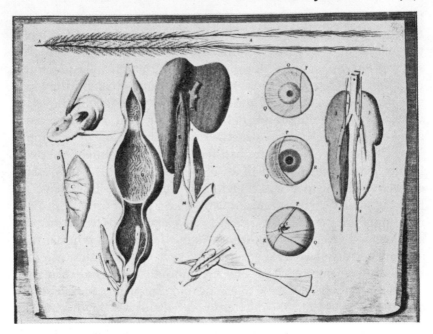

Fig. 178.—Perrault, 1676.   Cassowary.   Gut, biliary apparatus, renal organs,
Harderian gland, muscles of nictitating membrane of eye

from the optic nerve.   The motion of the membrane, in fact, is the
resultant of two forces, and not the expression of one.   These are
facts which have since been abundantly confirmed and exploited,
but the true merit of such a performance is rather the obligation it
imposes on posterity of precise and exhaustive observation.

It is unfortunate that the only part of *Manis* available to the
Parisians was the skin, which, on the evidence of its scales, they
concluded was that of a large lizard.   Could the head and internal
organs have been examined it is probable that this attribution would
have been reversed.   The fact that hairs as well as scales must
have been visible in their preparation would mean less to them
than it does to us, since at that time hair had not been precisely
defined, nor had it been shown that mammalian hair was not
found outside the mammalian series.

The most detailed and ambitious memoir produced by the
Parisians (1734) had for its subject the elephant, the dissection of
which was undertaken by Duverney himself.   The specimen was
an African elephant, twenty-seven years of age, which had lived

for thirteen years in the Royal Menagerie at Versailles. Owing to the external characters of the female resembling those of the male, the animal was recorded as a male during its lifetime, and its true sex was discovered only after dissection. A lengthy discussion did not suffice to determine the nature of the tusks, which, they say, some call teeth and others horns. The Parisians favour the latter view, because the tusks are hollow, are not shed, and arise from a bone which does not bear teeth — the " third bone of the upper jaw " [frontal].[1] They claim further that the structure of the tusks is more horn- than tooth-like. Garçias de Horto (1567) states correctly that in the Indian elephant the females either have no tusks at all or very short ones, but the Parisians show that this does not apply to the African species, as exemplified by their own specimen in which the tusks were two feet long. In proof that they are not deciduous and renewed every year they cite the case of a marked tusk being retained for four hundred years, and their own elephant had not shed its tusks during the whole period of its captivity. The duct of the temporal gland was known to the

---

[1] An error. The root of the tusk does not penetrate beyond the maxilla.

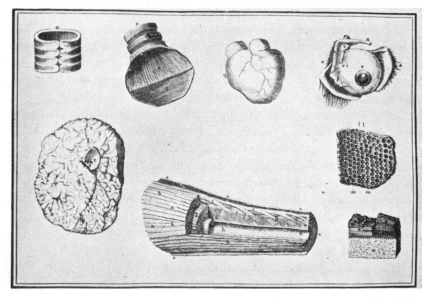

FIG. 179.—Perrault, 1734.   Elephant.   Temporal gland, heart, eye, structure of trunk

ancients, but the Parisians were the first to describe and figure the gland itself, and they were aware that it became active during the period of sexual heat. The structure of the trunk and its muscles were carefully examined and figured, and they were familiar with its varied functions and morphological status. The absence of a bony or cartilaginous support in the proboscis is noted, and the account and figures of the skeleton are specially good. They saw the nasal cartilages, but in the tarsus the navicular is neither mentioned nor figured. Attention is directed to the characters in which the elephant resembles man, such as the fusion of the mandibles at the symphysis, and the attachment of the pericardium to the diaphragm. Galen's errors in attributing a gallbladder and an os cordis to the elephant are corrected. In the viscera they found that the bile and pancreatic ducts unite to have a common opening into the duodenum, and the peculiar ileocolic valve was discovered and well figured. The heart was not so carefully examined, and they overlooked its bifid apex, which was nevertheless known to the ancients. Neither was the lobulated nature of the kidney observed, but this is not too obvious even in the foetus. The brain is briefly dismissed, but the unusual size of the " fifth pair " of cranial nerves [1] is explained as being due to the large number of fibres which this nerve is called upon to contribute to the large proboscideal nerve. The female genitalia are figured, and the penis-like appearance of the vulva is illustrated by a good drawing.

A remarkable vascular phenomenon is alleged to occur in the spotted deer, *Cervus axis* (1676), where the external and internal jugular veins are believed to possess sixteen valves disposed in six rows, the cavities of the valves being directed, not towards the heart, but *towards the head*. The Parisians fully recognize the unusual and incredible nature of this arrangement, which is explained as preventing " the too great impetuositie of the Blood which falls in its returne from the Brain into the Axillary Branches ". This statement does not appear to have ever been confirmed or denied, but without corroboration it is not possible to accept it. In the Barbary cow, *Bubalis boselaphus* (1676), valves are described in the primary hepatic branches of the portal vein, which are correctly interpreted as preventing the reflux of blood into the

---

[1] The trigemino-facial of modern usage.

parent vessel. These valves are not present in man. The independent blood supply of the cortex and medulla of the kidney was detected in the lion (1671), and an examination of the human kidney revealed a similar state of affairs, contrary to the statements of Vesalius. In the later edition of 1731 a good figure is added of the receptaculum chyli of the lion, together with the thoracic duct and its opening into the left subclavian. There is also introduced a figure of that " admirable structure ", the mechanism of movement of the claws. In the chapter on the leopard (1734) it is stated that the carotid does not give rise to a rete mirabile either in that species or in the lion. Under the names Chat-Pard (1671) and Panthére (1734) the Parisians describe a single species, *Felis serval*, as two different animals. A comparison of these two chapters is of some value as indicating the margin of error to be expected and allowed for in their observations and figures.

The completely lobulated kidney of the bear was first engraved by Eustachius in 1552, but the figure was not published until 1563. According to the Parisians (1669, 1671) it is composed of fifty-six lobules, each having its own artery, vein and efferent duct, and they compare it with the ten-lobular kidney of the otter and the superficially lobulated kidney of the porpoise and newly-born human infant, which last are stated to be essentially parenchymatous. In the common seal, *Phoca vitulina* (1676), they are again attracted by the lobulated nature of the kidney, which, with its superficial plexus of vessels, is satisfactorily described, but the investigation of the heart involves them in unexpected confusion. They are aware that the seal is not a fish, and is incapable of aquatic respiration, and they are led to assume that it must have an intranarial epiglottis to enable it to feed under water. They also understand that in the mammalian foetus blood is diverted from the right side of the heart to the left through the foramen ovale in order to avoid the lungs,[1] and they draw from this the fatal conclusion that the foetus does not respire. They profess to have found, and, indeed may actually have found, a persisting foramen ovale in the heart of the seal, and they believe that when the animal dives, and remains some time below water, the circulation follows the same course as in the intra-uterine embryo. The fact that the seal is below water only for a relatively short time, whilst the circulation

---

[1] This was known to Galen, and is discussed by Servetus and Harvey.

of the foetus remains the same throughout foetal life, should have warned them of the risk of assuming an interruption in the normal circulation every time the breathing organs are cut off from the atmosphere.

The Parisians display a marked capacity, unusual at a time when the Cetacea were still ranked as fishes, for detecting the natural relations of the animals they dissect. Thus, they say, the marmot and dormouse, judged by their size and appearance, appear to be very different animals. Nevertheless, both are species of rats, and they agree also in their habits of hibernation. The first *foreign* animal examined was the beaver (1669, 1671), and its skeleton was the first preparation to be placed in the " Salle des Squeletes " of the Royal Cabinet. The structure of the castoreum of this animal is accurately and minutely described.[1] It is distinguished from the scent glands of the civet-cat, which are regarded as secondary sexual characters. They establish that the castoreum is not the testis, as formerly believed. Their only failure of importance is to overlook the smallest subdivision of the second pair of glands, but on the other hand they note the difference in structure between the two sets of glands, and find also a difference in the character of the secretion. They point out, both in the beaver and in the civet-cat, that the contents of the glands have been produced *by the action of the gland tissue on the blood* — one of the earliest declarations of a physiological doctrine of supreme importance.

The ancients were induced by the apparent affinity of the hedgehog and porcupine to unite them under the one genus of *Echinus*, but the Parisians (1676), after anatomizing examples of both animals, conclude that they are " very different ", both as regards external characters and internal organs. But they go further than that, for they recognize the genetic relationship of the porcupine with such forms as the hare and the beaver, and the large rodent caecum of the porcupine is contrasted with the reduction of that organ in the hedgehog. By ligaturing the anterior extremity of the azygos vein, and inflating it backwards, they establish a posterior anastomosis with the iliac vein, as in man — a point of detail we hardly expect at this early period. The large bag-like and glandular

[1] The castor of the beaver was used in medicine as early as the seventh century B.C., and hence the gland secreting it was a structure of special interest to the Parisians.

vesiculae seminales of the hedgehog, however, naturally overtax their knowledge and experience, and they interpret them as vascular organs for the elaboration of the blood before it reaches the testes.

The chapter on the monkey, *Macacus cynomolgus* (1676), calls for comment only as regards the description of the laryngeal

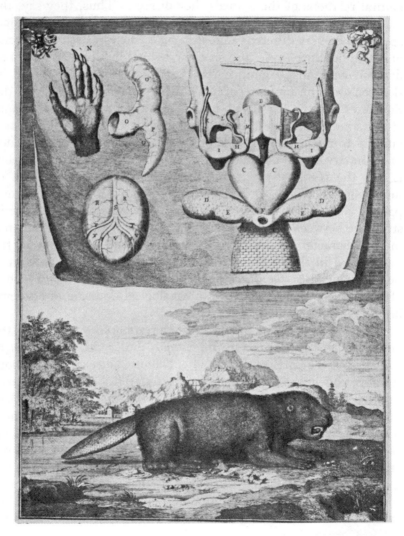

FIG. 180.—Perrault, 1671. Beaver, *Castor*. Os penis, male genitalia, castoreal glands

region. Its close resemblance to the speaking larynx of the human species is converted into an agreement so exact as to emphasize dramatically the vocal isolation of man, and to concentrate the philosophic genius of the company on the anthropomorphic but speechless monkey. They say : " For the *Ape* is found provided by Nature of all these Marvellous Organs of speech with so much exactness, that the very three small Muscles which do take

Fig. 181.—Perrault, 1676.   Comparative anatomy of porcupine and hedgehog.
Gut, male and female genitalia

their rise from the *Apophysis Styloides*, are not wanting, altho this *Apophysis* be extreamly small. This particularitie do's likewise shew that there is no reason to think that Agents do performe such and such Actions, because they are found with Organs proper thereunto : For according to these Philosophers, *Apes* should speake, seeing that they have the Instruments necessary for speech." Ray (1691), in commenting on this passage, claims that the larynx of the monkey closely resembles the human type, but that the resemblance is less marked in the case of the hands. Notwithstanding this, the monkey can use his hands like a man, but cannot speak. Again, the larynx of birds is wholly unlike that of man — nevertheless some birds can be taught to speak.[1] An opening such as this could hardly fail to provoke the ingenuous advocates of created man, and consequently we find Tyson, in confirming the statements of the Parisians, drawing an inference which he says the " Atheists can never answer ". Yet neither the Parisians nor Tyson could be expected to comprehend those functional refinements which alone can evoke the harmony of speech, and it was reserved for the more instructed vision of Camper to supply a profane but convincing explanation of the silence of the forest.

Apart from their contributions to the collected *Memoires* published under the aegis of the Academy of Science, Perrault and Duverney both produced independent works on comparative anatomy which are so intimately related to the *Memoires*, and are in themselves of such importance, that they fall to be noticed here. " La Mechanique des animaux " of Claude Perrault, first printed in 1680 in the *Essais de Physique*, is one of the few classics in which a speculative approach to the bare facts of comparative anatomy is even attempted. He himself vigorously justifies this attitude, and maintains that it is permissible to follow the example of those celebrated philosophers who do not hesitate to invoke the most fantastic imagination, provided that outstanding and convincing results are not otherwise possible. He fears that many lovers of science and letters will be scandalized by this declaration, and will treat as an outrage any defence of a philosophy the beauty and excellence of which is not rooted in factual verity. His own point

---

[1] Ray is here referring to the anterior or true larynx. The syrinx of birds was unknown to him.

of view is that an animal is a true machine, but a machine which can be activated only by a principle called the " soul ". It is, he says, like an organ which, although it is capable of producing all the varied sounds which the nature of its parts makes possible, is nevertheless condemned to perpetual silence except at the will of the organist. Perrault is therefore opposed to the Cartesian " new school ", which professes to understand and explain *all* the activities of animals on purely mechanical principles. Neither does he hold with the complete vitalist, but admits that mechanism is the more attractive and promising faith, and, like the modern critic, he attacks vitalism for glorying in its ignorance and stagnation. In his acknowledgments he refers with admiration to the excellent researches of Duverney, which had made their author famous in the learned world. Indeed, in the account of the anatomical investigations which follows, it is sometimes doubtful how much of it we owe to Perrault and how much to Duverney.

The Parisians completely ignored the invertebrates in their *Memoires*, and Perrault in this later work could hardly be expected to fill a gap of the existence of which he was none too conscious. He remarks unhappily that animals like insects possess only one sense — that of touch, in which they excel. No other sense organs appear to be present. The tracheae of insects, he says in another place, are related to the gills of fishes and the lungs of terrestrial animals, and he names them branchiae, although they extend over the whole body. There is a group of them or " little lung " in each joint, which has its own opening on to

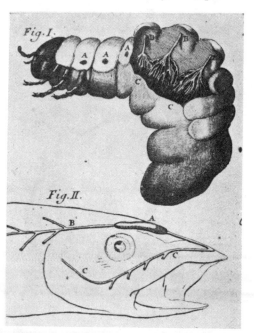

FIG. 182.—Perrault, 1680. Tracheal system of larva of *Lucanus* and lateral line canals of *Gadus*

the surface.   He adds a figure of a dissection of the "lungs" in the larva of the stag beetle, *Lucanus cervus*.

Perrault's methods are essentially comparative.   Thus he points out that the lens of the eye is spherical in fishes but lenticular in terrestrial species, which he explains on the basis of the difference between the refractive indices of air and water.   In fish, he says, the lens has to make good the loss of refraction due to the optical identity of the aqueous humour of the eye and water, in which medium, therefore, the humour ceases to operate.   In animals which combine terrestrial and aquatic habits, such as the seal and diving birds, the shape of the lens is midway between the spherical and the lenticular.   A comparison of the brains of diverse vertebrates reveals variations in size which have no relation to the bulk of the animal.   For example, in the large crocodile the brain is very small. In birds its external conformation is simple and the ventricles are small compared with those of other vertebrates, but they are very large in the o3trich and four in number as in the Vivipara.   The bird, however, has a ventricle not present in other animals — the sinus rhomboidalis of the spinal cord. Perrault found it first in the eagle and later in other birds.   He compares it with the fourth ventricle of the brain, and concluded that it was formed in the same way by the enlargement and partial divergence of the two halves of the cord.   Its presence, he says, may be associated with the high rate of tissue waste or muscle consumption inevitable in flight. The brain of fishes was

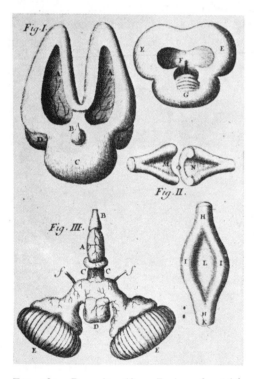

FIG. 183.—Perrault, 1680.   Brains of ostrich and *Scyllium*, sinus rhomboidalis of spinal cord of eagle

found to be much smaller than in birds, but the usual ventricles were present. In the cartilaginous species such as the dog-fish, *Scyllium catulus*, the most prominent part of the brain was the olfactory region. In his figure of this species the optic lobes are represented as if they formed a transverse ring in front of the cerebellum, and the restiform bodies are too small. A final example of Perrault's exploitation of the comparative method is his long discussion of the different types of valve found in the vascular system, the operation of which he illustrated by ingenious working models.

Fishes and birds are the groups from which Perrault selects the material for his investigations, mammals being employed only occasionally. We have from him the first published figure of the lateral line system of a teleostean fish, the species being *Gadus virens*. It shows the lateralis, supra- and infra-orbital sensory canals, not too accurately, the so-called "gland" being the supra-orbital commissure. He looks upon the canals as mucus-secreting organs, producing the oily substance which lubricates the surface of the body. Perrault displays great interest in the devices which produce a delaying action on the passage of the food down the gut, such as the spiral valve of *Alopias vulpes* and the scroll valve of *Carcharias glaucus*, both of which are figured, the latter for the first time. In this connection the spiral rectal

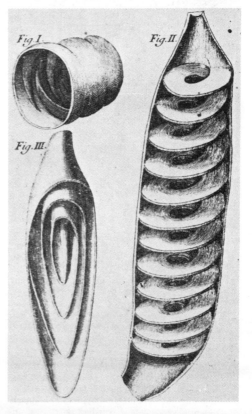

FIG. 184.—Perrault, 1680. Spiral valve of *Alopias vulpes*, scroll valve of *Carcharias glaucus*

caeca of the ostrich and the spiral caecum of the hare come in for consideration. Two kinds of swim-bladder are distinguished in fishes by Perrault. The first, as in *Clupea alosa*, has a small duct which connects it with the stomach, and by which " apparently " it receives the air which it contains. The second type, exemplified in the codfish, has no duct, but instead a glandular fleshy mass [gas gland] which, he says, is presumably responsible for the production and absorption of the gaseous contents of the bladder. He believes that this " air " is tenuous, and that when it is compressed by the muscles of the bladder its volume is diminished and the fish sinks. When, therefore, these muscles relax, the " air " expands to its former volume and the fish rises. In the aquatic tortoises, he adds, the lungs take the place of the swim-bladder of fishes, but with this difference — in the latter the quantity of air in the bladder remains the same, but in the tortoise it varies according to whether the animal is in or out of the water. The same, he says, applies to seals, dolphins and crocodiles. It is strange that Perrault, having got as far as deducing that the gas gland may increase or rarefy the contents of the bladder, should have thought it necessary to provide an additional, and erroneous,

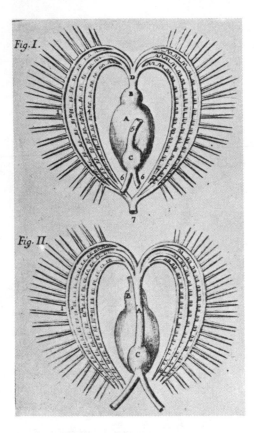

FIG. 185.—Perrault, 1680. Diagrams illustrating circulation of blood in a Teleostean fish, the carp. *A*, (Fig. I) ventricle, (Fig. II) inferior jugular vein; *B*, bulbus arteriosus; *C*, auricle +sinus venosus ; *D*, cardiac aorta ; 5, inferior jugular vein ; 6, precaval veins ; 7, dorsal aorta

explanation of the hydrostatic function of this organ.

Perrault, evidently inspired by his mentor Duverney, devised a scheme of the circulation of the blood in the carp. The first [cardiac] aorta, he says, divides into many branches in the gills, which rejoin to constitute the second [dorsal] aorta giving off branches to all parts of the body. As in *Gadus pollachius*, the heart has only one ventricle, but there are " what appear to be two auricles ", a superior or left auricle [bulbus arteriosus], from which the ventral aorta arises, and an inferior or right auricle, into which open two [precaval] veins. His right auricle must have been the true auricle +the sinus venosus, and it will be noted that the circulus cephalicus and cranial arteries are neither mentioned nor figured, but the structure and vascularity of the gills receive attention. Perrault's first figure comprises the afferent branchial system only, and is based on the highly improbable assumption that blood which was free to pass *direct* and unimpeded from the ventral to the dorsal aorta could be persuaded to traverse a capillary system in the gills on the way. He was aware that each branchial arch was accompanied by *two* vascular trunks, and his " veins of the gills " represented in the second figure are in fact the efferent branchial vessels, which, according to him, have no connection with the dorsal aorta, whilst ventrally they unite to constitute the inferior jugular vein, *A*, and so discharge into the heart. The latter error, as will be seen, was repeated in 1699 by Duverney, and it implies that the blood in the heart of the fish is mixed. Perrault appears to have discovered the inferior jugular vein before Duverney. He believes that the function of the gills of fishes is much the same as that of the lungs of terrestrial animals, but he did not perceive that the branchial circulation is maintained by the motion of the heart, and that the squeezing action of the operculum had nothing to do with it, nor did he understand that the respiratory air was *dissolved* in the water.

The mechanical explanation put forward by Perrault of the complex movements of the tongue of the woodpecker is based on a comparison with the controls of the copper dome of an astronomical observatory, but his dissections of the muscles of the tongue were not sufficiently careful or detailed to warrant the conclusions based on such a comparison. He may, however, claim to be the first to describe and figure with any accuracy the lower

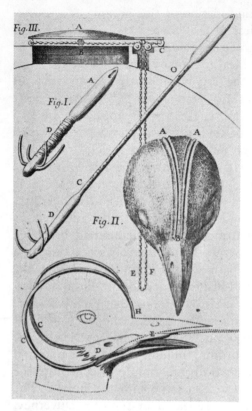

Fig.III.

Fig. 186 —Perrault, 1680.   Woodpecker.
Mechanism of movement of tongue

larynx of a bird, and to associate it with the production of voice. He holds that his experiments on the goose and duck prove that the glottis of birds is not the seat of the vocal organ, which is to be sought rather in some membranes situated in another larynx functioning at the bifurcation of the trachea. He tested this by cutting off the head and anterior larynx of a duck and then pressing the abdomen, whereupon a sound was produced similar to the voice of the bird when it was living *and possessed an anterior larynx.* In his earlier writings Perrault had already described the syrinx, and had discovered the semilunar membrane, but he had not recognized in these structures the vocal organ of birds. He now describes the internal tympaniform membrane (Fig. 187, *F*) and the pessulus (*H*). It is: he vibration of the former, he says, and in this he is not far wrong, which is responsible for the voice. He associates the long coiled trachea of some birds with the production of certain forms of sound, and compares the coils with the crooks of an orchestral trumpet. His search for mechanical parallels is in evidence again in the chapter on the air sacs of birds, the action of which he seeks to illustrate by comparing them with a farrier's bellows.

The avian eye has two peculiarities which attracted Perrault's attention. He points out firstly that a flying animal has need of perfect sight ; on which account, he adds, the eye of birds possesses a structure [pecten] which varies in different birds and is not found outside that group. Its high vascularity was known to him, which suggested a nourishing function, but further consideration resulted in the more subtle inference that the pecten, like the uvea, exercises a filtering or refining influence on the blood, and in that way mysteriously produces a more penetrating vision. He reaffirms a former statement that the pecten is wanting in the Numidian crane. The second feature is the nictitating membrane and its two muscles, which receive detailed consideration accompanied by notable illustrations.[1] This

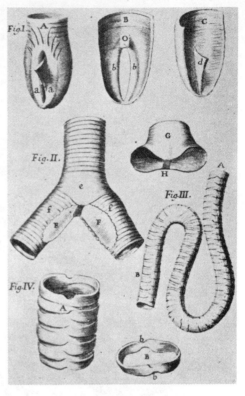

FIG. 187.—Perrault, 1680.   Structure of larynx and posterior larynx (syrinx) of birds

third eyelid, he says, is usually wanting in fishes, although present in most other vertebrates. In the shark, *Carcharias glaucus*, however, he rediscovered a true well-developed third eyelid and its muscle, which are described and figured.[2] He found it also in those " fishes " which are only partially aquatic, such as *Phoca vitulina*. The eyeball of *Rhina squatina* is described as being firmly attached to the optic peduncle [cartilago sustentaculum oculi], which, he says, is " osseous ",[3] and it is asserted that

[1] A comparison of the 1676 plate of the nictitating membrane and its muscles with the present figures of 1680 reveals a curious inaccuracy in the latter, whereby the tendon of the pyramidalis is drawn as if it were threaded through a *perforation* in the quadratus.   The earlier figure correctly represents the relations of these two muscles.

[2] Cf. Ridewood, " . . . the morphology of the nictitating membrane of sharks ". *Journ. Anat. Phys.*, 33, 1899.   Rondelet (1554) first described the membrane in sharks.

[3] This does not necessarily mean bony.   Perrault is guilty of such expressions as " cartilages osseux ".

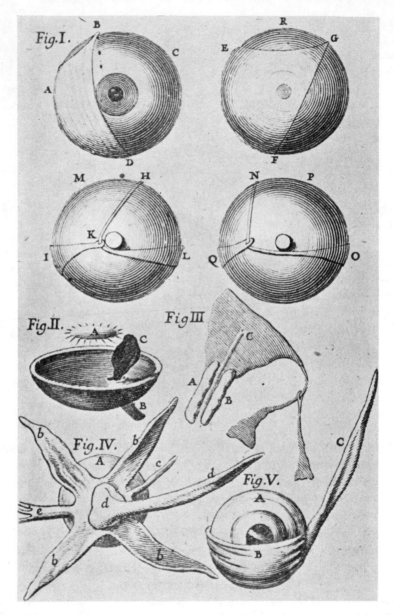

FIG. 188.—Perrault, 1680. Fig. I, muscles of nictitating membrane of bird (species not stated) ; II, III, ostrich, pecten of eye and Harderian gland with duct ; IV, angel fish, *Rhina*, *d*, optic peduncle ; V, shark, *Carcharias*, third eyelid

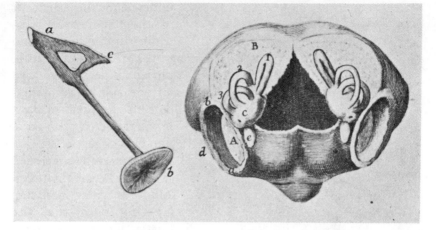

Fig. 189.—Perrault, 1680. Columella auris and extra-stapedial of bird ; bony labyrinth and lagena of Indian cock, *Crax alector*

the poise of such an eye makes it much more responsive to the play of the eye muscles than is the case in those animals where the eyeball is bedded in membranes or fatty substance. After completing a detailed examination of the ear of man, Perrault proceeds to compare it with the ears of other vertebrates. He recognizes that this sense organ presents a common structure in all the higher animals. In the curassow, *Crax alector*, the bony labyrinth and columella, including the extra-stapedial elements, are well figured. He holds that the three auditory ossicles of the mammal are reduced to one in birds, and that the labyrinth is also abbreviated, the spiral cochlea being represented by a simple, short projection like a small sac [lagena]. In fishes the drum and ossicles are wanting, and the only part of the ear they possess is the labyrinth, which, he wrongly adds, has no analogue of the cochlea. Many fishes have no visible external auditory meatus, and in " some " of them three semicircular canals are found, but in others only two. The last statement must be an error, as there is no reason to believe that Perrault had ever examined the ear of a lamprey.

Perrault gives a good account and figures of the mechanism of the claws in the Felidae. He compares it happily with the opening and closing of the shell valves of a mussel — the former movement being effected by an elastic ligament and the latter by muscular action. He has not, however, disclosed the complete story. He

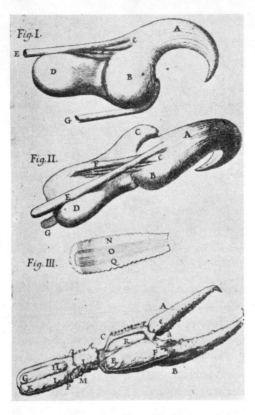

FIG. 190.—Perrault, 1680. Mechanism of claw movement in Felidae and in chela of crayfish

understands the behaviour of the flexor profundus digitorum muscle, and notes that the middle phalanx is deeply hollowed on one side to accommodate the terminal phalanx in the retracted position of the claw. But there are other muscles concerned of which he says nothing, and he is wrong in stating (and figuring) that the elastic ligament is reinforced by an " extensor muscle ". Perrault also compares this mechanism with the muscles of the chela of a crayfish, in which, however, as he himself is aware, *both* the occlusor and divaricator elements are muscular. Considerable attention is devoted to the structure and mode of action of the compound stomach of ruminants, and this section is a worthy introduction to the monograph by Peyer of 1685. The varied character of the lining of the four chambers is treated in some detail. He believes that the parrot ruminates, and he reaches the same conclusion in the case of the insect *Gryllotalpa*, which he figures as having a three-chambered stomach, the chambers in question being the crop, gizzard and enteric caecum of that quaint and attractive animal.

Duverney's classic researches on the comparative anatomy and physiology of the blood system in the lower vertebrates were concentrated largely on the tortoise, but he deals also with a bony fish, the frog and the viper (1699). The fish selected was the carp. He describes the sinus venosus (Fig. 191, *A*), and opening into it he found

three hepatic veins (*D*), right and
left cardinal veins (*B*), the unpaired
inferior jugular vein (*E*) and the
right and left jugular veins (*C*). An
accurate account follows of the
single auricle and ventricle and
their valves. In fishes, he says, the
gills are the respiratory organs, and
act as such by extracting the small
quantity of air held in water. His
idea as to how this is effected can
have a prophetic value only, but
he deduced that air was absorbed
into the blood stream, and that
water was concerned with the respiration of fishes *only in so far as
it was impregnated with air*. Hence
the ventral aorta is the pulmonary
artery, the gills are the lungs, and
the " veins of the gills " [efferent
branchial vessels] acquire thicker
walls and become the arterial factors of the systemic aorta. Even

FIG. 191.—Duverney, 1699.   Carp.
Heart and great vessels

the detailed vascular structure of the gills was not beyond him, and
he understood the anatomy of the blood system in fish better than
any of his predecessors and many of his successors, so that his
researches are the real basis of our modern knowledge of these
matters.   But for one unhappy and very serious blunder he
would have left little of first importance for posterity to discover.
His miniature figure illustrating the circulation of the blood through
the capillaries of the gills is a gem of its kind (Fig. 192).   It represents a transverse section through a double branchial filament,
and includes the surfaces of two lamellae.   The afferent branchial
trunk (*A*) gives off two afferent filamentar vessels (*B, B*) which
pass with diminishing calibre to the apices of the lamellae, where
they inosculate with their corresponding efferent partners.   Before
doing so, however, they give off numerous capillaries (*D, D*), which
discharge into the two efferent filamentar vessels (*E, E*).   The
latter open into the efferent branchial trunk (*F*).   Blood, therefore,

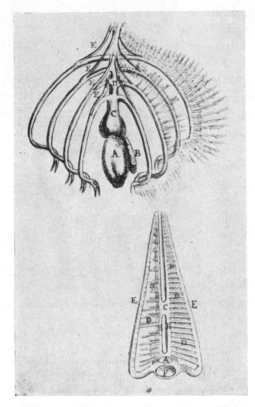

FIG. 192. — Duverney, 1699. Carp. Heart, afferent branchial system and scheme of circulation in gills

can reach the efferent from the afferent system *only by traversing a capillary plexus*, and hence *two* vascular trunks are postulated in each gill arch. This important result is illustrated by two figures. His Fig. (192) shows the auricle (*B*), ventricle (*A*), bulbus arteriosus (*C*), ventral aorta (*D*) and the afferent branchial arteries (*E*), "the aorta and its branches passing from the heart to the [dorsal] extremity of the gills, *where they end*". This is an important emendation of Perrault's luckless version. Duverney's Fig. (193) is very misleading. It represents the efferent branchial system only, and illustrates correctly the constitution of the circulus cephalicus and the dorsal aorta (*A*). The efferent trunks, he says, consist of dorsal muscular arterial portions and ventral membranous venous portions, the four latter joining up to form the vessel *F* opening into the sinus venosus.[1] This vessel is clearly the inferior jugular vein, which, needless to say, does *not* arise in the way described. On the left of the figure a small part of an afferent trunk is shown, to illustrate the relations of afferent and efferent systems.

Duverney fully recognizes that the two-chambered fish heart, which, he says, is "known to everybody", is a branchial or venous heart dispatching *all* its blood to the respiratory organs, from which it is collected as arterialized blood in vessels which unite to con-

---

[1] Duverney therefore believes that the heart of fishes contains a percentage of *arterial* blood derived from the efferent branchial vessels.

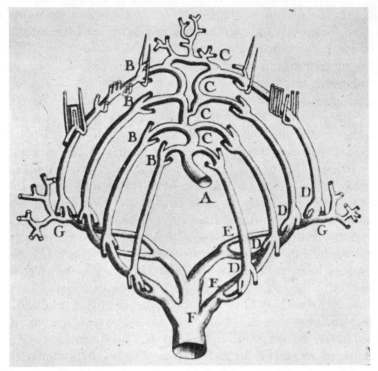

FIG. 193.—Duverney, 1699.   Carp.   *A*, dorsal aorta and circulus cephalicus.
Efferent branchial system wrongly connected up with inferior jugular vein, *F*

stitute the dorsal or " descending aorta ", and that such blood is
then distributed to the system generally, to be returned from it as
venous blood to the sinus venosus.[1]   He is the first to describe and
figure in a teleostean fish the circulus cephalicus and the carotids
or " ascending aorta ", and the first after Perrault to discover the
inferior jugular vein.   Duverney does not regard the sinus venosus
as a true chamber of the heart, but as being formed by the fortuitous
concourse of the caval veins.   It is, he says, still a vein in spite of
its muscular walls.   The bulbus arteriosus and its relations to the
ventricle and ventral aorta are correctly described, but he errs in
believing it to be muscular, and hence a kind of second ventricle.

In the paper dated 1701, but published in 1704, Duverney does
not add materially to his previous account of the blood system of
fishes — a " most singular " type of circulation, as he now says,
with its " 4320 twigs ".   He is sound and discriminating on the

[1] This was partly known to Needham in 1667.

gross anatomy of the mechanism of respiration and its mode of action in a bony fish, and he confirms the discovery of the breathing valves in the mouth, first described by Collins in 1685. In the gills he describes the blood as being spread out in very small volume over an extensive respiratory surface — a necessity, he says, due to the greater difficulty of extracting air from water. The blood leaving the gills has a brighter red colour than the blood in the ventral aorta, which indicates that during its passage through the gills it has become charged with particles of air, and so converted into arterial blood. In tortoises, frogs and similar animals one-third only of the blood passes through the lungs at each circulation. This statement is not as mistaken as appears at first sight, since it refers to one-third of the entire contents, venous and arterial, of a *single* ventricle, and not to one-third of the venous constituents only. None the less, as will be seen later, it hardly tallies with Duverney's own views on the passage of the blood through the heart of a tortoise.

Duverney's successful attack on the heart of the frog and the vessels associated with it is marred by the statement that the heart of that much investigated animal has but one auricle. He was doubtless led into this error by the contiguity of the pulmonary and sinu-auricular apertures, which may be separated only by the inter-auricular septum itself. He concludes, therefore, that the blood is mixed before it reaches the single ventricle. His description of the structure of the truncus arteriosus with its spiral and other valves, which he says is a " very curious mechanism ", is of outstanding merit, and he compares and contrasts it with the muscular conus arteriosus of the skate and other fishes. The latter is figured opened up to show the four rows of semilunar valves.[1]

Besides the two descriptions of the heart of the crocodile which have already been noticed, Duverney gives a correct account of the heart and great vessels of the viper, in which animal he distinguishes two auricles, and he is struck by the asymmetrical condition of the pulmonary arteries and veins — due to the unequal development of the two lungs. He found only a single ventricle in the

---

[1] The exact date when Duverney came into possession of the manuscript of Swammerdam's treatise on the frog is not known, but it must have been at about this time. The only important point of resemblance between the two accounts, however, is that both of them fail to reveal the existence of an inter-auricular septum in the heart of this species.

heart, the cavity of which was incompletely divided up into three, as in the tortoise. The sinus venosus was seen but not recognized as a separate chamber.

Duverney wrote four descriptions of the heart of the tortoise, the first of which was printed in the *Memoires* of the Parisians, to which reference may be made (p. 412). In the third [1] he found that the pulmonary veins opened separately, but very close together, into the left auricle. The sinus venosus is

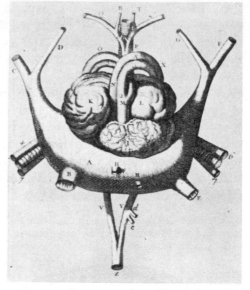

FIG. 194.—Duverney, 1699. Heart of tortoise. Ventral view

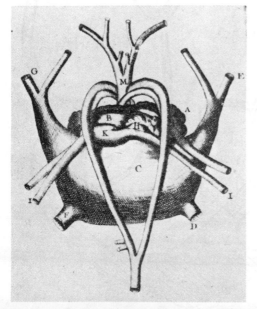

FIG. 195.—Duverney, 1699. Heart of tortoise. Dorsal view, showing pulmonary veins and sinus, *I, K*

now definitely mentioned. He discovered it under the right auricle as a venous trunk formed by the union of many vessels. It opened into the right auricle, at which point there was a crescentic valve encircling the aperture. In his fourth, last and by far the most important version, written in 1699 and published in 1702, he describes opening into the muscular sinus venosus the inferior vena cava (Fig. 194, *B*), the right and left

[1] Left by Duverney in manuscript, *c.* 1695. Abstracts from it were made and published by Mery in 1703.

"axillaries" (precavals — *C*), hepatic veins (*E, H, H*) and the coronary vein. Hence all the blood from the body is returned to the sinus venosus except the contents of the pulmonary veins. The latter unite to form a small sinus before opening into the left auricle (Fig. 195, *I, K*). The sinu-auricular valve consists of two muscular flaps like eyelids, and there is also a muscular pulmo-auricular semilunar valve, but the last statement is open to question. There are two muscular auriculo-ventricular valves, separated by the prolongation of the inter-auricular septum, which prevent blood regurgitating from the ventricle into the auricles. The single ventricle has *three* cavities :[1] the first (right dorsal or cavum venosum, Fig. 196, *C.V.*) receives blood from the

---

[1] Caldesi (1687), in his remarkable work on the anatomy of three tortoises, distinguishes the two auricles of the heart, but is less certain of the existence of three cavities in the ventricle.

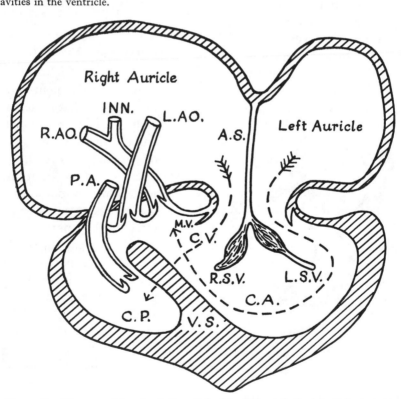

FIG. 196.—Diagram of the circulation of the blood through the heart of a tortoise based partly on Duverney's descriptions. *M.V.*, marginal valve.

right auricle; the second (left dorsal or cavum arteriosum, *C.A.*) receives blood from the left auricle ; and the third (right ventral or cavum pulmonale, *C.P.*) receives blood from the cavum venosum and gives rise to the pulmonary artery. The first and second cavities communicate with each other, and may be regarded as one, and the first also communicates with the third, so that all three constitute a single chamber, which, related as it is to the great vessels, and although both auricles discharge into it, does not represent the right and left ventricles of the higher animals, but the single ventricle of fishes and frogs and the left ventricle of the mammal. Similarly the three arteries arising from it must be compared with the undivided conus arteriosus of frogs. But, he adds, the blood which enters the ventricle from the two auricles is able to mix freely and to pass from one cavity to another, and he then proceeds to negative this statement by maintaining that the right auriculo-ventricular septal valve (*R.S.V.*) is so orientated that the venous blood leaving the right auricle is guided *to the right*, first into the cavum venosum and finally into the cavum pulmonale. At the same time the arterial blood entering from the left auricle is *at first* deflected to the left by the septal valve of that side (*L.S.V.*) and retained in the cavum arteriosum before passing into the cavum venosum on the right, where the venous blood is pushed before it into the cavum pulmonale. The blood stream in the ventricle, therefore, *flows from left to right*. Duverney admits that the right auriculo-ventricular valve hinders, but does not wholly prevent, blood passing from the cavum venosum into the cavum arteriosum. He describes an incomplete and not very definite ventricular septum (*V.S.*), and notes its relation to the opening of the left aortic arch, and how this septum more or less separates the cavum venosum from the cavum pulmonale. He saw also that *both* aortic arches arose from the cavum venosum, and that *no* artery was associated with the cavum arteriosum. The result of these arrangements, he says, is that the two aortic arches at the outset transmit venous blood, and afterwards largely arterial blood, the cavum pulmonale transferring the unmixed venous blood [1] to the lungs. The two bloods therefore mix in the ventricle, but it is not a *complete* mixture, which is prevented by the partitions (partly valvular) more or less isolating its three cavities. His final opinion

---

[1] " About one-third of the contents of the ventricle."

is that the right aorta, which supplies the important organs of the head, transmits a mixture, but one which carries a higher percentage of arterial blood.

Further details of the anatomy of the Chelonian heart noted by Duverney are that the aperture of each of the three arteries is guarded by a pair of sigmoid valves ; that the right aorta gives off the innominate artery or " ascending aorta ", the remainder of the arch detaching no further conspicuous branches before uniting with the left arch to form the " descending aorta " ; that previous to this union the left arch dispatches two arteries to the gut ; and that the coronary artery arises as a single vessel from the root of the right aorta and is distributed to the auricles and ventricle.

All this is a remarkably correct account of the heart of the tortoise anatomically and functionally, and as far as it goes differs little from modern descriptions. Duverney rightly concluded that the pulmonary artery transmitted venous blood, and that the left aortic arch was partly venous, but he did not emphasize sufficiently the mainly arterial function of the right aortic arch, although he had correctly deduced the course of the venous and arterial streams after entering the ventricle. Having got so far, and having a sound working knowledge of the anatomical relations of the parts, he might have hazarded a guess, but he was averse from taking so dubious a step. The course of the blood through the ventricle being from left to right, it was at least probable that the blood passing out by the pulmonary artery would be venous, that by the left aorta mixed but largely venous, and that by the right aorta mixed but largely arterial. We shall, however, exhibit a better understanding of the historical situation if we dwell rather on the merits than on the shortcomings of this pioneer research. It must not be forgotten that Duverney had to free his mind from the indifferent observation and worse reasoning of perverse and argumentative opponents — a task always difficult and often invidious. But despite all obstacles the problem of the reptilian heart became soluble after Duverney's final assault upon it, although, it must be added, the emergence of the solution did not follow until some two hundred years had passed !

VIII

THE ANATOMICAL MUSEUM

XXXII
FOUNDATION OF MUSEUMS

No history of comparative anatomy can afford to ignore the considerable part played by museums in the development of anatomical studies, and it is therefore necessary to enquire into the conditions which made possible the foundation of these extraneous aids to knowledge. We owe to Gesner, and especially to Aldrovandus whose collections may still be inspected at Bologna, the institution in the sixteenth century of the earliest and most notable biological museums. These museums, however, expressed the knowledge and purpose of founders whose major preoccupations were certainly not anatomical, nor did the anatomists themselves in these pioneer days display any enthusiasm for museums. It was therefore the field naturalist, attempting to reconcile the abstractions of philosophy with the realities of Nature, and ever seeking to alleviate the tedium of instruction, who finally devised the academic museum. He was well satisfied with his new creation. Here was something to illustrate, and occasionally embellish, the nakedness of truth, and to acclaim the victories of science over the mysteries of creation and the ailments of mankind. In all the centres of Europe museums shot up with the speed of fashion and the confidence of zeal. The spoils of the sea and the sweepings of the land were thrust indiscriminately into glass jars and mahogany cases of classic but unsuitable design. The public stamped through the galleries in heavy boots, and persons of quality, adorned with cocked hat, sword and quizzing-glass, lounged gracefully from case to case. The vogue of things created entered upon its long and fluctuating career.

The medieval museum, however, in the absence of efficient technique and fluid preservatives, could operate only within narrow limits. Dried objects of a non-perishable nature, such as skeletons, horns, corals, eggs, skins, shells, seeds, sponges, gorgonids, fossils

FIG. 197.—Levin Vincent, 1719. Ideal conception of a Natural History cabinet of the period

and minerals, were collected and mingled in heedless confusion. The spirit specimen was not available, but anatomical preparations were not excluded. The viscera could be inflated and dried, the vascular system injected, dissected out and dried, or isolated by corrosion, and indeed whole organs might be injected and afterwards mummified for exhibition. Even after the introduction of alcohol as a means to secure the permanency of the moist preparation, its use was considerably restricted by the expense of both spirit and glass, and technical works published as late as the nineteenth century counsel its avoidance so far as possible. The Hunterian Museum, for its time unusually rich in spirit material, contained, as left by John Hunter, only 4829 moist preparations, as against 8636 not requiring a fluid preservative. Therefore the first use of spirit in biological museums is an historical event of no small interest.

XXXIII

PRESERVATION OF SPECIMENS IN SPIRIT

At a meeting of the Royal Society held on May 28, 1662, " Mr. Croune [Croone] produced two embryos of puppy-dogs, which he had kept eight days, and were put in spirit of wine in a glass-vial sealed hermetically ".[1] This is one of the earliest records, if not the first,[2] of the use of alcohol as a preservative of animal tissues. Some years later (1672), in his paper on Preformation in the Chick,[3] Croone states that he had preserved embryos of this species in " spiritus vini ", and he describes the effect of the alcohol on the appearance of the tissues. In June 1663 Boyle published the following important observations : " Nor were it amiss that diligent Tryal were made what use might be made of Spirit of Wine, for the Preservation of a humane Body : For this Liquor being very limpid, and not greasy, leaves a clear prospect of the Bodies immers'd in it ; and though it do not fret them, as Brine, and other sharp things commonly employ'd to preserve Flesh are wont to do, yet it hath a notable Balsamick Faculty, and powerfully resists Putrefaction, not onely in living Bodies . . . but also in dead ones. And I remember that I have sometimes

---

[1] Birch, **1**, p. 84.
[2] It seems certain that Croone got the idea from Boyle, who was experimenting with alcohol before 1662, and who is quoted by Grew as the inventor of the method.
[3] Written, however, in or soon after 1670.

preserv'd in it some very soft parts of a Body for many Moneths (and perhaps I might had done it for divers Years, had I had opportunity) without finding that the consistence or shape was lost, much less, that they were either putrifi'd or dry'd up. . . . Nay, we have for curiosity sake, with this Spirit, preserv'd from further stinking, a portion of Fish, so stale, that it shin'd very vividly in the dark."

In 1664 Boyle exhibited to the Royal Society a linnet and a small snake which had been kept for four months in spirit of wine, during which time they had undergone no change. They were ordered to be placed in the Society's Museum and were still there in 1681, when the collection was catalogued by Grew. In 1665 and 1666 Boyle continued to experiment with spirit of wine, which he successfully employed to preserve the head of a colt [1] and chick embyros of various stages. He affirms that he had " long since " used the method to control putrefaction, and emphasizes the importance of *changing* the spirit after some days' immersion of the specimen to ensure freedom from discoloration. In 1665 Coxe presented to the Society's museum a human foetus " kept in spirit of wine well rectified ",[2] and in the following year a dissection of a human foetus preserved in spirit of wine was demonstrated to the Society by King.[3]

According to Boerhaave, Swammerdam (*c.* 1670) kept insects from putrefaction for some time by immersing them in spirit of wine,[4] and in another place brandy is the vehicle mentioned. In 1678 the French Jesuit Barbillart was exhibiting a large bottle of spirit of wine in which the body of a child had been kept for several years without exhibiting any signs of decay.

The museums left by Robert Hubert (catalogued in 1664) and the Hon. Robert Boyle found their way into the Royal Society's cabinet, of which they formed the nucleus. Unhappily there is no separate catalogue of Boyle's museum, but two of its moist preparations were described during his lifetime by Grew in 1681. One is a " male humane foetus ", of which " The Skin hath been kept white and smooth for so long a time, *scil.* above fifteen years, by being included with rectified spirit of Wine in a Cylindrical Glass ;

---

[1] Birch, **2**, p. 50. [2] Birch, **2**, p. 45. [3] Birch, **2**, p. 129.
[4] So the Latin version. The Dutch word used by Boerhaave is *voorloop*, which means the first runnings of a distillate — not necessarily alcohol.

to the middle of which the *Foetus* is poised, by means of a Glass Buble of an Inch diametre, the Neck whereof is fastned to the *Anus* of the *Foetus* by a wyer ". The other is " A young LINET which being first embowel'd, hath been preserved sound and entire in rectified Spirit of Wine, for the space of 17 years. Given by the Honourable Mr. *Boyl.* Who, so far as I know, was the first that made trial of preserving Animals this way. An Experiment of much use. As for the preserving of all sorts of Worms, Caterpillars, and other soft Insects in their natural bulk and shape, which otherwise shrink up, so as nothing can be observed of their parts after they are dead. So also to keep the Guts, or other soft parts of Animals, fit for often repeated Inspections. And had the Kings or Physitians of *Egypt* thought on't, in my Opinion, it had been a much better way of making an everlasting Mummy."

In 1681 and 1683 Tyson [1] produced at the Royal Society a [human] embryo " not bigger than a bee " and a tapeworm preserved in spirit of wine, and a few years later the " incomparable Museum " of Squire Courtine,[2] ultimately taken over by the British Museum, included a great variety of plants and animals mounted in brandy. He recommends re-distilling the liquid when it becomes discoloured. In 1710 Leeuwenhoek was employing spirit of wine, in which he kept his young oysters,[3] but there is no evidence as to date, except that his first experiments with alcohol could hardly have been later than 1695. Finally, in 1698, Merrie [Mery], as reported by Martin Lister, was exhibiting the dissected heart of a land tortoise " preserved in Spirit of Wine ".

These records suffice to show that the early naturalists found spirit of wine adequate to their purpose, and it was therefore almost invariably adopted. Daubenton, writing in 1749, suggests that this was " perhaps only because it happened to be more plentiful ". He himself thinks that any other fermented spirit would be just as efficient, and he does not appear to have recognized that in all these products the anti-putrefactive principle is the same. Other substances tested were salt and alum, but it was soon evident that they did not *fix* the tissues at all satisfactorily, nor were their effects lasting. Daubenton (1749) mentions that the moist preparations

---

[1] Birch, **4**, pp. 96, 176.
[2] Otherwise William Courten or William Charleton (1642–1702). No catalogue published. [3] Dobell, p. 66.

in the Paris Museum of Natural History were mounted in rum, which was maintained in a transparent condition by frequent distillation. He also discloses the nature of the preservative used in Ruysch's museum, the first catalogue of which was printed in 1691. In this museum, which was universally considered at the time to be unsurpassed in Europe, and " the eighth wonder of the world ", there were a large number of the finest anatomical exhibits beautifully displayed in a liquid of unknown composition. Ruysch obstinately declined to reveal his methods on the ground that his preparations were unique, and they would cease to remain so if he explained how they were made. After his death the secrets were divulged to the French Academy of Science, and Geoffroy was instructed by the Academy to experiment with the liquid preservative. The receipt was as follows : Pulverize 1 oz. 6 drams of black pepper, $\frac{1}{2}$ oz. of the seeds of the small cardamomum, and an equal quantity of cloves. Place the resulting powder in a flask with 12 lb. of spirit of wine, in which a bag containing 2 oz. of camphor is suspended. The whole must then be distilled, when it should give about 11 lb. 3 oz. of clear liquid. Dilute before use. This disclosure, as in the case of so many jealously guarded secrets, is not very enlightening, and Ruysch's fluid would have been no less effective had it consisted of the spirit of wine alone.

It has been stated that the development of the anatomical museum was restrained by the cost of preservatives and glass. In spite of many attempts to discover an efficient substitute, alcohol is still the only medium which fixes and permanently preserves an anatomical preparation. In the financial year 1938–9 the scientific and medical institutions of Great Britain purchased 42,998 gallons of alcohol, the cost of which must have represented no negligible fraction of an economized laboratory expenditure. It hence concerns the historian to weigh the influence of economic and fiscal conditions on the production of this auxiliary material. The importance of these conditions can be gauged by the fact that in 1905 the cost of a gallon of ethyl alcohol in Germany was 10*d.*, whilst in England it was £1 : 1 : 6, the difference being explained by, on one hand, a State-aided industry selling below cost price, and on the other a heavily taxed article sold at a profit.

The invention of flint glass in the seventeenth century is an event of first-rate importance. The commercial glass of the period was unsuited for museum purposes, but the transparency and colour of the flint-glass jar made possible for the first time a satisfactory exhibition of moist preparations. In the second half of the eighteenth century Hunter was buying about 5000 museum jars suitable for the display of his spirit material. It is improbable that these jars were manufactured and stocked in the ordinary course of business routine, for it was only after about 1830 that the demand for laboratory and museum glass began to justify such a practice. Hunter's jars must therefore have been made to order at a special price. Repeated enquiries have failed to recover any details of these transactions, many of the older firms having been extinguished by the introduction of free trade, and the early files of the survivors destroyed. During Hunter's time, however, the duty on flint glass rose considerably. In 1745 it was 9s. 4d. a cwt., and in 1803 it was £1 : 12 : 8, this period, as it happens, coinciding with a perceptible diminution in the number of museums founded in England. The tax then dropped to 9s. 6d. in 1845, and on the advent of free trade it was reduced to one-half in 1846 and to one-quarter in 1848.

In the matter of spirit, a product so closely affecting the comfort of the individual and the prosperity of the State, we naturally have information of a more detailed character. A duty on spirits was first imposed in England in 1643, when " strong waters " were rated at 6d. a gallon. This example was at once followed by Scotland, where every pint of the national beverage bore an impost of 2s. 8d. Scots (about 2½d.). But these halcyon days were not to last. In 1736, owing to the increasing popularity of gin in England, 20s. per gallon were added to the duty, and the licence stamp was raised to £50. The result was so great an increase in illicit distillation that in 1743 the tax was reduced to 3d. and the licence to £1. From that time to 1791 the rates fluctuated considerably, but the public, especially in Scotland and Ireland, invariably defeated a high duty by the double evasion of smuggling and secret distillation, so that in 1821, when the tax had reached a culminating point, the duty was not paid upon more than half the spirit consumed. This introduces an unexpected and a complicating feature. Between 1739 and 1800 thirty-nine museums were founded in England, mostly by private individuals, and the anatomist, who had to choose between the law and his work, may have added to body-snatching the perils of unlawful distillation. We can well imagine that John Hunter, whose scruples vanished where the welfare of the museum was concerned, and whose dramatic abduction of the body of the giant was an unfailing source of satisfaction, foregathered as blithely with the spirit-runners as he did with the resurrectionists. It is therefore difficult to estimate precisely the influence of the cost of spirit, and the more so, for example, as no less than thirty-one different rates are given in the Act of 1803. In England the duty in 1791 was 3s. 4d. per gallon ; it rose steadily to 11s. 8d. at the close of the Napoleonic wars, after which it was dropped to 7s. in 1826, from this figure rising in 1941 to the unprecedented level of 97s. 6d. per gallon at proof. In Scotland and Ireland the duties were different — in Ireland much lighter, and in parts of Scotland so heavy as to prohibit the legal production of spirit altogether. From 1800, however, the Scots duties were generally lowered so as to be well below the English scale, and in 1859 a uniform rate of 10s. was imposed throughout the whole of the United Kingdom. At the beginning of the eighteenth century spirit began to be generally used as a museum preservative, but it was not until about 1740 that the museum movement commenced to make headway in England, and its slow progress was at least partly due to the taxes, which by that time were making themselves severely felt, on the already costly glass and spirit. Thus no moist specimen could be added to a collection without weighing heavily in the balance the cost of mounting it, and additions which had no special interest or importance could not be entertained. The modern museum is much less handicapped. The introduction of the Coffey still considerably reduced the cost of production, and permission to use duty-free spirit, made non-potable by some noxious adulterant, has eliminated the tax.[1]

XXXIV
EUROPEAN MUSEUMS (1528–1850)

The propagation of biological museums may be studied by graphic methods embodying the dates of foundation of the museums of Europe and America between the dates of 1528 and 1850. In some such way it is possible to trace in a comprehensive

---

[1] This is the methylated spirit of commerce, first produced in 1855.

graph the origin and progress of the natural science museum in all countries and at all times, and we can detach from it, for separate consideration, those collections which have an anatomical bias. Data relating to 445 general and 92 anatomical museums have been available for this attempt, but although these small numbers represent a large proportion of the maximum possible, the risk of using them as the basis of a statistical enquiry is not thereby lessened. The conclusions, therefore, must be stated with some reserve, although general tendencies can scarcely be mistaken. It is impossible to be informed of every museum, especially of the earlier private collections, of which no printed record was ever made, and of whose existence we are only aware from the chance visits of distinguished travellers. Cabinets are broken up and change hands to the complete bewilderment of the historian. The life of individual museums is in most cases so inscrutable that the attempt to incorporate this important factor had to be abandoned. Therefore, dates of foundation only can be considered, and even these are often the subject of active speculation. Generally speaking, the date of foundation may be deduced from (*a*) the publication of the first edition of the printed catalogue ; (*b*) the extent and nature of the collection ; and (*c*) the birth or death of the founder. But no printed catalogue may have been published, the contents of the museum may be little known, and details of the founder's life entirely wanting. This notwithstanding, the maximum error in the average case should not exceed ten years, and, subject to this qualification, the following conclusions may, in the main, be stated.

For about the first century, that is, to 1630, foundations occur with fair regularity, but there is no significant presence or absence at any stage. Judging from the general graph, no current depression, in spite of a break of six years, can be ascribed to the Thirty Years War — in fact, some progress is even shown. The period between 1646 and 1680 includes forty-four foundations, and corresponds very closely with that access of peaceful activity which followed the signature of the Treaty of Westphalia in 1648. England, Germany and Holland all share in it, but France, although under the stimulus of national aggrandizement, is represented only by three entries between 1644 and 1713, and the record of Italy is almost as meagre. The period in question, moreover, is interesting in other ways. The Royal Society of London was instituted in 1662 and the French Academy of Science in 1666, and both exercised an influence which extended considerably beyond their immediate environment. Nevertheless they must be regarded as the result, and not as the cause, of the interest in natural science which precipitated this group of museums. The authority of these corporate bodies may be traced rather in the works of Boyle, Malpighi, Hooke, Redi, Swammerdam, Leeuwenhoek and Grew, published between 1660 and 1680.

The War of the Spanish Succession (1701–14), affecting as it did France, Spain, England, Holland, Belgium, Germany and Italy, coincides with an almost complete cessation of museum foundation in all the centres of Europe except Germany, a country not fundamentally affected by the war ; but after the signature of the Treaty of Utrecht in 1713, we have the inception of a period of activity which reached its highest point in 1747 and did not begin to decline before 1770. In it all European countries participated except Russia, where the revival of scientific learning had not at that time extended. In Germany, Holland, France and Sweden it is the most fruitful age of all, but not so in England, although the foundation of the British Museum in 1753–9 belongs here. There can be no doubt, judging from the character of the museums founded at the time, that the prevailing influence was the publication of two important works in systematic zoology — the *Systema Naturae* of Linnaeus, the various editions of which range from 1735 to 1768, and the *Histoire naturelle* of Buffon, which was first issued between 1749 and 1804. It is, on the other hand, highly interesting to observe that museums of an anatomical cast, so far from increasing at this period, show a slight falling-off. With natural history in the ascendant, and under the guidance of two such masters, anatomy must give ground. A later and more important development, as we shall see, reveals the influence of the anatomical type of mind.

The Napoleonic Wars (1794–1815) are responsible for another check to progress, and a break between the years 1800 and 1810 is of obvious significance. Here, however, England, adversity and panic notwithstanding, has relatively overtaken her rival Germany. Holland and Belgium are beginning to retire from the contest, France, from the first revolution in 1789, was severely shaken, and

Italy, so sensitive to external dominance, almost abandons her science during that stricken period. But the conclusion of peace in 1814–15 results at once in a general revival, this time evident in the United States. It reached its limit in 1835, and was followed by the usual reaction. This second great advance, however, derives its inspiration from another source. It is the time of peace and of the Industrial Revolution. The powers of the mind are bent on the abridgement of distance, the advancement of science, and the expansion of commerce. Between 1800 and 1817, Cuvier, whose work was continued in Germany by Meckel, was publishing treatises which acquired a European vogue and influence. In England and Germany a number of museums of a more scientific and anatomical type may be directly traced to his writings. In the United States many general and anatomical museums are instituted, France and Italy begin to revive, Holland puts forth a final effort, but we witness the temporary extinction of Denmark and Belgium.

The achievements of individual countries between the sixteenth century and 1850 may now be briefly examined. Germany alone, developing early, has maintained from start to finish a steady and growing interest in scientific museums, but the record of France, though almost as lengthy, is not so continuous. Italy, under the unique and protracted influence of Vesalius, was greatest in the sixteenth and seventeenth centuries, she then suffered various relapses, but took some part in the nineteenth-century revival. Of those entering later, England, if irregular at first, has been growing in strength all the time. Holland was greatest between 1660 and 1670, the activity of Sweden naturally centred round Linnaeus, and Denmark hardly survived the moderate interest she displayed between 1619 and 1672. The earliest anatomical museums were founded in Denmark, France, Holland, Italy and Germany. Holland has been more or less continuously interested in anatomy, but the final honours rest with Germany and England. The relative interests of anatomy and general biology may be ascertained by working out the proportion of anatomical foundations to the total number of museums instituted in each country, and on this basis the order of merit would be : Holland, 28 per cent ; Germany, 20 per cent ; Italy, 16 per cent ; England, 13 per cent ; and France, 9 per cent. Germany was at first interested rather in the general than in the anatomical museum, but with the growth of anatomical traditions this policy is relatively, if not absolutely, reversed. In England, on the other hand, interest in the general collection has rapidly and continuously gathered, but we find only a slight increase in the number of anatomical foundations — almost the opposite of the trend of events in Germany.

The actual numbers of museums associated with each country up to the year 1850 may be assembled in the following table ; but too much importance should not be attached to figures which express fluctuating values only :

	Total Number of Museums founded	Anatomical Museums founded
Germany . .	156	32
England . .	140	19
France . .	64	6
Holland .	50	14
Italy . .	44	7
Sweden .	14	1
Switzerland .	11	1
Austria .	10	3
Denmark .	10	2
Russia .	10	2
Belgium . .	8	0

XXXV

MEDIEVAL, RUYSCH'S AND THE HUNTERIAN MUSEUMS COMPARED

We may now compare three collections, representing the most important stages in the evolution of the anatomical museum — first the medieval cabinet, of which that instituted by the Royal Society may be taken as an example, then the early anatomical museum of Ruysch, and finally the great foundation of John Hunter.

(1) The Royal Society was fortunate in the choice of a cataloguer for its museum. Nehemiah Grew, the friend of Dugdale, Ashmole and John Gibbon, will always be distinguished as the author of a little book, published in 1672, bearing the modest title *The Anatomy of Vegetables Begun*. The museum catalogue is dated 1681. Grew indeed shares with Malpighi the honour of having initiated the morphological study of plants. Such a man might be

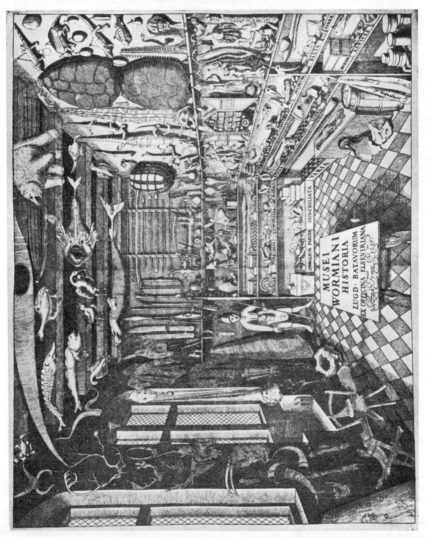

FIG. 198.—Ole Worm, 1655.   Early figure of a cabinet of " Natural Rarities "

expected to make the most of the museum, and his year's work is not wholly unprofitable. He respectfully classifies this adventitious collection of oddments and monstrosities, and, as in the case of the shells, he seizes the few opportunities it affords of exercising an ingenious and philosophic mind. In criticism he is a shrewd and healthy sceptic, but is not invariably sound. He recognizes indeed that the older zoologists are not altogether averse from the wiles of fiction, for the story of the barnacle goose he says is " fabulous ", and the " stupendious power of the shiphalter ", or sucker fish, which was credited with the ability to stop a ship under full sail, is dismissed in the irony of a single sentence. But his task is an impossible one, and the catalogue seldom rises above the level of : " a humane skull cover'd all over with moss " ; the windpipe of a crocodile ; a powder said to be taken out of a serpent's head ; the swaptail lizard ; the egg of a swan with another within it ; " a bone said to be taken out of a maremaid's head " ; the frog-wilk ; " the chaps (perhaps) of the greenland needlefish " ; the palmer worm which " pilgrims up and down everywhere, feeding upon all sorts of plants ".

It is to Grew's credit that he is only partially captured by the empiric medicine of his time, the history of which is at once the most pathetic and astounding in the records of natural science. The barefaced lack of connection between a disease and the means proposed for its reduction does occasionally assail his critical faculty. And yet the necessity of rifling the universe in the attempt, of using materials of the most varied and often offensive character in the hope of casually lighting upon a specific, attaches his sympathies and beguiles his reason. To his generation an important function of the museum was that it assembled that fortuitous assortment of natural objects which the craft and subtlety of the physician had ground up or boiled down in a frantic effort to alleviate the distresses of mankind. Boyle tells us that he had seen " life itself almost disgorged together with a potion ", and this callous disregard for the sufferings of the patient made him "apprehend more from the physician than from the disease ". Galen mentions a prescription which contained some hundred ingredients. At times Grew's credulity is obviously strained when he cautions the patient to read Fienus' *Of the Power of Phancy*, and a draught of a scruple of soft alabaster taken in milk is only recommended in the absence of a more relevant remedy. On the other hand, the ingenuous appeal of powdered crabs' claws worked up with a jelly extracted from the skins of snakes subdues his scepticism, nor does the cure of baldness appear difficult if a " wilk, being burnt, powdered and mixed with old oil to the consistence of glew " be well rubbed into the refractory scalp. Indeed the magic of the " wilk shell " declares itself in the minor and vague respect of " doing good " if only it be used as a drinking vessel, whilst the stomach of the ostrich sustains its high reputation by dissolving the stone which is incapable of afflicting its digestion. The prevailing taint of Carolean cynicism, rather than his own humour, is perhaps responsible for the passage in which he says that the " stag's tears " are " a thicken'd excretion from the inward angle of his eye. In colour and consistence almost like to mirrh ; or ear-wax that has been long harden'd in the ear. . . . They are generally affirmed to be sudorifick, and of an alexi-pharmick nature. And if they were as easie to be had, as some womens, it were worth the trying." Even many years later than the period of the Royal Society's museum the " Rarities " in the Anatomy Hall of the University of Leyden, one of the most famous Universities of the period, was just such another collection of odds and ends.

So much for the character of the medieval museum. A super-fluous institution, it may be thought, but useful in keeping alive the interest of the public, and in preparing the world for better things.

(2) Of the pre-Hunterian museums of anatomy, that founded by Frederik Ruysch, professor of anatomy at Amsterdam, is clearly the most important. Ruysch was born in 1638 and died in 1731, and therefore lived a century before Hunter. The first catalogue of the " Musaeum Ruyschianum Anatomicum " was published in 1691, the zoological collections were described in 1710, and the anatomical in a series of quartos ranging from 1701 to 1715. The museum contained in 1710 more than 1300 anatomical preparations mounted in liquid. It was purchased by Peter the Great on his second visit to Amsterdam, and removed to St. Petersburg in 1717. Ruysch was then in his eightieth year, but his mind was unclouded, and the " silence and darkness of declining years " found his energies still intact. With fine courage he endeavoured to complete

another collection, of which short descriptions were published in 1724 and 1728, and might even have succeeded had not an accident crippled his movements and shortened his life.

As a *préparateur* Ruysch occupies a unique position both in his own time and in ours, but a jealous and suspicious disposition preserved the secrets of his methods to the end. His son, who assisted him in the work, and might in his own time have divulged the technique, unhappily predeceased him. In his publications Ruysch confines himself to such statements as that the preparation is preserved in balsam, which presumably in some cases is turpentine, or " in liquore limpidissimo ", which appears to be spirit of wine. In an access of confidence he tells us in one place that the liquid has been prepared from spirit of wine, frumentum, sugar and arrack, but does not give either the quantities or the method of procedure. His unique reputation as an injector is tarnished by similar reserves, and he mentions only that vessels may be filled with a " coloured material ", red wax (massa ceracea subtilissima) or mercury. His passion for injection discovers itself in weird and unexpected fashion. The skeletons in the museum are thrown into dolorous attitudes, and provided with anatomical pocket-handkerchiefs of injected omentum, and even the bladder used in sealing the mouths of the jars has been carefully injected. The results are naturally a source of pride and congratulation. He says : " Sunt mihi parvula cadavera, à viginti annis balsamo munita, quae tam nitidè sunt conservata, ut potius dormire videantur, quam exanimata esse corpuscula ".

Ruysch's museum includes human and comparative preparations of all kinds — in fluids, dried and inflated, but the lesson it conveyed was beyond the comprehension of his objective mind. To him an animal was only the *corpus vile* on which to exercise unrivalled powers of dissection and display. No serious attempt is made towards a scientific classification of animals or organs, or to build up a system of philosophic anatomy. In place of this the preparations are arranged, not to illustrate any principle of biological science, but to produce a picturesque effect. A skeleton balances an injected spermatic plexus in one hand and a coil of viscera in the other ; minatory assortments of calculi of all sizes and shapes occupy the foreground ; in the rear a variety of injected vessels backed by an inflated and injected tunica vaginalis

FIG. 199.—Ruysch, 1703. Pictorial museum exhibit in an anatomical setting

combine to form a grotesque arboreal perspective; another skeleton *in extremis* is grasping a specimen of that emblem of insect mortality, the may-fly, and a third is performing a composition " expressing the sorrows of mankind " on a violin symbolized by a bundle of injected arteries and a fragment of necrotic femur. Bones are arranged to represent a cemetery, wrists are adorned with organic and injected frills, and human, comparative and pathological exhibits are indiscriminately mingled as the exigencies of space required. Of those vast philosophic conceptions which discriminate the Hunterian collection, no suggestion can be traced. Instead of the joy and stimulus of scientific speculation, the museum only reminds him of the sorrows of this world and the perils of the next. Quotations from the Latin poets of a gloomy and despairing nature, insistent in big type, inspire the hope that the diffusion of the Latin tongue was more restricted then than it is to-day. " Ab utero ad tumulum ", " communis ad lethum via ", " ah fata, ah aspera fata ", are examples of the cheerful subjects for moral reflection which restrained the levity of the medieval student ; and " mundus lachrymarum vallis " was considered an appropriate introduction to the skeleton of a woman.

(3) It was in 1763, when he was thirty-five years of age, that John Hunter laid the foundations of his own museum. Up to that time he had been working with his brother William, not always harmoniously, and the preparations of those early years may still be seen in the Hunterian Museum at Glasgow. He died in 1793. For thirty years, therefore, he laboured without intermission at a task the perfect accomplishment of which is the wonder and inspiration of the anatomical world. His ambition was to investigate, and crystallize in the museum, the anatomy and physiology of the whole animal kingdom. He is alive even to the importance of embryology, and that much-debated speculation known as the recapitulation theory, wrongly attributed to von Baer, was affirmed in unmistakeable language by John Hunter. Several attempts are made to base a natural system of classification on structural detail. He divides his material into the following sections, in each case arranged in order from the simplest to the most complex : (*a*) preparations illustrating the general or monographic anatomy of selected animals, to serve as an introduction to comparative anatomy ; (*b*) twelve series of preparations expressive of the general life of the individual as apart from its reproductive activity, such as animal motion, digestion and nutrition, heart and circulation, respiration, excretion, nervous system and organs of sense, connective tissues and skin, and anatomical peculiarities of particular species ; (*c*) ten series of preparations demonstrating animal reproduction.

As examples of Hunter's method, the history of the vascular system is traced by means of beautiful injections (for he lived in an

FIG. 200.—Seba, 1761. Pictorial museum exhibit in a zoological (conchological) setting

age when the syringe was pushed into every crevice an animal presented) from the radial canals excavated in the jelly of a medusa, via the specific circulatory system of an annelid, in which the dorsal vessel begins to assume the contractile function, and the definite pulsating heart of the arthropod and mollusc, to the complex dynamic pump of fishes, amphibians, reptiles, birds and mammals. The transition from the aquatic to the terrestrial animal is followed step by step in the breathing and circulatory organs, and illustrated by a complete and convincing series of preparations. Here we have no assemblage of uncouth and unrelated fragments, nor, like the collections of Swammerdam and Ruysch, is it composed of heterogeneous dissections, however beautiful. We find instead series of exhibits arranged to illustrate the fundamental truths of anatomical science, so that we may compare organs rather than animals. Thus we may trace every separate organ in different species, under varying conditions, in young and old, summer and winter, male and female. No stage or variety is omitted which can throw any light on morphological standards, and for the first time in its history comparative anatomy acquires the status of an ordered and constructive branch of science. The medieval museum either addressed itself frankly to the curiosity of the vulgar, or it aimed at an admirable and even complete presentment of anatomical detail. Such a museum was manifestly deficient in philosophic interest. But Hunter's collection embodied living principles and ideas, and on that basis must it be judged. That his ideas no longer command unqualified assent is not the point. They call for recognition rather because they challenged enquiry, and imposed a standard of research which might be accepted or amended but could not be ignored. From his contemporaries, however, this recognition did not come. His methods and opinions involved a revolution far beyond the knowledge and vision of his time. His museum was of no more use, they said, than so many pig's pettitoes. Indeed his friend Sir Joseph Banks, himself a distinguished naturalist, writes after the death of Hunter as follows : " Had I thought my friend John Hunter's collection an object of importance to the general study of natural history, or indeed to any branch of science except to that of medicine, I hope that two years would not have elapsed without my taking an active part in recommending to the public the measure of purchasing it ". On the other hand it is a pleasure to record that his enemy Jessé Foot, whose implacable jealousy even the death of Hunter could not disarm, should sound the only note of genuine and unreserved approval. He says : " I know of no museum similar to this ; it may be said to be unique, or *sui generis* ; nor do I think that the aggregation or consolidation of any former museums would have produced anything like this ; and I believe that the idea of forming such a collection originated with John Hunter ".

It is illuminating to trace the subsequent history of the anatomical museum. Hunter's collection, it is true, has no rival, but then it has had no imitators. Modern ideals discover a narrower outlook, and a relaxing hold on anatomical verity. The museum of the present day is designed to illustrate, first the general principles of classification, and afterwards the elements of *taxonomic* anatomy. It is above everything zoological, and is in striking and almost painful contrast to the big-scale and rigorous science of the Hunterian conception. Upon this museum the reputation of John Hunter as a man of science must ultimately depend. It is a foundation which has undergone no decay, and requires no restoration.[1]

[1] Written before the destruction of the museum in the late war.

CONCLUSION

THE progress of our knowledge of comparative anatomy illustrates the three stages through which such a pilgrimage commonly passes. First comes the recognition that the less conspicuous animals and the minutiae of anatomy, or, in other words, smallness as such, are not *ipso facto* contemptible, but repay protracted and careful study. The dictum that man was not made to contemplate midges was exploded by the revealing studies of Swammerdam, Leeuwenhoek and Malpighi, but nevertheless it was a long time in dying. Even in the present century Rádl, in discussing the mechanism of reproduction, dismisses the contributions of the " highly learned who can tell us only of centrosomes and chromosomes ; their exact science has discovered nothing more. Tant pis pour elle ! "

The second stage is dependent on the development of the skill and technique necessary for the investigation of the finer details, and the accumulation of a body of comparative data. As usual, skill came first. The old masters produced the most spectacular results with the simplest of tools, and Swammerdam's dissections of the body louse still continue to amaze us. Technique is after all a concession not so much to our physical infirmities and limitations as to a desire for ease, speed and comfort in working. If Swammerdam could have commanded a modern binocular dissecting microscope he would doubtless have done so, and thereby saved himself much toil and strain, but whether he would have *discovered* more is not so certain.

In the third stage the data are integrated into a logical pattern or co-ordinating principle, which gives a new impetus and direction to the enquiry. And until this stage is reached, the observer, as Mery once remarked to Fontenelle, is in the position of the porters of Paris, " who know all the streets, even the smallest and least accessible of them, but who do not know what goes on inside the houses ". The vitality and expansion of research is unquestionably determined by speculation, which in its turn must be sanctioned by observation. If the speculation is wholly false, not only is progress halted but reaction may supervene, in proof of which witness the

464

evil influence of the preformation doctrine in embryology. Unhappily it cannot be said that speculation is ever entirely or even largely sound. Nevertheless it has to embody but the smallest nucleus of truth to provoke the production of piles of literature, in perplexing disproportion to the slight advance that is registered. This hopeful tendency to accept and overwork novel ideas results in nullifying criticism, and in delaying a just verdict on their merits. They are in fact often hailed as triumphant solutions of the difficulties they set out to solve. Nor does a damping succession of failures suffice to warn the younger generation that its own solution awaits a similar inexorable fate. " Drelincourt ", says Blumenbach, " collected from the writings of his predecessors no less than 262 groundless hypotheses on generation, and nothing is more certain than that his own theory was the 263rd." And if the genius of the seventeenth century deserved such wholesale condemnation, perhaps some Blumenbach yet unborn will have occasion to exercise his wit on the speculations even of our own time.

It has been said in haste, if not in jest, that the energies of the scientific worker are employed in correcting the mistakes of his contemporaries, and in making new ones of his own. But the errors of the past are condoned, even if they are perceived, and what should prove a chastening experience fails to encourage that unassuming and circumspect attitude of mind which permeated the scientific labours of Steno. Thus he says (1669) : " Instead of promising to satisfy your curiosity touching the anatomy of the brain, I hereby frankly and sincerely avow that I know nothing about it ". Much confusion of thought would be avoided if the historical approach to scientific learning were permitted to occupy a more prominent place in the curriculum. The boundless optimism of the research worker would be suitably curbed, and the prodigal exploitation of popular and ambitious speculations would be reduced to less wasteful proportions. It is difficult to be satisfied with the final state of the Mendelian doctrine when we contemplate the mountain of research of which it is the modest product, and future generations will view with amusement the involved jargon and mechanical elaboration which condemn the chromosome theory of heredity, and remind us of the evanescent frenzy of the nineteenth-century transcendentalists.

There is much to be learnt from a consideration of two works on comparative anatomy separated by an interval of time sufficiently ample to justify expectations of striking advances in methods of observation and inference. We might select Eustachius' famous little essay on the azygos vein (1563), which, however, though comparative in treatment, is restricted in scope, and concerned only with matters of detail. Duverney's comparison of the hand of man with the fore-foot of the lion (1693) would in itself be more suitable, but it is too late in point of time, and was never published in full. The earliest and most important example is clearly Belon's analysis of the skeleton of the bird (1555), with which we may contrast Goethe's views on the skull published some 250 years later.

Belon tells us that he never missed an opportunity of dissecting a bird, and had examined the internal parts of two hundred diverse species. Hence, he says, it is appropriate that he should describe the skeleton of these animals, and be able to do so with accuracy. If, he adds, you examine the wing or leg of a bird and compare them with the limbs of a quadruped or man, it will be found that the bones are strictly comparable in the two cases. In order to simplify this exercise " so that even a peasant can understand it ", he proposes to consider each bone separately, and to formulate comparisons with other animals and man. He does not undertake to discuss human anatomy, except in so far as it is necessary to establish the truth that, whilst Nature is highly variable in her individual works, *she is ever mindful of the structural unity of the whole*. Hence it can be shown that birds and man are fundamentally identical, in spite of their apparent divergence. He adds that although he restricts himself to two types,[1] he intends to establish similar comparisons with other animals in his Commentary on Dioscorides — a work which was never published. In spite of the apparent absence of sutures in the skull of birds he firmly believes that they can be demonstrated, since in the boiled and disarticulated skull the usual sutures and cranial bones present in man are distinguishable. There are, however, differences to record. Thus in birds he found twelve cervical vertebrae,[2] but only seven in mammals. Again, the sternum is much larger, to

---

[1] His selection of the bird for one could not have been bettered.
[2] The number of cervical vertebrae varies considerably in birds.

provide attachment for the powerful muscles of flight. The
" clavicle " [coracoid] articulates with the scapula to assist in the
support of the wing bones, which latter are no other than the bones
of the arm of man and of the fore limbs of quadrupeds. The ala
spuria or " little wing " corresponds with the thumb of man and
the dew claw of other animals. He compares in detail the bones of
the bird's wing with those of the human arm, although he was
unavoidably ignorant of the fusions which had occurred. He saw
and identified the two free carpals of the bird, but was puzzled
by the elongated metacarpus, which, without further comment, he
likens to a weaver's shuttle. He seems, however, to distinguish
three digits in the wing, and to imply that they are comparable with
the first, second and third of man. On the foot he is more certain
of himself. He claims to have discovered that what is commonly
regarded as the shin of the bird is in fact the heel or tarsus (as
distinct from the toes) of man. This is not far from the truth, and
indeed is as near to it as he could be expected to get without a
knowledge of the development of the skeleton. He implies further
that the proximal ossicle of the first digit of the foot is not a member
of the phalangeal series, but he does not clearly assert this, nor does
he claim that it belongs to the metatarsal region.

Belon's remarkable demonstration, however, did not provoke
in his own mind, or in that of his readers, the larger issues implicit
in the comparison. On the contrary, he admits that the homologies
which he had succeeded in establishing did not disturb his conception of the living world as a unified whole. He was therefore
a teleologist, with leanings towards transcendentalism, of which
ingenious and perfervid school he was one of the founders. Hence
evolution had no place in his philosophy of Nature. Some two
and a half centuries later Goethe was facing a similar problem, and
if we contrast the views of these two distinguished observers, we
discover with surprise how little change the lapse of time had
wrought in the attitude of the biologist towards the question of
the mutability of species.

It was in an essay written in 1795, but not published until
1820, that Goethe focused attention on the unity of plan or organic
constant which was supposed to underlie the organization of all
living beings. Since there was only one plan or idea, the world
was regarded as something akin to a vast musical symposium, in

which the poverty of a solitary theme was enriched by an endless and expanding series of variations. All forms were interpreted in terms of this ideal simulacrum,[1] and even man himself must be studied comparatively, working downwards from higher to lower types. This is the so-called law of reduction, or, to paraphrase the words of Geoffroy, written after his famous debate with Cuvier in 1830, " there is but a single intangible being which becomes patent to our senses under diverse forms ". Goethe was not the first to adopt this attitude, but he was the first to convert it into an ordered scientific theory. The plan was put into operation by a controlling dynamic principle or *Bildungstrieb*. It is obvious at the outset that the theory is a scheme of development and not a principle of evolution, and that its working-out must depend on the institution of a system of anatomical homologies.

Goethe's eager pursuit of homologies, of uniformity in diversity, was bound to lead to the conception that there was only one animal. Further, this single animal type was itself constituted by linking up a number of similar, complete subordinate entities, like a string of beads. He extended the same speculation to plants, and interpreted, as others had done before him, the leaves, sepals, petals, stamens and pistils as variants of a common structure, just as the skull of the animal represented an efflorescence of the vertebral column. Leaf and vertebra were the elemental structures of animal and plant. He was not unmindful of the fact that the working-out of the plan demanded modifications which could not be introduced without affecting the status of the whole, and his law of balance, that " in order to spend on one side, Nature is forced to economize on the other side ", is closely related to the famous deduction made by Cuvier in 1812 of the law of correlation, which was exploited with such dramatic effect by himself and later comparative anatomists. Goethe was not the only biologist to become obsessed with this idealistic morphology. Even Johannes Müller embodied it in his " Inaugural Dissertation " — a lapse which this sensitive man lamented so deeply in later years that he bought up and destroyed all copies of the offending thesis.

[1] Goethe would perhaps not have disapproved this term, since the " reality " of the idea is that of a mental abstraction projected on to paper, like Owen's Archetype. He himself warns us that the type is not a visible reality, but " the generalized expression of that which really exists ".

It is not difficult to understand why Haeckel should have been led to credit Goethe with a share in the triumph of evolution, and to regard him as a forerunner of Darwin. Both Haeckel and Goethe were convinced upholders of the comparative method, and Haeckel was powerfully influenced by the genius of the poet. In Goethe's work on the metamorphosis of plants published in 1790, which is considered to be the beginning of speculative morphology,[1] there are passages which give some support to the claim. Everything depends on his interpretation of what is called " metamorphosis ". Is it an objective historical reality, as any principle of organic evolution is considered to be, or is it a philosophical conception expressing nothing more real than a generalized subjective plan or idea, which itself is not incompatible with Nature philosophy and special creation ? He hesitates between these alternatives, nor does it appear that they were ever sharply contrasted in his own mind, and he has no consistent views with regard to either. His contemporaries included him with the Nature philosophers, and Goethe, who did not hesitate to express the liveliest dissatisfaction when his scientific works were misunderstood, never thought it worth while to correct them. In his later writings, however, he appears at times to have a clearer perception of the evolutionary alternative, but again, at the close of his life, in 1831, he was supporting Geoffroy in his contest with Cuvier, which shows how far he was from comprehending the historical implications of evolution. There is no evidence that he ever specifically discussed the problem of species, regarding such speculations as " a useless occupation which we may well leave to those who are fond of busying themselves with insoluble problems".

The main point of contact between Goethe and the evolutionist lies, however, in the fact that the latter, like the Nature philosopher, also endeavours to find uniformity in diversity, but he looks for it only where it may reasonably be expected to exist, nor is he in any way constrained to mould the facts or stretch the argument in those cases where the quest has been unsuccessful. Failure must be expected and even accepted, since it does not affect the essential truth of the principle of evolution, but merely its mode of operation. Whilst, therefore, Goethe's methods were partly

[1] Goethe introduced the term morphology in 1817, but his definition of it is not retained in modern usage.

those approved by the evolutionist, he goes, and must go, considerably beyond the permitted limits of scientific induction. Having *assumed* the existence of a single animal type, it follows that every species can and must be manipulated to conform to it. Such was the error of Goethe and his successors the Nature philosophers, and it must be held definitely to exclude them from the evolutionary circle, although Goethe himself was never guilty of the ludicrous extravagances of his disciples. In one respect, however, he was remarkably sound. He realized that the *complete* organism only can give us understanding of its inner meaning or idea.

The attitude adopted by Goethe towards the theoretical applications of comparative anatomy may be profitably explored by a comparison of his two researches on the intermaxillary bone and the vertebral theory of the skull. The former was written in 1784 and bears the date 1786 (*sic*),[1] but it was not published until 1831, although an abstract without illustrations appeared in 1817. The latter paper on the skull was drafted and communicated to friends in 1790, but was actually published in 1820, this belated publication being responsible for the priority dispute with Oken. The alleged absence of the premaxilla in man was considered to separate him from the apes — a conclusion which ignored the inference that if the bone is absent in man it is also wanting in the higher apes. Goethe sought to establish its presence in man by methods not dissimilar from those of the modern comparative anatomist. He examined its condition in a number of mammals, noted variations in its state of development and correlated them with feeding habits, investigated aberrant cases such as that of the elephant, and succeeded in finding the bone in some species in which it had not been described. Finally, by comparing human adult and foetal skulls, he demonstrated its occurrence in man.[2] He thus established the importance of embryology in the interpretation of adult structure. His work, in fact, is an admirable essay in comparative anatomy.[3]

The research on the skull, however, belongs to another category. It has no concern with experience, but is an exploitation of the

---

[1] This date is usually but erroneously quoted as the date of publication.

[2] The bone was not first discovered in man by Goethe. Fernel, Vesalius and Fallopius saw indications of it in 1542, 1543 and 1561, and Vicq-D'Azyr described it in the human foetus in 1786.

[3] Cf. Wood Jones, *Nature*, 159, 1947.

Idea. Whether the skull is vertebral in nature or not cannot be tested by observation. If, however, we concede the point, the skull may have been formed on the vertebral pattern as the result of the operation of a creative force working according to plan, or it may have been produced from modified vertebrae by a process of evolution. In either event, the investigator is confronted with an abstraction beyond the means of objective verification. It is otherwise with the case of the intermaxillary bone. Here the " metamorphosis " may be tested and observed in all its stages, and an evolutionary conclusion becomes not only possible but almost inevitable. But how did Goethe regard these two problems ? To him their implications and significance were *identical,* and the only inference he drew from them was a confirmation of his theory of the common plan or idea. To such an outlook deduction and induction are interchangeable. Had the history of the intermaxillary bone suggested to his mind any conception of genetic relationships, he would surely have said so. Goethe therefore was a forerunner of Darwin only in the sense that before the historical continuity of species could be established it was necessary to formulate a doctrine of *homologies,* and in this important work Goethe played a leading part.

We thus see that Belon demonstrated the importance of comparative studies as early as 1555, but the significance of the homologies then disclosed was not recognized by himself, or by Goethe, pondering similar facts, over two hundred years later. The lapse of time therefore had left the situation almost unchanged. The rationale of this prolonged lag is sufficiently obvious. Descriptive anatomy had served its purpose, and could do no more. Until an evolutionary *principle* was demonstrated, further random research could but swell the accumulation of data which awaited integration into a science. In this inchoate state anatomy remained until it was quickened by the publication of the *Origin of Species,* which forthwith raised this uneventful record of factual competence to the dignity of a learned discipline. It is to be regretted that certain modern commentators, who chastise error and belittle accepted discoveries (with some disregard for consistency), should have spoken of the great naturalist who demonstrated the doctrine of evolution in terms of patronizing condescension. This is to neglect the plain historical fact that the publication of the *Origin* put an end to centuries of stagnation, and translated the study of anatomy to a philosophical plane not otherwise attainable.   Moreover, discoveries must be digested before they can be absorbed into the body politic of science, and hence the world rightly honours the men who accomplish this secondary but formidable task.

APPENDIX

BIOGRAPHICAL NOTES

Abbatius (Abati), Baldus Angelus. Date of birth and death unknown. Physician to the Duke of Urbino *c.* 1530. Born at Gubbio in Umbria, Italy. He wrote the first account of the anatomy of the viper.

Aldrovandi, Ulisse. Born at Bologna, 1522, and died there, 1605. He studied at Rome and Padua, and graduated M.D. at Bologna in 1553. He settled in his native town and taught logic, philosophy and botany. He was the most prolific encyclopaedist of an encyclopaedic age, and was renowned as a teacher and practising physician.

Ambrosini, Bartholomeo. Born at Bologna *c.* 1588 (? 1600), and died there in 1657. He collaborated with Aldrovandus in writing his Natural History, and succeeded him as director of the Botanic Gardens of the University. He was the Professor of Philosophy, Botany and Medicine.

Bartholin, Caspar. Son of Thomas (1616–80) and grandson of the first Caspar (1585–1629). Born at Copenhagen in 1655, and died there in 1738. At the age of thirteen his name appears as editor on the title-page of his father's work on the anatomy of the swan. The *ductus Bartholinianus* and the *glandulae Bartholini* are named after him. His first interests were medicine and anatomy, but he abandoned his scientific work to enter the Civil Service, in which he became so distinguished that a patent of nobility was conferred on him in 1731.

Bartholin, Thomas, sen. Born at Copenhagen in 1616. One of three generations of University professors of medicine and science, and founder of the *Acta Medica Hafniensia*. He studied medicine, and more particularly anatomy, in the University of Copenhagen, and also for ten years in Holland, France and Italy. He was an illustrious teacher, and brought fame to the University of Copenhagen. He discovered in 1652–4 the thoracic duct and lymphatic system in man. He retired early in life from his chair of anatomy, and died on his estate out from Copenhagen in 1680.

Belon, Pierre. Pioneer of the comparative method in anatomy. He was born in the hamlet of Soultière in the French province of Maine in 1517. He placed himself under the protection of the Bishop of Le Mans, who aroused his interest in gardening and botany. He studied botany under Valerius Cordus at Wittenberg, and medicine at Paris in 1542. He visited England three times, and travelled for three years in Italy, Greece, Asia Minor, Palestine, Egypt and Arabia, working at the natural history of those countries and the fauna of the Mediterranean. He was assassinated in Paris in 1564.

Bidloo, Govert. Born at Amsterdam, 1649. Professor of anatomy first at The Hague and afterwards at Leiden. For a time he was physician to William III, King of England, and also inspector of military hospitals in England. His imposing atlas of anatomy was plagiarized by the English anatomist W. Cowper. He died at Leiden, 1713.

Blair, Patrick. Born at Boston, Lincolnshire, c. 1670, and died there, 1728. He practised as a physician, and was imprisoned as a Jacobite in 1715. He dissected an elephant at Dundee in 1706.

Blasius (Blaes), Gerard. Born in Kadzand, Holland, *c.* 1625, and died at Amsterdam, 1692. He studied medicine at Copenhagen and Leiden, graduating M.D. at the latter University, 1646, and was elected Professor of Medicine at Amsterdam. He was the author of bibliographical and original works on comparative anatomy.

Borrichius, Olaus, Ole Borch. Born at Nörre Bork, Jutland, 1626, and died at Copenhagen, 1690. His name is taken from his birthplace. He studied medicine at Copenhagen, and afterwards travelled in Holland, England, France and Italy. He returned to Copenhagen to occupy the chair of medicine, and became librarian to the University in 1680. He worked at anatomy under Steno, and discovered the lachrymal duct.

Boyle, Hon. Robert. Fourteenth child of the " great " Earl of Cork, and one of the most famous Englishmen of his time. He was born at Lismore, Ireland, 1627, and died in London, 1691. He promoted the " Invisible College ", which developed into the Royal Society, of which he was one of the original Fellows. He is important as a protagonist of *experimental* science, and also for his practical interest in anatomical methods. He completed his air pump, the " machina Boyleana " or " pneumatical engine ", in 1659, with which he carried out a number of physiological experiments in respiration.

Brown, Edward. Eldest son of Sir Thos. Browne. He was born at Norwich, 1644, and died at Northfleet, Kent, 1708. He travelled extensively in Europe in furtherance of his scientific and medical interests. Comparative anatomy was his favourite study, and he was one of the first to dissect an ostrich. Only brief notes have survived of his numerous dissections of other vertebrate types in 1664 and 1665.

Buffon, Comte de. Born at Montbard, near Dijon, in 1707, and died at Paris, 1788. Buffon was not himself an anatomist, and called in L. J. M. Daubenton to write the comparative anatomy of his *Histoire Naturelle*.

Camper, Petrus. Anatomist, artist and physician. Born at Leiden, 1722, and died at The Hague, 1789. Occupied chairs of anatomy, botany, surgery and medicine at Franeker, Amsterdam and Groningen. One of the greatest teachers of his century. He discovered the pneumaticity of the bones of birds, and made valuable contributions to the comparative anatomy of the Mammalia.

Casserio, Giulio. Born at Piacenza, Italy, 1552 (? 1561), of humble parents. Died at Padua, 1616. He became man-servant to Fabricius at Padua, and was at first self-taught, but was afterwards instructed by his master. Fabricius delegated to him first the teaching of surgery, and later nominated him as his successor to the chair of anatomy, but Casserius did not long hold this office. He is the most skilled and thorough of the early comparative anatomists, and he was the first to construct an accurate figure of the circle of Willis.

Charas, Moyse. Apothecary, physician, chemist and botanist. Born at Uzès in France, 1619 (? 1618), and died in Paris, 1698. Apothecary to Charles II. He was at one time imprisoned by the Spanish Inquisition. His only anatomical publication in his own name is his memoir on the viper, which secured for him admission to the company of the " Parisians ".

Charleton, Walter. Born at Shepton Mallet in Somerset, 1619, and died in London, 1707. He graduated M.D. Oxon., and became President of the College of Physicians and Harveian Orator. He was physician to Charles I and II, and one of the first elected Fellows of the Royal Society. Charleton was a naturalist and a voluminous writer, but only incidentally an anatomist.

Coiter, Volcher. Born at Groningen, Holland, 1534. According to Nuyens he died in 1576, but place unknown. He studied anatomy at Padua, Bologna, Rome and Montpellier under Fallopius, Eustachius and Rondelet. He returned to Bologna as Professor of Anatomy in 1564. In 1569 he became medical officer of Nürnberg, and in 1576 a military surgeon in the German army. He was the first to study comparative anatomy on a large scale, and his work is detailed and accurate.

Collins, Samuel. Born at Rotherfield, Sussex, 1618, and died in London, 1710. He graduated M.D. Padua in 1654, and later became Censor and President of the College of Physicians. For a time he was physician to Charles II. He devoted his leisure to the study of human and comparative anatomy, but his *Systeme of Anatomy* is his only published work.

Columbus, Realdus. Born at Cremona, 1516, and died at Rome, 1559. He studied anatomy under Vesalius at Padua, and succeeded him in the chair of anatomy, 1544. In the following year he accepted a call to Pisa, and finally, in 1549, he was summoned to Rome as Professor of Anatomy. In Rome he made the acquaintance of Michelangelo, and carried out the post-mortem examination on the body of Ignatius Loyola. The name of Columbus is associated with the discovery of the pulmonary circulation, and his work on human anatomy indicates that he had given some attention to comparative anatomy.

Cowper, William. Born at Alresford, Hants, 1666, and died in London, 1709. He practised surgery and studied human anatomy. Cowper's gland is named after him, but he did not discover it. He was one of the first to dissect a marsupial, and was severely criticized for his production of a pirated edition of Bidloo's anatomical plates.

Daubenton, Louis Jean Marie. Born at Montbard, France, the birthplace of Buffon, in 1716, and died in Paris, 1799. He was brought to Paris by Buffon in 1742 to take up an appointment in the Muséum d'Histoire Naturelle, of which he became Professor-Conservator in 1793. He was a good zootechnician, but little more. His chief work was his contribution to Buffon's *Histoire Naturelle*, in which he was solely responsible for the comparative anatomy. His contribution, however, was only printed in full with adequate illustrations in the first edition of that work.

Duverney, Guichard Joseph. Born at Feurs en Forez, France, in 1648, and died at Paris, 1730. He studied medicine at Avignon, where he graduated in 1667. He then went to Paris to improve his anatomy, and was elected into the Academy of Science in 1674, and to the chair of anatomy at the Jardin du Roi. Finally he devoted himself entirely to comparative anatomy and physiology, producing an important work on the organ of hearing, and making still more important contributions to the anatomy of the vascular system and the circulation of the blood in the lower vertebrates.

Ent, Sir George. Born at Sandwich, Kent, in 1604, and died in London, 1689. He studied medicine at Padua, where he graduated M.D. in 1636. Returning to London he became President of the College of Physicians, and was knighted by Charles II in 1665. He was one of the original Fellows of the Royal Society, and the friend and supporter of Harvey, whose work on generation was sponsored by Ent. He wrote a memoir on the anatomy of *Lophius*, *Galeus* and *Rana*, in which he describes the bursa Entiana.

Eustachius (Eustacchi), Bartholomaeus. Born at San Severino Marche, Italy, in 1524, and died at Rome, 1574. He was city physician and Professor of Anatomy in Rome. His lesser works, which are comparative, appeared during his lifetime, but his important anatomical plates, executed in 1552, were not printed until 1714.

Fabricius ab Aquapendente, Hieronymus. On the statue erected to him in Acquapendente the name is spelt Girolamo Fabrizio. Born at Acquapendente, Italy, 1533, and died on his estate near Padua, 1619. He commenced his studies at Padua in 1550, first of all taking letters subjects, but finally devoting himself to anatomy and surgery under Fallopius, whom he succeeded as professor in 1565. Fabricius built the old anatomy theatre at Padua in 1594, which still survives. He saw the valves of the veins in 1574, and was one of the earliest workers to study comparative anatomy on an extensive scale. As the teacher of William Harvey, who went to Italy *c.* 1598, Fabricius provided the stimulus which led to the discovery of the circulation of the blood. His fame attracted students to Padua from all parts of Europe.

Fallopius, Gabriel, Gabriele Falloppia. Born at Modena, Italy, 1523, and died at Padua, 1562. His first profession was the Church, but he forsook it to study medicine at Ferrara under Brassavola, following which he taught anatomy at Pisa. In 1551 he was called to succeed Columbus as Professor of Surgery and Anatomy at Padua, to which Botany was added in his own case. Amongst his important discoveries is that of the bony labyrinth of the ear.

Frederick the Second, Emperor of the Romans and King of Germany. Born at Jesi, Ancona, Italy, 1194, and died at his Italian hunting lodge near Lucera, 1250. That Frederick himself was the author of the celebrated treatise on Hawking is well established and it was composed between 1244 and 1250. The anatomical part of it was first printed in 1596. Apart from Aristotle's brief observations on the anatomy of birds, Frederick's monograph is the first comprehensive attempt to describe the structure of these animals. Nor is it confined to the hawking species, but may claim to be a system of *comparative* ornithology. Amongst other points, Frederick described the uropygial oil gland, the pectoral region and muscles of flight, the sinus rhomboidalis of the spinal cord, the distinction between the stomachs of carnivorous and graminivorous birds, the absence of an epiglottis, and the disappearance in some species of the gall-bladder. The most important point established by him, however, is the true homologies of the bones of wing and leg, or rather as far as that was possible at the time. He says that the bone usually called the shin or tibia [tarso-metatarsus] is in fact a part of the foot, and that in the leg of a bird the same three regions are present as in other animals, that is, thigh, shin and foot. The bones of the wing and leg are comparable in structure, and both are modifications of the generalized vertebrate limb.

Galen, " Claudius ". According to Klebs the Claudius is a misreading for Cl.[arissimus].[1] Born at the ancient city of Pergamum in N.W. Asia Minor, A.D. 130, and died, place unknown, A.D. 200. Pergamum still possessed a medical school in Galen's time, and he was therefore able to study in his native city the subjects of philosophy and medicine, which continued to be his main interests throughout life. Later he added to his anatomical knowledge at Smyrna, Corinth and Alexandria.

Glisson, Francis. Born at Rampisham, Dorset, 1597. He studied at Cambridge, where his first appointment was a College lectureship in Greek. He then moved to London and changed over to medicine, becoming Regius Professor of Physic in Cambridge, but he treated this appointment as a sinecure and spent the greater part of his time in London. He was elected President of the College of Physicians, and was one of the original Fellows of the Royal Society. He died in London, 1677.

Goethe, Johann Wolfgang von. Born at Frankfort-on-the-Main, 1749, and died at Weimar, 1832. The poet's contributions to natural science were many and important, and he completed an article on Geoffroy's views on comparative anatomy in the month that he died.

Gouye, Thomas. Born at Dieppe, 1650, and died in Paris, 1725. He was elected a member of the French Academy of Science in 1699, and joined the Parisian group of comparative anatomists, dissecting some of the material sent to Paris by the mission of the Jesuit Fathers in Siam.

Graaf, Reinier de. Born at Schoonhoven in Holland, 1641, and died at Delft, 1673. He studied under de le Boë (Sylvius) at Leiden, and practised medicine at Delft. He was the author of important works on the comparative anatomy and functions of the generative organs.

Grew, Nehemiah. Born at Coventry, 1641, and died in London, 1712. He studied letters at Cambridge and medicine at Leiden, where he graduated M.D. He returned to Coventry, but soon settled in London

---

[1] If so, the error goes back at least to 1528.

to practise medicine. In 1677 he succeeded Oldenburg as Secretary of the Royal Society. He was one of the first to recognize the value of the comparative principle in anatomy.

Haller, Albrecht von. Born at Bern, Switzerland, 1708, and died there, 1777. He studied philosophy at Tübingen and anatomy at Leiden. After travelling in Belgium, England and France, he became Professor of Anatomy, Botany and Medicine at Göttingen, but returned to Bern in 1753 in order to spend the remainder of his life in his native town. He was a man of vast erudition and unfailing industry, and the author of a bibliography of anatomy which has never been surpassed.

Harder, Johann Jacob. Born at Basel, Switzerland, 1656, and died there, 1711. He studied medicine in Basel, Geneva, Lyon and Paris, and returned to Basel in 1687 as Professor of Anatomy and Botany. He was admitted to the Academia Naturae Curiosorum, and contributed papers on comparative anatomy to their Transactions. The Harderian lachrymal gland is named after him, and he described the *glandulae Pacchionii* in his *Apiarium* of 1687 before Pacchioni himself drew attention to them in 1705.

Hartmann, Philipp Jacob. Born at Stralsund, Pomerania, 1648, and died at Königsberg, 1707. He studied letters, theology and medicine in Königsberg from 1669. He then travelled in France, Holland and England to improve his knowledge and gain experience. Returning to Königsberg, he was appointed successively Professor of History and Medicine. In 1685 he was elected into the Academia Naturae Curiosorum under the pseudonym of Aristotle II. In addition to his memoirs on comparative anatomy he wrote a useful history of anatomy.

Harvey, William. Born at Folkestone in Kent, 1578, and died at London, 1657. He studied anatomy under Fabricius at Padua, where he graduated M.D. in 1602. He settled in London to practise medicine, and became physician to St. Bartholomew's Hospital and to King Charles I. The first public statement of his discovery of the circulation was made in 1616. In 1645 he was the Warden of Merton College, Oxford. Harvey's notes on comparative anatomy were destroyed in his lifetime during the civil war, but his works on the Circulation and Generation include numerous valuable observations on the subject.

Heide (Heyde), Anton van der. Dates of birth and death unknown. He practised medicine in Middleburg, Holland, about the middle of the seventeenth century. He saw striped muscle and cilia soon after Leeuwenhoek, and described the anatomy of *Mytilus* in 1683, and of the liver fluke, tapeworm and the medusa *Aurelia* in 1686. He also studied the anatomy and circulation of the frog.

Heroard, Jean. Born at Montpellier, 1561, and died at the siege of La Rochelle, 1627. He studied medicine at his native town, and became physician to four French sovereigns, the first of which was Charles IX. His only anatomical work, but a noteworthy one, is his osteology of the horse, which seems to have been written before Ruini's treatise, and to be the only surviving part of a general work on the anatomy of the

Appendix 479

horse, the remainder of which, he says, " was shipwrecked during the recent troubles ".

Hodierna, Gianbattista. A Sicilian, born at Ragusa, 1597, and died at Palermo, 1660. He investigated the compound eyes of insects, and the poison apparatus and tongue of snakes.

Hooke, Robert. Born at Freshwater, Isle of Wight, 1635, and died in London, 1703. He went to Christ Church, Oxford, as a chorister in 1653, and came under the influence of Wilkins, Willis and Boyle. This settled the trend of his own researches, and he took up residence in London as Curator of Experiments to the Royal Society in 1664, becoming Secretary to the Society in 1677. His only works covering biological observations are the *Micrographia* of 1665, the first important work on the microscope, and the *Microscopium* of 1678. No portrait of him exists.

Hunter, John. Born at Long Calderwood, near Glasgow, 1728, and died in London, 1793. He joined his brother William in 1748, and studied anatomy and surgery at St. Bartholomew's and St. George's hospitals. An attempt to transfer his studies to Oxford in 1755 was abandoned. He served as a naval surgeon in 1761–3, and on returning to London he started private practice. For twenty-five years he was surgeon to St. George's Hospital. Most of his unpublished observations in comparative anatomy were destroyed after his death, but some survived, and they were published between 1833 and 1840 in the Catalogues of the Royal College of Surgeons.

Jacobaeus, Oligerus, Holger Jacobsen. Born at Aarhus, Denmark, 1650, and died in Copenhagen, 1701. He studied anatomy at Copenhagen under Steno, whose " dearest pupil " he was, and in that University he became successively Professor of History, Geography, and Medicine, and University librarian. He read medicine at Leiden and continued his studies in Florence. He was a leading member of Thomas Bartholin's group of comparative anatomists in Copenhagen, and a frequent contributor to the *Acta Medica Hafniensia*.

Jonston, Johann. Of Scots descent, but born at Samter, Great Poland, 1603. When he retired he took up residence in Silesia, and died near Liegnitz in 1675. As an encyclopaedist he was " a weak successor to Aldrovandi " (Miall). He visited England and Scotland, and studied at St. Andrews, London and Cambridge. He travelled also in Germany and Holland, where he read medicine, anatomy and botany at Franeker and Leiden. He was a prolific writer on Natural History, and his works passed through several editions, but they include little anatomy.

King, Sir Edmund. Born in 1629, but place unknown. Died in London, 1709. Anatomist, surgeon and chemist. He maintained the vascular autocracy of the tissues, and worked at comparative anatomy for Willis, who owed much to his " most dexterous dissections ". He was appointed physician to Charles II, whom he attended in his last illness, and by whom he was knighted in 1676.

König, Emanuel. Born at Basel, Switzerland, 1658, and died there, 1731. He studied philosophy and medicine in his native town, and sought further experience in France and Italy. Returning to Basel, he was successively appointed Professor of Greek and Natural Philosophy, and finally on the death of Harder in 1711, Professor of Medicine. He was a prominent member of the Academia Naturae Curiosorum under the pseudonym of Avicenna.

Lachmund, Fridericus. Born at Hildesheim, Germany, 1635, and died there, 1676. He practised medicine at Osterwieck, and was the author of one of the earliest monographs on a single bird, in this case the petrel *Procellaria*, published in 1674. It includes some anatomical details. Lachmund was a member of the Academia Naturae Curiosorum.

La Hire, Philippe de. Born at Paris, 1640, and died there, 1718. He was primarily an applied mathematician and only incidentally interested in biology, and then from the mechanistic point of view. His early training in art was useful to the Parisians when preparing their anatomical illustrations.

Leeuwenhoek, Antony van. Born at Delft in Holland, 1632, and died there, 1723. He was an amateur microscopist, and had had no scientific training. His researches were communicated (in Dutch) to the Royal Society and to friends in the form of letters, the first of which was written at the age of forty, and the last shortly before he died in his ninety-first year. He followed the occupation of draper, and for thirty-nine years was the City Chamberlain of Delft. He visited England in 1668, and was elected a Fellow of the Royal Society in 1680. He used only simple microscopes constructed throughout by himself and employing a single biconvex lens. Several hundred were made, but only a very few have survived.

Leonardo da Vinci. Born at the village of Vinci, near Empoli and Florence in 1452, and died at Cloux, near Amboise in France, in 1519. His anatomical studies were undertaken in the conviction that such a training was indispensable to the work of an artist. Three only of his scientific manuscripts, which include many important observations on comparative anatomy, are dated, the years being 1489, 1510 and 1513. He must have studied the anatomy of the horse before 1493, by which time his model of the Sforza horse had been completed and was on exhibition. It was never cast.

Libavius (Libau), Andreas. Born at Halle, Germany, *c.* 1546, and died in Coburg, 1616. He was a chemist and physician and studied at Jena, where he was elected Professor of History and Poetry in 1588. Later he became Gymnasiarch at Rothenburg and Coburg. His biological researches are reported in his *Singularia* of 1599–1601.

Lister, Martin. Second son of Sir Martin Lister, M.P. Born at Radclive, Bucks, 1638, and died at Epsom, 1712. He was brought up under the care of his great-uncle Sir Matthew Lister, M.D., physician to James I and Charles I, whose influence sufficed to commit his nephew to the career of medicine. After resigning his fellowship of St. John's College,

Cambridge, he began to practise in 1669 with much success in York, and later in London. In 1698 he attended the English ambassador to Paris as his physician, and described his own scientific and medical experiences in his *Journey to Paris* of 1698. In 1709 he was appointed Physician to Queen Anne. Lister wrote on natural history, medicine, antiquities and cookery, and was the first after Swammerdam to study the comparative anatomy of the Mollusca.

Lorenzini, Stefano. Dates of birth and death unknown. His work on the torpedo of 1678 is the first monograph on the general anatomy of a single fish. It includes also the first mention of red and white muscle.

Major, Johan Daniel. Born at Breslau, Germany, 1634. He studied medicine at Wittenberg and Leipzig, and graduated M.D. at Padua in 1660. He practised medicine in his native town and also in Hamburg, and later became Professor of Medicine and Botany at Kiel. He entered the Academia Naturae Curiosorum under the pseudonym of Hesperus, and was the author of a few noteworthy contributions to comparative anatomy in their Transactions. He died in 1693 at Stockholm, where he was attending professionally the Queen of Sweden.

Malpighi, Marcello. Born near Bologna, 1628. He studied philosophy and medicine at Bologna, where he was appointed Professor of Medicine. In 1656 he joined the University of Pisa as Professor of Medicine, Borelli being elected to the chair of mathematics at the same time. In 1662 he was called to the chair of medicine at Messina, but returned to Bologna in 1666. He was elected a Fellow of the Royal Society in 1668, and maintained a regular correspondence with the Society, which published most of his important works. Much unpublished material was destroyed by fire in 1684. In 1691 he removed to Rome to become physician to the Pope, and died there in 1694, but was buried at Bologna in accordance with his wishes.

Mentzel (Menzel), Christian. Born at Fürstenwalde in Brandenburg, 1622, and died there, 1701. He studied botany and medicine at Frankfort-on-the-Oder and at Königsberg, continuing his studies in Poland, Holland, Spain and Italy. He graduated M.D. at Padua in 1654. He returned to Germany as physician to the Elector of Brandenburg, and was admitted a member of the Academia Naturae Curiosorum under the pseudonym of Apollo.

Mery (Merry), Jean. Born at Vatan, France, 1645, and died in Paris, 1722. He studied medicine in Paris, and was appointed surgeon to the Queen in 1681. On his election as first surgeon to the Paris Hospital he retired from private practice, and occupied himself with anatomical studies. He engaged in a prolonged dispute with Duverney on the circulation of the blood in the lower vertebrates and in the mammalian foetus. He was one of the " Parisians ", and a member of the French Academy of Science.

Moulin (Moulen, Mullen, Molines), Allen. Born in Ireland, date unknown. He studied medicine at Dublin, where he graduated M.D. in 1684. He then carried out researches in human and comparative anatomy, the most important of which was his memoir on the anatomy of the elephant. He went to the West Indies to improve his fortunes, but died soon after landing in 1690.

Muralt (Muralto), Johannes von. Of Italian origin but born at Zurich, 1645, and died there, 1733. Zurich was the centre of scientific and medical learning in Switzerland in the seventeenth century, and Muralt took a leading part in that revival. He studied medicine in Germany, Holland, France and England, and was appointed medical officer of his native town. In 1691 he became Professor of Physics to the Cathedral Chapter and also to the Gymnasium. In 1681 he was elected a member of the Academia Naturae Curiosorum under the pseudonym of Aretaeus, and he contributed numerous papers on comparative anatomy to their Transactions.

Needham, Walter. Born in Shropshire *c.* 1631, and died in London, 1691. He graduated M.D. at Cambridge, 1664, and studied anatomy in Oxford under Willis, Lower and Millington. He wrote only one work containing important observations on comparative anatomy—his *De formato Foetu* of 1667. In this work he assigns a tonal function to the air sacs of birds, and compares the diversity in size of the sacs with the similar diversity in the pipes of an organ.

Nicholls, Frank. Born in London, 1699, and died at Epsom, 1778. He studied at Oxford and graduated M.D. in 1729. Having specialized in anatomy, he was appointed reader in that subject in the University, but continued his anatomical researches and lectures in London, and extended his knowledge by travels in France and Italy. He was physician to George II between 1753 and 1760, and an early *préparateur* of corroded anatomical preparations, being also one of the first to study and teach *minute* anatomy.

Peiresc, Nicolas Claude Fabri de. Senator of the Parliament of Aix. Born at Beaugensier in Provence, 1580, and died at Aix, 1637. He studied at Padua, Rome and Paris, and travelled in England and Holland, examining scientific collections. He built up a vast correspondence with men of learning in all the states of Europe and parts of Asia, with the result that a great accumulation of literary and scientific manuscripts and material was assembled, which has proved invaluable for purposes of research. It included 10,000 letters, most of which were destroyed after his death. Peiresc published no scientific works himself, but his collections were made available to others. He was one of the first to accept Harvey's discovery of the circulation, and to provide verification of it. He was also responsible for numerous valuable observations on comparative anatomy, which were printed by his distinguished friend P. Gassendi, himself, however, an opponent of the Harveian circulation.

Perrault, Claude. Leader of the "Parisians", an early and an active member of the French Academy of Science, and a naturalist who did more to promote the study of comparative anatomy in the seventeenth century than any other worker. He was born in Paris, 1613, and died there, 1688. He studied medicine, anatomy and mathematics at Paris, but abandoned a successful medical practice for the profession of architecture, in which he greatly distinguished himself. His work on the façade of the Louvre, commenced in 1666, was considered at the time to be the *chef-d'œuvre* of French architecture, and the most beautiful building in Paris. The Arc de Triomphe of 1670 was also designed by Perrault, but it is known to us now only from the admirable engraving of Leclerc. Perrault's interest in comparative anatomy grew as his architectural commitments declined, and his mechanistic attitude towards animal structure is but another expression of the technical bias of the architect.

Peyer, Johann Conrad. Born at Schaffhausen, Switzerland, 1653, and died there, 1712. He studied medicine at Basel and Paris — in the latter case under Duverney. He was elected successively to the chairs of rhetoric, logic and physic in the Schaffhausen College, and figures as Pythagoras in the Academia Naturae Curiosorum. He was not the discoverer of Peyer's patches, which were, however, well described and figured by him in 1677. He first found them in 1673, but Severino had already seen them in 1645. Peyer is the author of an important monograph on the compound stomach of Ruminants.

Portius, Lucas Antonius, Lucantonio Porzio. Born at Positano, Italy, 1639, and died at Naples, 1723. He studied medicine in Rome, afterwards travelling through Italy to Vienna, where he investigated problems of the preservation of health in a community at war. He was called to the chairs of anatomy at Rome and Naples in 1670 and 1687. He made only one contribution to comparative anatomy, his memoir on *Astacus*, but it was a memorable one. He wrote also on the history of medicine.

Power, Henry. Born 1623, place unknown. Entered Christ's College, Cambridge, 1641, and graduated M.D., 1655. In 1662 he was received into the Royal Society as one of the first *elected* Fellows. He practised medicine in Yorkshire, and died at New-Hall near Halifax in 1668. His only published work is the "Experimental Philosophy", the main title of which bears the date 1664, but the volume was completed in 1661 and printed in 1663. This treatise is famous for its opulent literary style and many microscopical discoveries. It is the prototype of Hooke's "Micrographia" and other similar works. Apart from Power's poem "In Comendation of ye Microscope" (1661) none of his numerous other writings have been printed.

Ranby, John. Born in London, 1703, and died there, 1773. He studied surgery in London, and was appointed surgeon to George II in 1740. He appears as one of the characters in Fielding's *Tom Jones*. His only contributions to comparative anatomy are his observations on the poison apparatus of a rattlesnake and on the dissection of an ostrich.

Ray (Wray), John. Born at Black Notley, Essex, 1627, and died there, 1705. He was the son of a blacksmith, and entered Cambridge with a scholarship. Finally he secured a Fellowship at Trinity College, and indulged his interest in natural history. In this pursuit he was associated with Francis Willughby, whose Ornithology and Ichthyology were completed and edited by Ray after the death of their author in 1672. In consequence of conscientious scruples, he lost his fellowship, and was forced to leave Cambridge. The remainder of his life was overcast by poverty, isolation and ill-health, his numerous writings, and the pension received under Willughby's will, failing to provide more than a bare livelihood. Ray travelled on the Continent with Willughby, and made valuable natural history collections. He was primarily a systematist, and his most notable contribution to comparative anatomy was his memoir on the dissection of a porpoise.

Réaumur, René-Antoine Ferchault de. A modern Leonardo, who achieved eminence in natural history, physiology, mathematics, physics, technical science and arts, and invention. He was born at La Rochelle, 1683, and died at La Bermondière, Maine, 1757. He was admitted to the Academy of Science at the early age of twenty-five, and was an active member for nearly fifty years. His observations on comparative anatomy are scattered but important, and include an account of the electric organ of the torpedo.

Redi, Francesco. Born at Arezzo, Italy, 1626, and died at Pisa, 1697, but 1696 and 1698 have also been given. Naturalist, physician and poet. He studied philosophy and medicine at Florence and Pisa, and settled in Florence, where he was appointed first physician to the Duke of Tuscany. He successfully attacked by experiment the universal belief in spontaneous generation. His most important anatomical work was published in 1684, and it includes accounts of dissections of the following types : *Aphrodite*, leech, earthworm, lobster, snails and slugs, three Cephalopods, the simple Ascidian *Microcosmus*, various fishes, and one bird, the owl. Redi was a member of the Accademia del Cimento — the first organized scientific society.

Rondelet, Guillaume. Born at Montpellier, 1507, and died there, 1566. He studied the arts in Paris and medicine at his native University, where he became Professor of Medicine and Anatomy in 1545, and Chancellor of the University in 1556. His great reputation attracted able students to Montpellier from all parts of Europe. He devoted himself particularly to the study of marine animals, and for more than two centuries his work on the Mediterranean "fishes" was unrivalled. Although responsible for many important anatomical observations, he was rather a systematist than a comparative anatomist.

Ruini, Carlo. Born *c.* 1530, but place and date unknown. Died at Bologna, 1598. He was the son of an eminent lawyer, and himself studied law. At the time of his death Ruini was a man of considerable wealth and a Senator of Bologna. Nothing is known of his life. In a family record it is stated that shortly before his death he had written two books, both very much praised — one treating of the anatomy, and the other of the diseases of the horse. The complete work was published twenty-eight days after his death, and it is one of the outstanding achievements in the history of comparative anatomy. Ruini's father was assassinated, and he himself is said to have been poisoned, but no record to this effect can be traced in the criminal archives of the city.

Ruysch (Ruijsch), Frederik. Born at The Hague, 1638, and died at Amsterdam, 1731. He studied medicine at Leiden, and in 1665 published the first adequate description of the valves of the lymphatics, which, however, he did not discover. He was called to Amsterdam in 1666 as Professor of Anatomy, where he established a European reputation as a *préparateur* of anatomical injections, and the founder of one of the earliest anatomical museums. It is significant of the high, but mistaken, regard in which Ruysch was held at the time that he should have been selected to fill the vacancy in the French Academy of Science left by the death of Isaac Newton.[1]

Salviani, Hippolyte. Born at Città di Castello, Italy, 1514, and died in Rome, 1572. Ichthyologist, poet, playwright, and physician to three Popes. His great work on " Fishes " was published in parts from 1554 to October, 1557, and it was printed in his own house. The plates are good early examples of line engravings on copper, and closely follow nature. Salviani was primarily a systematist, but there are anatomical notes in the text. He does not confuse the Cephalopoda with the true Fishes.

Schelhamer, Günther Christoph. Born at Jena, 1649, and died at Kiel, 1716. He studied medicine at Leipzig and Leiden, and took part in the defence of the latter city when it was besieged by the French. After the customary academic travels, during which he met Robert Boyle, he was called to Helmstädt as Professor of Botany, and then to Jena to profess anatomy, surgery and botany, finally settling down at Kiel in 1695 as Professor of Medicine. His interest in comparative anatomy arose out of his election into the Academia Naturae Curiosorum, of which he was an active member under the pseudonym of Theophrastus.

Seger, Georg. Born at Nürnberg, 1629, and died at Danzig, 1678. He studied medicine in Germany, and afterwards spent five years in Copenhagen under Thos. Bartholin. He then proceeded to Basel, where he took his doctor's degree in 1660. After holding a medical appointment at Thorn, West Prussia, the birthplace of Copernicus, he was appointed medical officer to the city of Danzig in 1675. His studies in comparative anatomy were stimulated by the Academia Naturae Curiosorum, of which, however, he was apparently not a member.

Severino, Marco Aurelio. Born at Tarsia in Cosenza, Italy, 1580, and died at Naples, 1656. He quitted law to study medicine at Naples, and graduated M.D. at Salerno in 1606. In 1610 he was elected Professor of Anatomy and Medicine at Naples, and retained the chair until his death. His reputation as a surgeon attracted students to Naples from all parts, but this migration ceased with his death. At one time, as the result of the intrigues of his enemies and the threats of the Inquisition, he was driven from Naples, only to be recalled by general consent. The

---

[1] Newton possessed one of Ruysch's works presented by the author, who had autographed it as follows : Illustri Viro ! Isaaco Neuton Equiti reverentiae ergo ; misit auctor.

Zootomia of Severino is one of the most curious and attractive works in the history of comparative anatomy, and deserves careful study.

Snape, Andrew, jun. Born in 1644, but place of birth and death and date of death unknown. Veterinarian to Charles II. For two hundred years, Snape says, his family had served the Crown " in the Quality of *Farriers*". Snape himself was the father of Andrew Snape, a distinguished divine, chaplain to Queen Anne and George I, and at times Master of Eton and Provost of King's College, Cambridge. Nothing is known of Snape's own life, beyond his wholesale appropriation, and publication under his own name, of the plates of Ruini's *Anatomy of the Horse*.

Stelluti, Francesco. Born at Fabriano, Italy, 1577, and died there, 1646 (or ? 1651). Naturalist and poet, and an original member of the Accademia dei Lincei, 1603, taking the pseudonym of Tardigrado. The founder of the Academy, Duke Federico Cesi, died in 1630 and the Academy was dissolved, in spite of Stelluti's vigorous efforts to preserve it. Stelluti was the first to use Galileo's microscope in research, and in 1630 he described and figured the external morphology of the honey bee and the grain weevil *Calandra*. In the latter insect he seems to have understood the nature of the rostrum, and to have seen something of the mouth parts before Leeuwenhoek.

Steno, Nicolaus. The Danish form of the name is Niels Steensen, which Steno himself latinized into Stenonis. The version Steno is an error based on the assumption that Stenonis is the genitive of Steno, but it has passed into general use and may be retained. Steno was born at Copenhagen, 1638, and died at Schwerin, Germany, 1686. He was an anatomist, physiologist, geologist, physician and priest, and one of the most gifted observers of Nature in the seventeenth century. He studied medicine at Copenhagen, where he was befriended by Thos. Bartholin. When the city was besieged by the King of Sweden, Steno and his fellow students took part in its defence. He left before peace was ratified to continue his studies at Amsterdam, where he was associated with Blasius and Swammerdam, and in 1660 he discovered the duct of the parotid gland (*ductus Stenonianus*). He next visited Leiden, where he worked under de la Boë Sylvius and van Horne, and stayed there nearly four years, Swammerdam, Borrichius, Jacobaeus and Spinoza being his companions and friends. He returned to Copenhagen in 1664, but failing to secure a professorship there he proceeded to Paris, and in the following year to Florence. It was during his residence in the latter city that his most important scientific work appeared. Steno now became friendly with Redi and Malpighi, and in 1667 he was converted to the Catholic faith. In 1672 he was again in Copenhagen, this time as Anatomicus Regius, and it was then that he contributed to the *Acta Hafniensia*. He returned to Florence in 1674, and decided to abandon his scientific work. In the following year he took Holy Orders, and in 1677 he was appointed to a titular bishopric, with heavy obligations in the missionary field. At this difficult task he laboured unceasingly in Hanover, Munster, Hamburg and Schwerin, subjecting himself to cruel privations, in consequence of which, according to Maar, he died " in unspeakable misery, forty-eight years old ".

Swammerdam, Jan. Born at Amsterdam, 1637, and died there, 1680. The house still exists, although the original façade has vanished as the result of repairs. Swammerdam was intended for the Church, but he decided in 1661 to study medicine at Leiden, which brought him into contact with Steno and de Graaf. He continued his studies at Saumur, France, and there discovered the valves of the lymphatics in 1664, but Ruysch had anticipated him in this, at all events as regards publication. Swammerdam next visited Paris, where he was befriended by Thévenot, but returned to Amsterdam in 1665 to study anatomy and anatomical methods. In 1667 he graduated M.D. at Leiden, presenting as the thesis his famous work on respiration. The naturalist was now attacked by malaria, which returned from time to time, and his health became very uncertain. In 1673 he came under the influence of Anthoinette Bourignon, and from this time his interest in science began to decline, and his life was devoted more and more to religious exercises. Finally he was afflicted with melancholia and a form of religious mysticism, and died at the early age of forty-three. No portrait of him has been discovered. The decennial Swammerdam medal, instituted in 1880 in celebration of the bicentenary of his death, has been awarded to Siebold, Haeckel, Gegenbaur, de Vries, Max Weber and Spemann.

Sylvius, Jacobus, Jacques Dubois. Born at Louville, near Amiens, 1478, and died in Paris, 1555. At first he studied the classic languages, in which he became very proficient, and then applied himself to medicine in Paris, where he interrupted his course to examine in detail the works of Hippocrates and Galen. Later he proceeded to Montpellier, and graduated there at the mature age of fifty-one. On returning to Paris his genius for exposition attracted to his lectures large numbers of students, including Vesalius, and in 1550 he was elected to a chair in the Royal College. Although he professed to base his anatomy on man rather than on the pig, he was one of the first to dissect and to describe a series of animal types, and hence has a place in the history of comparative anatomy. An excessive admiration for Galen, however, was a bar to the prosecution of his own research.

Tyson, Edward. Born at Bristol, Somerset, 1651, and died in London, 1708. He graduated at Oxford and Cambridge, and then settled in London, where he practised medicine and lectured on anatomy. He abandoned human for comparative anatomy, and was the first Englishman to investigate animal structure on an extensive scale. Much of his work was published, but he left behind a number of unpublished manuscripts on comparative anatomy which are now preserved in the libraries of the British Museum and the Royal College of Physicians. Tyson also sponsored a partial English translation of Swammerdam's monograph on the may-fly, and contributed notes to Willughby's " Fishes " on the anatomy of *Mustelus* and *Cyclopterus*. The section on the anal scent glands of mammals in Plot's *Oxford* is likewise by him.

He was a classical scholar and a bibliophile, which was so little to the liking of Garth that Tyson appears in " The Dispensary " as

> With lumber of vile Books besieg'd around
> . . . retriev'd from *Cooks* and *Grocers*.

Valentini, Michael Bernhard. Born at Giessen, Germany, 1657, and died there, 1729. He studied medicine first in German Universities, and later in Holland, England and France, finally settling down in Giessen as Professor of Physics and Medicine. He was the author of a useful compendium of Zootomy and an exhaustive folio work on Museums.

Vesalius, Andreas. Born at Brussels, 1514, and died in the Greek island of Zante, 1564. He studied letters and philosophy at Louvain, and medicine at Montpellier. For three years from 1533 he was the pupil at Paris of Jacobus Sylvius, with whom he worked at human anatomy and also dissected some animals. He spoke very highly of Sylvius, but not as an original anatomist. At Paris also he attended the lectures of Fernel and John Guinther of Andernach, but again did not value Guinther's anatomy. In Italy he studied anatomy at Padua and Venice, and at Bologna he made the skeleton of a monkey. Elected to the chair at Padua in 1537, he built up a famous cosmopolitan school of anatomy, and conducted public dissections in Padua, Bologna and Pisa. His first anatomical plates were published in 1538, and his great work on human anatomy in 1543. When visiting Basel in 1543 he articulated a human skeleton for the medical school, parts of which may still be seen. Vesalius was bitterly attacked by Sylvius and others for his criticisms of Galen, which so incensed him that he destroyed his own unpublished commentary on Galen, and abandoned his anatomical work. Finally he became court physician to Philip II of Spain, and accompanied him to Madrid in 1559. In 1563, for reasons which are conjectural, Vesalius undertook a pilgrimage to Jerusalem. On his way back he was taken ill and put ashore on the island of Zante, where he died. Vesalius' knowledge of comparative anatomy may be deduced from numerous scattered observations in the *Fabrica* and the China Root Epistle. He seems to have renewed his interest in zootomy after his return to Louvain from Paris, when he began work on an osteological collection of monkeys, quadrupeds and birds. He continued these studies in Italy, and it was his investigations into the anatomy of monkeys that led him to doubt Galen, and finally to distrust those anatomists who owed all that they knew to Galen. " Thus he formed the proud resolve to bring anatomy to life again *by his own efforts* " (Boerhaave and Albinus). Of the 289 woodcuts which illustrate the works of Vesalius, 227 of the blocks still exist in the library of the University of Munich. They were reprinted in a special volume issued in 1934.

Vesling, Johann. Born at Minden in Westphalia, 1598, and died at Padua, 1649. He studied philosophy, anatomy and medicine at Vienna and Venice. He then spent some time in study-travel in Egypt, returning to Europe via Jerusalem. In 1628 he was back in Venice, where he lectured privately on anatomy and botany with such success that he was elected to the chair of anatomy at Padua in 1632. Vesling was the author of a popular textbook on anatomy, and he also made some notable observations on the anatomy of snakes.

Volckamer, Johann Georg. Born at Nürnberg, 1616, and died there, 1693. He studied philosophy and mathematics at Jena, and medicine at Altdorf, Padua and Naples. After travelling in France, he returned to Altdorf to take his medical degree. In 1676 he was elected into the Academia Naturae Curiosorum under the pseudonym of Helianthus, and became its president in 1686. He prepared for publication the *Zootomia* of his old master Severino, which appeared at Nürnberg in 1645.

Waller, Richard. Birth unknown, but *c.* 1650. He died at Northaw in Hertfordshire, 1715. Nothing is known of his early life, and he was probably a business man in the city of London. For twenty-eight years he officiated as Secretary to the Royal Society, and seems to have had a wide knowledge of the sciences. He published translations of the scientific memoirs of the Italian and French Academies, and edited the *Posthumous Works of Robert Hooke*, to which he added a biography of Hooke. His publications show that he was actively interested in comparative anatomy.

Wepfer, Johann Jakob. Born at Schaffhausen, Switzerland, 1620, and died there, 1695. One of the most celebrated anatomists, experimentalists and physicians of the seventeenth century. He studied at Basel and Strasbourg, and after touring the Universities of Italy for two years, he returned to Basel in 1647 to take his doctor's degree. Much of his life was spent as a state and military physician. He was Machaon III of the Academia Naturae Curiosorum, and contributed to its Transactions several memoirs on the anatomy of animals which had not before been dissected.

Willis, Thomas. Born at Great Bedwyn, Wiltshire, 1621, and died in London, 1675. He was buried in Westminster Abbey. His first studies at Oxford were theology and letters, but eventually he changed over to medicine. After the restoration of the monarchy in 1660, Willis was rewarded for his loyalty to the royal cause with the chair of natural philosophy at Oxford, and he remained there for some years practising medicine and engaged in anatomical researches. Also he took a prominent part in the Oxford deliberations which contributed to the foundation of the Royal Society, of which he was one of the first elected Fellows. He removed to London in 1666, and established a medical practice of such dimensions as to draw from the King the epigram that Willis had accounted for more of his subjects than a hostile army would have done. It is difficult to decide how much of the important research in comparative anatomy associated with the name of Willis we owe to himself. It is probable that all the dissections were the work of others, and even the text is perhaps not wholly his own.

BIBLIOGRAPHY

ABBATIUS, BALDUS ANGELUS. *De admirabili viperae natura.* Urbini, 1589. 4to.
Academia Naturae Curiosorum. See under Anonymous, Baier, T. Bartholin, Clauder, Gahrliep, Hannaeus, Harder, Hartmann, König, Lachmund, Major, Mentzel, Muralt, Nebelius, Peyer, Polisius, Portius, Reisel, Schelhamer, Schröck, Seger, Velschius, Volckamer, Waldschmid, Wedel, Wepfer, Wolfstrigel, Zambeccari.
Acta Medica Hafniensia. See under T. Bartholin, sen., C. Bartholin, Borrichius, Jacobaeus, Paulli, Steno, Willius.
ADELMANN, H. B. See Coiter, V.
The Embryological Treatises of Hieronymus Fabricius of Aquapendente. Ithaca, New York, 1942. 8vo.
ALBERTI, S. (1540–1600). *Tres Orationes.* Norimbergae, 1585. 8vo.
ALDROVANDUS, U. (1522–1605). *Ornithologia.* Lib. I, " De Aquila ". Lib. XIX, " De Cygno ". Bononiae, 1599, 1603. Fol.
De animalibus insectis. Bononiae, 1602. Fol.
Quadrupedum omnium bisulcorum historia. Bononiae, 1621. Fol.
Serpentum, et Draconum Historiae libri duo. Bononiae, 1640. Fol.
Monstrorum historia. Pp. 87-8. Bononiae, 1642. Fol.
See addenda by B. Ambrosinus.
AMBROSINUS, B. (*c.* 1600–1657). *Paralipomena accuratissima historiae omnium animalium, quae in voluminibus Aldrovandi desiderantur.* Bononiae, 1642. Fol.
Amsterdammers. *Observationes anatomicae selectiores Collegii privati Amstelodamensis. Observationum anatomicarum Collegii privati Amstelodamensis, Pars altera.* Amstelodami, 1667, 1673. 12mo. Reprinted, University of Reading, 1938. 12mo.
Anonymous. *Hist. Acad. Roy. Sci.* (1666-86). T. I, p. 117. Obs. dated 1670. Paris, 1733. 4to.
" Anoymi [*sic*] curiosi observationes quaedam circa animalia." *Misc. Med.-Phys. Acad. Nat. Curios.* Dec. I, Ann. II. Jenae, 1671. 4to.
" A. I." "·A Conjecture concerning the Bladders of Air that are found in Fishes." With addendum by Hon. R. Boyle. *Phil. Trans.* X. London, 1675. 4to.
ARANTIUS (ARANZI), J. C. (1530–89). *De Humano Foetu Opusculum.* Romae, 1564. 8vo.
ARISTOTLE (384–322 B.C.). *Historia Animalium : De Partibus Animalium : De Generatione Animalium.* Oxford, 1910, 1911, 1910. 8vo.
De historia animalium. Latin trans. by T. Gaza. Venetiis, 1476. Fol.
ATHENAEUS (II-III cent. A.D.). *The Deipnosophists.* English trans. by C. B. Gulick. 7 vols. London, 1927–41. 8vo.
BACON, FRANCIS, Viscount St. Albans (1561–1626). *De dignitate et augmentis Scientiarum.* London, 1623. Fol.
BAIER, J. J. (1677–1735). " De pisce praegrandi Mular [*Physeter*]." *Acta Phys.-Med. Acad. Nat. Curios.* III, p. 2. Norimbergae, 1733. 4to.

BARTHOLIN, C., Thomae fil. (1655–1738). Papers on comp. anat. in *Acta Med. Hafn.* I (1671–2), II (1673), III + IV (1674–5), V (1677–9).
Specimen historiae anatomicae partium corporis humani. Hafniae, 1701. 4to.
BARTHOLIN, T., Sen. (1616–80). *Historiarum anatomicarum rariorum Centuria I-VI.* Hafniae, 1654–61. 8vo.
Dissertatio de Cygni Anatome, ejusq., Cantu. [Hafniae], 1668. 8vo. 1st ed., 1650. 4to.
" De Sirene Danica." *Misc. Med.-Phys. Acad. Nat. Curios.* Dec. I, Ann. I. Lipsiae, 1670. 4to.
Papers on comp. anat. in *Acta Med. Hafn.* I (1671–2), II (1673).
Acta medica & philosophica Hafniensia. 5 voll. in 4. Ann. 1671–9. Hafniae, 1673–80. 4to.
De Anatome Practica Consilium. Hafniae, 1674. 4to.
BAYON, H. P. " The authorship of Carlo Ruini's *Anatomia del Cavallo.*" *Jour. Comp. Path.* XLVIII, p. 138. Edinburgh, 1935. 8vo.
BELLINI, L. (1643–1704). *De structura et usu renum.* Florentiae, 1662. 4to.
BELON, P. (1517–64). *L'histoire naturelle des estranges poissons marins.* Paris, 1551. 4to.
De aquatilibus. Parisiis, 1553. Obl. 8vo.
L'Histoire de la Nature des Oyseaux. Paris, 1555. Fol.
Portraits d'oyseaux, animaux, serpens, herbes, arbres, hommes et femmes d'Arabie & Egypte. Paris, 1557. 4to.
BENEDICTUS, A. (*c.* 1470–1525). *Singulis corporum morbis a capite ad pedes, generatim, membratim, atque remedia, causas, eorumque signa.* Venetiis, 1508. Fol.
BERENGARIO DA CARPI, J. (*c.* 1470–1530). *Commentaria cum amplissimis additionibus super anatomiam Mundini.* Bononiae, 1521. 4to.
BIDLOO, G. (1649–1713). *Anatomia humani corporis.* Amstelodami, 1685. Fol.
Observatio, de animalculis, in ovino aliorumque animantium hepate detectis. Leidae, 1697. 4to.
BIRCH, T. (1705–66). *The History of the Royal Society of London.* 4 vols. London, 1756–7. 4to. Covers the period 1660–87 only.
BLAIR, P. (?–1728). " Osteographia Elephantina." " A description of the organ of hearing in the Elephant." *Phil. Trans.* XXVII, XXX. London, 1710 1718. 4to.
BLASIUS (Blaes), G. (?–1692). *Miscellanea anatomica, hominis brutorumque variorum, fabricam diversam magna parte exhibentia.* Amstelodami, 1673. 8vo.
Observata Anatomica in Homine, Simia, Equo, Vitulo, Ove, Testudine, Echino, Glire, Serpente, Ardea, variisque animalibus aliis. Lugd. Batav. & Amstelod., 1674. 8vo.
Anatome Animalium. Amstelodami, 1681. 4to.
See also under Amsterdammers.
BLUMENBACH, J. F. (1752–1840). *Ueber den Bildungstrieb.* Göttingen, 1780. 8vo.
BORELLI, G. A. (1608–79). *De motu animalium.* 2 voll. Romae, 1680–81. 4to.
BORRICHIUS (Borch), O. (1626–90). Papers on comp. anat. in *Acta Med. Hafn.* I (1671–2), II (1673), V (1677–9).
BOTALLO, L. (1530–?). *Commentarioli duo.* Lugduni, 1565. 16mo.

[BOYLE, Hon. R.] (1627–91). *Some Considerations touching the Usefulnesse of Experimental Naturall Philosophy*. Oxford, 1663. 4to.

" Of preserving Birds taken out of the Egge, and other small Faetus's." *Phil. Trans.* I. London, 1666. 4to.

BRESCHET, G. (1784–1845). *Histoire anatomique et physiologique d'un organe de nature vasculaire découvert dans les Cétacés*. Paris, 1836. 4to.

British Museum. *The history of the collections contained in the natural history departments.* Vol. II. London, 1906. 8vo.

BRONZERIO, G. H. (1577–1630). *Dubitatio de principatu jecoris ex anatome Lampetrae.* Patavii, [1626]. 4to.

BROWN, E. (1644–1708). " An Account of the Dissection of an Oestridge sent to the Royal Society." Hooke's *Phil. Coll.* No. 5. London, 1682. 4to.

BUFFON, COMTE DE (1707–88) and L. J. M. DAUBENTON. " Le Sarigue ou L'Opossum " : " Description du Sarigue." *Histoire Naturelle*, X. Paris, 1763. 4to.

BULWER, J. (*fl.* 1650). *Pathomyotomia, or a Dissection of the significative Muscles of the Affections of the Minde.* London, 1649. 12mo.
Variants of this title, bearing the same date, have been recorded.

CALDESI, G. B. *Osservazioni anatomiche intorno alle Tartarughe Marittime, d'Acqua dolce, e Terrestri.* Firenze, 1687. 4to.

CAMPER, P. (1722–89). " Over het gehoor van den Cachelot, of Pot-Walvisch." *Nat. Verh. Maatsch.* IX. Haarlem, 1767. 8vo.

" Verhandeling over het zamenstel der groote Beenderen in Vogelen." *Verhand. Bataaf. Genoot. Wijsbeg.* Rotterdam, 1774. 4to. Paper dated 2 March 1771.

" Account of the Organs of Speech of the Orang Outang." *Phil. Trans.* LXIX. London, 1779. 4to. Letter to Royal Society dated 2 Dec. 1778.

Natuurkundige verhandelingen over den Orang Outang ; en eenige andere Aapsoorten. Amsterdam, 1782. 4to.

Description anatomique d'un Éléphant mâle. Paris, an XI, 1802. Fol.

CASSERIUS, J. (1552 ?1561–1616). *De Vocis Auditusque Organis.* Ferrariae, 1601–1600. Fol.

Pentaestheseion, hoc est de quinque sensibus liber. Venetiis, 1609. Fol.

CHARAS, M. (1619–98). *Novvelles Experiences sur la Vipere [V. aspis].* Paris, 1669. 8vo.

CHARLETON, W. (1619–1707). *Onomasticon Zoicon.* With *Mantissa Anatomica* by G. Ent. Londini, 1668. 4to.

De Differentiis & Nominibus Animalium. With extended version of the *Mantissa Anatomica* by G. Ent. Oxoniae, 1677. Fol.

Enquiries into Human Nature. London, 1680. 4to.

CHIAIE, S. DELLE (1794–1860). *Memorie sulla storia e notomia degli animali senza vertebre del regno di Napoli.* II. Napoli, 1827. 4to.

CLARK, G. N. *The Seventeenth Century.* Oxford, 1929. 8vo.

CLAUDER, G. (1633–91). " Cricetus (der Hamster) providus sibi contra famem hyemalem promus-condus, singularibus à Natura organis ad haec instructus." *Misc. Cur. Acad. Nat. Curios.* Dec. II, Ann. V, 1686. Norimbergae, 1687. 4to.

COITER, V. (1534–76). *De ossibus et cartilaginibus corporis humani tabulae,* Bononiae, 1566. Fol.

Externarum et internarum principalium humani corporis partium tabulae. Noribergae, 1572. Fol. The less rare 1573 issue is a title page edition.

Lectiones Gabrielis Fallopii . . . ex diversis exemplaribus a Volchero Coiter summa cum diligentia collectae. Noribergae, 1575. Fol.

" The ' De ovorum Gallinaceorum generationis primo exordio progressuque, et pulli Gallinacei creationis ordine ' of Volcher Coiter." Trans. and edited by H. B. Adelmann. *Ann. Med. Hist.*, N.S. V. New York, 1933. 4to.

COLE, F. J. " History of the Anatomical Museum." *Mackay Miscellany.* Liverpool, 1914. 8vo.

" The History of Comparative Anatomy. A Statistical Analysis of the Literature." *Science Progress*, XI. London, 1917. 8vo. With N. B. Eales.

" The History of Anatomical Injections." Singer's *Studies in the History and Method of Science*, II. Oxford, 1921. 8vo.

Early Theories of Sexual Generation. Oxford, 1930. 8vo.

" Goethe as Biologist." *Nature*, 129. London, 1932. 4to.

" Leeuwenhoek's Zoological Researches." *Annals of Science*, II, pp. 1 and 185. London, 1937. 8vo.

" Bibliographical Reflections of a Biologist." *Proc. Bibl. Soc.* V, p. 169. Oxford, 1939. 4to.

COLLINS, S. (1618–1710). *A Systeme of Anatomy.* 2 vols. London, 1685. Fol.

COLLINSON, P. (1694–1768). " Some observations on a sort of Libella or Ephemeron." *Phil Trans.* XLIV. London, 1746. 4to.

COLUMBUS (Colombo), R. (1516–59). *De re anatomica.* Venetiis, 1559. Fol.

COPHO. See G. W. Corner.

CORNER, G. W. *Anatomical texts of the earlier middle ages. . . . With a revised Latin text of Anatomia Cophonis.* Washington, 1927. 8vo.

COSTA, O. G. (1787–1867). " Note sur le prétendu parasite de l'*Argonauta Argo.*" *Ann. Sci. Nat.* XVI (Zool.). Paris, 1841. 8vo.

COWPER, W. (1666–1709). *Myotomia reformata.* London, 1694. 8vo.

" An account of two glands, and their excretory ducts, lately discovered in human bodies." *Phil. Trans.* XXI. London, 1699. 4to.

See also Tyson and Cowper, 1704.

CRAIGIE, D. (1793–1866). *Elements of Anatomy, General, Special, and Comparative.* Edinburgh, 1831. 4to. Cf. *Edin. New Phil. Jour.*, 1830–31, pp. 146 and 291, and 1831, pp. 42 and 355. No more published.

CRIÉ, L. (1850–1912). " Pierre Belon du Mans et l'anatomie comparée." *Rev. Sci.* IV. Paris, 1882. 4to.

CROONE, W. (1633–84). " De formatione pulli in ovo." Birch's *History of the Royal Society of London*, III, pp. 30-40. London, 1757. 4to. Written in 1672.

CRUIKSHANK, W. C. (1745–1800). *The Anatomy of the Absorbing Vessels of the Human Body.* 2nd ed. London, 1790. 4to.

CUVIER, G. (1769–1832). " Mémoire sur le larynx inférieur des Oiseaux." Millin's *Magasin encyclopédique*, I. Paris, 1795. 8vo.

" Mémoire sur l'organe de l'ouie dans les Cétacés." *Bull. Sci. Soc. Philomath.* I. Paris, 1796. 4to.

" Recherches sur les Ossemens Fossiles de Quadrupèdes." T. I. Paris, 1812 4to.

CUVIER et A. VALENCIENNES (1794–1865). "Histoire Naturelle des Poissons."
T. I, Liv. I. Paris, 1828. 8vo.
"Mémoire sur un ver parasite d'un nouveau genre (*Hectocotylus octopodis*)."
Ann. Sci. Nat. XVIII. Paris, 1829. 8vo.
DAREMBERG, C. V. (1817–72). *Histoire des Sciences Médicales.* 2 voll. Paris,
1870. 8vo.
DAUBENTON, L. J. M. (1716–99). "Pièces d'anatomie conservées dans des
liqueurs." Buffon's *Histoire Naturelle,* III. Paris, 1749. 4to.
"De la Description des Animaux." *Ibid.* IV. 1753.
"Description du Pecari." *Ibid.* X. 1763.
See also Buffon and Daubenton.
DE BURE, G. F. (1731–82). *Bibliographie Instructive . . . Volume . . . des
Sciences et Arts,* p. 404. Paris, 1764. 8vo.
DENT, P. *The Correspondence of John Ray.* Letter by Dent dated June 21,
1675, p. 118. London, 1848. 8vo.
DERHAM, W. (1657–1735). *Philosophical experiments and observations of the
late eminent Dr. Robert Hooke.* London, 1726. 8vo.
DESCARTES, R. (1596–1650). *De Homine.* Lugd. Bat., 1662. 4to.
DOBELL, C. *Antony van Leeuwenhoek and his "little animals".* London,
1932. 8vo.
DRAKE, J. (1667–1707). *Anthropologia Nova.* 2 vols. London, 1707. 8vo.
DRYANDER, J. (1748–1810). *Catalogus Bibliothecae Historico-Naturalis Josephi
Banks.* 5 voll. Londini, 1796–1800. 8vo.
DUERDEN, J. E. "A Stalked Parapineal Vesicle in the Ostrich." *Nature,* 105.
London, 1920. 4to.
DUNCAN, D. (1649–1735). *Explication nouvelle et mechanique des actions ani-
males.* Paris, 1678. 12mo.
DUVERNEY, G. J. (1648–1730). *Description anatomique d'une grande tortuë des
Indes.* Paris, 1676. Fol.
Histoire de l'Académie Royale des Sciences depuis . . . 1666 jusqu'à 1686,
[Ann. 1679]. I. Paris, 1733. 4to.
"Description anatomique d'un Crocodile." *Mem. Acad. Roy. Sci.,* 1666–99,
III, Pt. III. Paris, 1734. 4to. Written *c.* 1681. Passage on heart ex-
tracted from the MS. and published by Mery, 1703.
Traité de l'organe de l'ouïe. Paris, 1683. 12mo.
"Description anatomique de trois crocodiles." In Gouye's *Observations,*
1688 (*q.v.*). Reprinted *Mém. Acad. Roy. Sci.,* 1666–99, III, Pt. II. Paris,
1732. 4to.
"Observation sur l'endroit ou se forme la voix du cocq." *Mém. Acad. Sci.,*
1686–99, II. Paris, 1733. 4to.
MS. written *c.* 1695. Partly transcribed and published by Mery, 1703 (*q.v.*).
"Observations sur la Circulation du Sang dans le Foetus : et description du
Coeur de la Tortue et de quelques autres Animaux." *Mém. Acad. Roy.
Sci.* Ann. 1699. Paris, 1702. 4to. Read Dec. 23, 1699.
"Memoire sur la Circulation du Sang des Poissons qui ont des Ouïes, et sur
leur respiration." *Mém. Acad. Roy. Sci.* Ann. 1701. Paris, 1704. 4to.
Oeuvres Anatomiques. 2 voll. Paris, 1761. 4to.
DUVERNOY, G. L. (1777–1855). [On Ruysch's injections.] *Ann. Sci. Nat.,
Zool.* Ser. III, T. VII. Paris, 1847. 8vo.

ELOY, N. F. J. (1714–88). *Dictionnaire historique de la Médecine.* 4 voll. Mons, 1778. 4to.

ENT, SIR GEORGE (1604–89). See W. Charleton, 1668 and 1677.
ANTIΔIATPIBH, *sive animadversiones in Malachiae Thrustoni diatribam de respirationis usu primario.* Londini, 1679. 8vo.
Opera omnia medico-physica. Lugduni Batavorum, 1687. 8vo.

EUSTACHIUS, B. (1524–74). *De renum structura : De auditus organis : De vena, quae ἄζυγος Graecis dicitur.* Venetiis, 1563. 4to.
Tabulae Anatomicae. Romae, 1714. Fol.

FABER, J. (1570–1640) and F. COLUMNA (1567–1650). Comments on Stelluti's Bee Plate of 1625 in Hernandez, 1648–9 (*q.v.*).

FABRICIUS, H.[1] (1533–1619). *De formato foetu.* Venetiis, 1600 [?1604]. Fol.
De visione, voce, auditu. Venetiis, 1600. Fol.
De locutione et eius instrumentis. Venetiis, 1601. 4to.
De brutorum loquela. Patavii, 1603. Fol.
De venarum ostiolis. Patavii, 1603. Fol.
De musculi artificio, de ossium dearticulationibus. Vicentiae, 1614. 4to.
De respiratione, et ejus instrumentis. Patavii, 1615. 4to.
De Totius Animalis Integumentis. [Sub nomine Hieronymi Senis.] Patavii, 1618. 4to.
De gula, ventriculo, intestinis tractatus. Patavii, 1618. 4to.
De motu locali animalium. Patavii, 1618. 4to.
De formatione ovi, et pulli. Patavii, 1621. Fol.

FAITHORNE, H., and J. KERSEY. *Weekly Memorials for the Ingenious.* London, 1682–3. 4to.

FALLOPIUS, G.[2] (1523–62). *Observationes anatomicae.* Venetiis, 1561. 8vo.

FERRARI, G. B. (?–*c.* 1569). *Trattato utile . . . per guarir cavalli, bovi, vacche, cani, asini, muli . . .* Bologna, [*c.* 1560].

FIENUS (Feyens), T. (1567–1631). *De viribus imaginationis tractatus.* Lovanii, 1608. 8vo.

FITZHERBERT, SIR ANTHONY (1470–1538). *A newe Tracte or Treatyse moost profytable for all Husbandemen.* London, [1523]. 4to.

FOLIUS (Folli), C. (1615–60). *Nova auris internae delineatio.* Venetiis, 1645. 4to.

FONTENELLE, B. LE B. DE (1657–1757). " Éloge de M. Mery." *Hist. Acad. Roy. Sci.* Ann. 1722. Paris, 1724. " Éloge de M. Ruysch." *Ibid.* Ann. 1731. Paris, 1733. 4to.

FOOT, JESSÉ (1744–1826). *The Life of John Hunter.* London, 1794. 8vo.

FRANCK VON FRANCKENAU, G. (1643–1704). *Bibliotheca parva Zootomica.* Heidelberg, 1680. 4to.

FRANKLIN, K. J. *De venarum ostiolis 1603 of Hieronymus Fabricius of Aquapendente. Facsimile edition with introduction, translation and notes.* Springfield and Baltimore, 1933. 8vo.
" A survey of the growth of knowledge about certain parts of the foetal cardio-vascular apparatus . . . in man and some other mammals. Part I : Galen to Harvey." *Ann. Sci.* V. London, 1941. 8vo.

FRIDERICUS II. (1194–1250). *De arte venandi cum auibus.* Avgvstae Vindelicorvm, 1596. 8vo. Composed *c.* 1244–50. English translation, 1943.

GAHRLIEP, G. C. (1630–1717). " De Organo auditus Lupini." *Misc. Med.-*

---

[1] Girolamo Fabrizio.   [2] Gabriele Falloppia.

496   *A History of Comparative Anatomy*

Phys. Acad. Nat. Curios. Dec. II, Ann. IX, 1690. Norimbergae, 1691. 4to.

GALEN (130–200). *Opera Omnia.* Ed. C. G. Kühn. Leipzig, 1821–33. 8vo.

GALEN. BRASAVOLA, A. M. (1500–1555). *Index refertissimus in omnes Galeni libros.* Venetiis, 1556. Fol.

GASSENDUS, P. (1592–1655). . . . *the life of The Renowned Nicolaus Claudius Fabricius Lord of Peiresk* . . . London, 1657. 8vo. 1st ed. in Latin : Paris, 1641. 4to.

GAUTIER D'AGOTY, J. (1717–85). *Observations périodiques, sur la physique, l'histoire naturelle, et les beaux arts.* Paris, 1756. 4to.

GAYANT, L. (?–1673). *Hist. Acad. Roy. Sci. depuis* . . . *1666 jusqu'à 1686* [for 1667]. I. Paris, 1733. 4to.

GAZA, T. (1398–1475). See Aristotle, 1476.

GEER, C. DE (1720–78). "Observations sur les Ephémères, dont l'accouplement a été vu en partie." *Mém. Acad. Roy. Sci.* II. Paris, 1755. 4to.

GEOFFROY SAINT-HILAIRE, E. (1772–1844). *Principes de Philosophie Zoologique.* Paris, 1830. 8vo.

GIBBON, E. (1737–94). Memoirs of his Life and Writings composed by himself. London, 1796. 4to.

GIRARDI, M. "Saggio di osservazioni anatomiche intorno agli organi della respirazione degli Uccelli." *Mem. Soc. Italiana,* II. Verona, 1784. 4to.

GLISSON, F. (1597–1677). *Anatomia hepatis.* Londini, 1654. 8vo.
De Ventriculo et Intestinis. Londini, 1677. 4to.

GOETHE, J. W. VON (1749–1832). *Versuch die Metamorphose der Pflanzen zu erklären.* Gotha, 1790. 8vo.
Zur Naturwissenschaft überhaupt, besonders zur Morphologie. 2 Bde. Stuttgard & Tübingen, 1817–24. 8vo. In this work the essays on Comparative Anatomy (1820) and on the Skull (1817, 1820) were first published.
"Über den Zwischenkiefer des Menschen und der Thiere." *Acad. Caes. Leop. Nova Acta,* XV. Jena, 1831. 4to.
"Réflexions sur les débats scientifiques de Mars 1830, dans le sein de l'Académie des Sciences." *Ann. Sci. Nat.* XXII. Paris, 1831. 8vo.

GOODSIR, J. (1814–67). "On the structure of the Intestinal Villi in Man and certain of the Mammalia." *Edin. New Phil. Jour.* XXXIII. Edinburgh, 1842. 8vo.

GOUYE, T. (1650–1725). *Observations physiques* . . . *envoyées de Siam à l'Acad. roy. Sci.* . . . *par les pères Jésuites.* Paris, 1688. 8vo. Reprinted *Mém. Acad. Roy. Sci.,* 1666–99, III, Pt. II. Paris, 1732. 4to.

GRAAF, R. DE (1641–73). *De natura et usu succi pancreatici.* Lugd. Bat., 1664. 12mo.
De Virorum organis Generationi inservientibus, . . . *et de usu Siphonis in Anatomia.* Lugd. Batav. et Roterod., 1668. 8vo.
"Letter dated July 25. 1669. at Delft : accompanied cum Testiculo Gliris dissoluto, & transmisso in Spiritu vini." *Phil. Trans.* IV. London, 1669. 4to.
De mulierum organis generationi inservientibus. Lugduni Batav., 1672. 8vo.

GREENHILL, T. (1681–c. 1740). ΝΕΚΡΟΚΗΔΕΙΑ : *or, the Art of Embalming.* London, 1705. 4to.

GREW, N. (1641–1712). *The Anatomy of Vegetables Begun.* London, 1672. 8vo.

The Comparative Anatomy of Trunks. London, 1675. 8vo.

Musaeum Regalis Societatis. . . . *Whereunto is Subjoyned the Comparative Anatomy of Stomachs and Guts.* London, 1681. Fol.

GRISELINI, F. *Discorso sopra l' utilità della Zootomia.* Venezia, 1749: 8vo. 1751 : 4to.

GUENELLON, P. *De genuina Medicinam instituendi Ratione.* Amstelodami, 1680. 12mo.

HALLER, A. VON (1708–77). *Elementa physiologiae corporis humani.* 8 voll. Lausannae, 1757–66. 4to.

Bibliotheca Anatomica. 2 voll. Tiguri, 1774–7. 4to. First issue of Vol. II dated 1776.

HANNAEUS, G. (1647–99). " Xiphias adumbratus." · *Misc. Med.-Phys. Acad. Nat. Curios.* Dec. II, Ann. VIII, 1689. Norimbergae, 1690. 4to.

HARDER, J. J. (1656–1711). *Examen anatomicum cochleae terrestris domiportae.* Basileae, 1679. 8vo.

Papers on comp. anat. in *Misc. Med.-Phys. Acad. Nat. Curios.* Dec. II, Ann. IV, VI and IX, 1685, 1687, 1690.

Apiarium Observationibus Medicis centum . . . *refertum* Basileae, 1687. 4to.

HARTMANN, P. J. (1648–1707). " Vermes vesiculares sive Hydatŵdes in caprearum Omentis." *Misc. Curios. Acad. Nat. Curios.* Dec. II, Ann. IV, 1685. Norimbergae, 1686. 4to.

Paper on comp. anat. in *Misc. Med.-Phys. Acad. Nat. Curios.* Dec. II, Ann. VII, 1688.

HARVEY, W. (1578–1657). *De motu cordis et sanguinis in animalibus.* Francofurti, 1628. 4to.

De Circulatione Sanguinis. Cantabrigiae, 1649. 12mo.

De Generatione Animalium. Londini, 1651. 4to.

Anatomical Exercitations, Concerning the generation Of Living Creatures. English trans. by M. Lluelyn. London, 1653. 8vo.

HASKINS, C. H. (1870–1937). *Studies in the history of mediaeval science.* Cambridge, U.S.A., 1927. 8vo.

HAUPTMANN, A. (1607–74). *Uralter Wolkensteinischer warmer Badt- und Wasser-Schatz.* Leipzig, 1657. 8vo.

HAZEN, A. T. " Johnson's Life of Frederic Ruysch." *Bull. Hist. Med.* VII. Baltimore, 1939. 8vo.

HEIDE, A. DE. *Anatome Mytuli, Belgicè Mossel,* . . . *nec non Centuria observationum medicarum.* Amstelodami, 1683. 8vo.

HERNANDEZ, F. (1515–87). *Rerum medicarum Novae Hispaniae Thesaurus.* 4th ed. 2 voll. Romae, 1648-9. Fol. 1st Latin ed. by N. A. Recchus : Romae, 1628. Fol.

HEROARD, J. (1561–1627). *Hippostologie.* Paris, 1599. 4to.

HEWSON, W. (1739–74). *A description of the lymphatic system in the human subject, and in other animals.* London, 1774. 8vo.

HOBOKEN, N. (1632–78). *Anatomia Secundinae Vitulinae.* Ultrajecti, 1672. 8vo.

HODIERNA, J. B. (1597–1660). *Opuscoli. 3. L'occhio della mosca.* Palermo, 1644. 4to.

De dente viperae virulento. Panorini, 1646. 4to. Reprinted in Severino, 1651 (*q.v.*).

HOFMANN, M. (1622–98). *De Nutritione.* Altdorfiae, 1648. 4to.

HOME, SIR E. (1756–1832). " Observations on the Structure of the Stomachs of different Animals." *Phil. Trans.* XCVII, p. 139. London, 1807. 4to.

HOOKE, R. (1635–1703). Letter to Boyle dated Nov. 10, 1664 describing respiration experiment on living dog. Boyle's *Works*, V, p. 541. London, 1744. Fol.
Micrographia. London, 1665. Fol.

HOPSTOCK, H. " Leonardo as Anatomist." In Singer's *Studies*, II. Oxford, 1921. 8vo.

HORTO, GARÇIAS DE. *Aromatum et simplicium aliquot medicamentorum apud Indos nascentium historia.* Antverpiae, 1567. 8vo. First edition, 1563.

HOTTON, P. (1648–1709). " On the late M. Swammerdam's Treatise de Apibus." *Phil. Trans.* XXI. London, 1699. 4to.

HUBERT, R. *A Catalogue of many natural rarities . . . collected by Robert Hubert, aliàs Forges.* London, 1664. 8vo.

HUNTER, J. (1728–93). " Account of certain receptacles of air in birds . . . in the hollow bones of those animals." *Phil. Trans.* LXIV. London, 1774. 4to.
" Observations on the Structure and Oeconomy of Whales." *Phil. Trans.* LXXVII. London, 1787. 4to.
Essays and Observations on Natural History. 2 vols. London, 1861. 8vo.

HUNTER, W. (1718–83). *Two introductory lectures . . . to his last course of anatomical lectures.* London, 1784. 4to.

Hunterian Museum. *Descriptive and illustrated catalogue of the physiological series of comparative anatomy contained in the Museum of the Royal College of Surgeons in London.* 5 vols. London, 1833–40. 4to.

HUTCHINSON, B. *Biographia Medica.* 2 vols. London, 1799. 8vo.

HUXLEY, T. H. (1825–95). " On certain errors respecting the structure of the Heart attributed to Aristotle." *Nature*, XXI. London, 1879. 4to.

IMPERATO, F. (1550–1625). *Dell' Historia Naturale di Ferrante Imperato Napolitano.* Napoli, 1599. Fol.

JACOBAEUS (Jacobsen), O. (1650–1701). Papers on comp. anat. in *Acta Med. Hafn.* II (1673), III + IV (1674–6), V (1677–9).
De Ranis et Lacertis Observationes. Hafniae, 1686. 8vo. Revised ed. of *Anatome Salamandrae*, 1676.

JACOPI, G. (1779–1813). *Elementi di Fisiologia, e Notomia comparativa.* 2 voll. Milano, 1808–9. 8vo.

JAMES, R. (1705–76). *A Medicinal Dictionary.* 3 vols. London, 1743–5. Fol.

JONSTON, J. (1603–75). *Historia naturalis de piscibus et cetis.* Francofurti ad Moenum, [1649]. Fol.
Historiae naturalis de Quadrupedibus libri. Ibid., 1650. Fol.
Historiae naturalis de Insectis libri III. Ibid., 1653. Fol.

KING, SIR EDMUND (1629–1709). " Some Observations [Dec. 17, 1668] Concerning the Organs of Generation." *Phil. Trans.* IV. London, 1669. 4to.

KING, T. W. " *Moderator* " band of the Heart in Ungulata. *Guy's Hospital Reports*, II. London, 1837. 8vo.

KLENCKE, H. (1813–81). *Swammerdam oder die Offenbarung der Natur.* 3 Bde. Leipzig, 1860. 8vo.

KNOX, R. (1791–1862). *The Races of Men.* 2nd ed. London, 1862. 8vo.

KÖLLIKER, R. A. VON (1817–1905). "Some Observations upon the Structure of two new Species of Hectocotyle." *Trans. Linn. Soc.* XX. London, 1846. 4to. Abstract : *Proc. L.S.*, 1845. 8vo.

"Hectocotylus Argonautae D. Chiaie und Hectocotylus Tremoctopodis Köll." *Berichte könig. zootom. Anstalt Würzburg . . . f. d. Schuljahr 1847/8.* Leipzig, 1849. 4to.

KÖNIG, E. (1658–1731). Papers on comp. anat. in *Misc. Med.-Phys. Acad. Nat. Curios.* Dec. II, Ann. IV and V, Dec. III, Ann. II, 1685, 1686, 1694.

LACHMUND, F. (1635–76). Papers on comp. anat. in *Misc. Med.-Phys. Acad. Nat. Curios.* Dec. I, Ann. IV and V, 1673–4.

LA HIRE, P. DE (1640–1718). " Nouvelle découverte des yeux de la Mouche." *Jour. d. Sçavans.* Paris, 1678. 4to.

" Explication mécanique du mouvement de la langue du Pivert ", Prop. cxi, in *Traité de Mécanique.* Paris, 1695. 8vo.

LANKESTER, E. Ray (1847–1929). " A Contribution to the Knowledge of Haemoglobin." *Proc. Roy. Soc.* XXI. London, 1872. 8vo.

LEEUWENHOEK, A. VAN (1632–1723). Numbers and dates of letters quoted. For identification and bibliography of these letters cf. Cole, 1937.

14 (1676) ; 22 (1679) ; 30 and 33 (1680) ; 37 and 38 (1683) ; 45 (1685) ; 51 and 52 (1686) ; 56, 57 and 58 (1687) ; 64 (1688) ; 67 (1689) ; 71 and 75 (1692) ; 76 and 77 (1693) ; 83 (1694) ; 90 and 95 (1695) ; 98 and 104 (1696) ; 111 (1698) ; 121 (1699) ; 130, 133 and 134 (1700); 138 (1701); 146 (1702) ; 156 (1704) ; 187 (1711) ; IX (1713) ; XI (1714).

The Collected Letters of Antoni van Leeuwenhoek. Parts I and II. All published. Amsterdam, 1939–41. 8vo.

LEONARDO DA VINCI (1452–1519). Cf. Hopstock, 1921 ; McMurrich, 1930.

LEYDIG, F. (1821–1908). "Ueber Organe eines sechsten Sinnes." *Nova Acta Acad. Nat. Curios.* XXXIV. Dresden, 1868. 4to.

LIBAVIUS, A. (*c.* 1546–1616). " *Singularium Pars secunda.*" Includes " Bombycum domesticorum historia." Francofurti, 1599. 8vo.

LIEBERKÜHN, J. N. (1711–56). *De fabrica et actione villorum intestinorum tenuium hominis.* Lugduni Batavorum, 1745. 4to.

" Sur les moyens propres a decouvrir la construction des visceres." *Mém. Acad. roy. Sci.* Ann. 1748. Berlin, 1750. 4to.

LISTER, M. (1638–1712). *Historiae animalium angliae tres tractatus.* Londini, 1678. 4to. Appendices, 1685.

Appendix ad Historiae conchyliorum Librum IV. Includes " Tabularum anatomicarum explicatio." Londini, 1692. Fol.

Exercitatio Anatomica. In qua de Cochleis, Maximè Terrestribus & Limacibus, agitur. Londini, 1694. 8vo.

Exercitatio Anatomica altera, In qua maximè agitur De Buccinis Fluviatilibus & Marinis. Londini, 1695. 8vo.

Conchyliorum Bivalvium Utriusque Aquae Exercitatio Anatomica Tertia. Londini, 1696. 4to.

" Anatome Pectinis." *Phil. Trans.* XIX. London, 1697. 4to.

LONES, T. E. *Aristotle's Researches in Natural Science.* London, 1912. 8vo.

LORENZINI, S. *Osservazioni intorno alle Torpedini.* Firenze, 1678. 4to.

LOWER, R. (1631–91). *Tractatus de Corde.* London, 1669. 8vo.

LYONET, P. (1706–89). *Traité anatomique de la Chenille, qui ronge le bois de saule.* La Haye, 1760. 4to.

MACKAY, J. Y. (1860–1930). "The arteries of the head and neck and the rete mirabile of the porpoise (*Phocaena communis*)." *Proc. Phil. Soc.* XVII. Glasgow, 1886. 8vo.

McMURRICH, J. P. *Leonardo da Vinci the Anatomist (1452–1519)*. London, 1930. 8vo.

MAJOR, J. D. (1634–93). "De anatome Phocaenae." *Misc. Med.-Phys. Acad. Nat. Curios.* Dec. I, Ann. III, 1672. Lipsiae, 1673. 4to.

"De Respiratione Phocaenae." *Ibid.* Ann. VIII, 1677. Vratislaviae & Bregae, 1678. 4to.

MALPIGHI, M. (1628–94). *De Pulmonibus.* Bononiae, 1661. Fol.

Epistolae Anatomicae. Bononiae, 1665. 12mo.

De viscerum structura. Bononiae, 1666. 8vo.

De Bombyce. Londini, 1669. 4to.

De Formatione Pulli in Ovo. Londini, 1673. 4to.

De Ovo Incubato. Londini, 1675. Fol.

De Structura Glandularum conglobatarum consimiliumque partium. Londini, 1689. 4to.

Opera Posthuma. Londini, 1697. Fol.

Opera Omnia. 2 voll. Londini, 1686. Fol.

MARKHAM, G. (*c.* 1568–1637). *Markhams maister-peece.* London, 1610. 4to.

MARSHALL, J. (1818–91). "On the Brain of a Young Chimpanzee." *Nat. Hist. Review*, N.S. I. London, 1861. 8vo.

MENTZEL, C. (1622–1701). Papers on comp. anat. in *Misc. Med.-Phys. Acad. Nat. Curios.* Dec. II, Ann. I, II, III, 1682, 1683, 1684.

MERY (Merry), J. (1645–1722). "Observations sur la peau du Pelican [1686]." *Mém. Math. Phys. Acad. Roy. Sci.* [II]. Paris, 1693. 4to.

"Traité Physique." *Mém. Acad. Roy. Sci.* Ann. 1703. Paris, 1705. 4to.

"Observations sur les mouvemens de la langue du Piver [Green Woodpecker]." *Ibid.* Ann. 1709. Paris, 1711. 4to.

"Remarques faites sur la moule des etangs." *Ibid.* Ann. 1710. Paris, 1712. 4to.

MOLYNEUX, SIR THOS. (1661–1733). "Account of a not yet described Scolopendra marina." *Phil. Trans.* XIX, XXI. London, 1697, 1699. 4to.

MONCONYS, B. DE (1611–65). *Journal des Voyages de M. de M.* 3 vols. Lyon, 1665-6. 4to.

[MONRO, A., primus] (1697–1767). *An Essay on Comparative Anatomy.* London, 1744. 8vo.

MOUFET, T. (1553–1604). *Insectorum sive Minimorum Animalium Theatrum.* Londini, 1634. Fol.

M[OULIN], A. (?–1690). *An Anatomical Account of the Elephant Accidentally Burnt in Dublin.* London, 1682. 4to. Variants of name : Moulen, Mullen, Molines.

MOULIN, A. "Anatomical observations in the heads of Fowl." *Phil. Trans.* XVII. London, 1693. 4to.

MÜLLER, H. (1820–64). "Ueber das Männchen von Argonauta Argo und die Hectocotylen." *Zeits. wiss. Zool.* IV. Leipzig, 1852. 8vo.

MÜLLER, J. (1801–58). *Diss. inaug. physiologica sistens commentarios de Phoronomia animalium.* Bonnae, 1822. 4to.

"Über den glatten Hai des Aristoteles." *Abhandl. Akad. Wiss.*, 1840. Berlin, 1842. 4to.

Acad.

MURALT, J. VON (1645–1733). Papers on comp. anat. in *Misc. Med.-Phys. Acad. Nat. Curios.* Dec. II, Ann. I, II and V, 1682, 1683, 1686.

Dissertatio physica de insectis eorumque transmutationibus. Tiguri, 1718. 8vo.

l. Nat.

MURRAY, D. *Museums Their History and their Use.* 3 vols. Glasgow, 1940. 8vo.

. 4to.

MUSSCHENBROEK, P. VAN (1692–1761). *Elementa physicae.* Lugd. Bat., 1734. 8vo.

NEBELIUS, S. " De glandula lachrymali Harderiana." *Misc. Med.-Phys. Acad. Nat. Curios.* Dec. III, Ann. III, 1695. Lipsiae & Francofurti, 1696. 4to.

NEEDHAM, W. (1631–91). *De formato Foetu.* Londini, 1667. 8vo.

NESBITT (Nisbet), R. (?–1761). *Human osteogeny explained in two lectures.* London, 1736. 8vo.

NICHOLLS, F. (1699–1778). " An Account of the Hermaphrodite Lobster presented to the Royal Society." *Phil. Trans.* XXXVI. London, 1730. 4to.

NUYENS, B. W. TH. *Het ontleedkundig onderwijs en de geschilderde anatomische lessen van het chirurgijns gilde te Amsterdam, in de jaren 1550 tot 1798.* Amsterdam, [1928]. 8vo.

OMURA, S. " Structure and function of the female genital system . . . and the reproductive system of the male of *Bombyx mori.*" *J. Fac. Agric. Hokkaido Imp. Univ.* XL. Sapporo, 1938. 8vo.

OWEN, R. (1804–92). *On the Archetype and Homologies of the Vertebrate Skeleton.* London, 1848. 8vo.

Parisians. See under Charas; Duverney; Gouye; La Hire, de; Perrault.

PARSONS, J. (1705–70). *Philosophical observations on the analogy between the propagation of animals and that of vegetables.* London, 1752. 8vo.

PASTEUR, L. (1822–95). *Études sur la Maladie des Vers à Soie.* 2 voll. Paris, 1870. 8vo.

PAULLI, S. (1603–80). " Equi Regii Anatome." *Acta Med. Hafn.* Ann. 1671 & 1672. Hafniae, 1673. 4to.

PEIRESC, N. C. F. DE (1580–1637). See P. Gassendus.

PERRAULT, C. (1613–88). . . . *Observations qui ont este faites sur un grand Poisson dissequé dans la Bibliotheque du Roy, le vingt-quatriéme Iuin 1667 : . . . sur un Lion dissequé dans la Bibliotheque du Roy, le vingt-huictiéme Iuin 1667.* Pp. 27, 2 Pls. Paris, 1667. 4to.

Description anatomique d'un Cameleon, d'un Castor, d'un Dromadaire, d'un Ours, et d'une Gazelle. Paris, 1669. 4to.

Memoires pour servir a l'histoire naturelle des animaux. Suite des memoires pour servir a l'histoire naturelle des animaux. Paris, 1671, 1676. Fol.

Memoires pour servir a l'histoire naturelle des animaux. Dressez par M. Perrault, de l'Academie Royale des Sciences. Paris, 1676. Fol.

Essais de Physique. 4 voll. Paris, 1680–88. 12mo.

Memoir's for a Natural History of Animals. English trans. by A. Pitfeild. London, 1688. Fol. Engraved title dated 1687.

" Memoires pour servir a l'histoire naturelle des animaux." *Mém. Acad. Roy. Sci.,* 1666–99, III, 3 Pts. Paris, 1732–4. 4to.

The History, and Anatomical Description of a Cassowar, from the Isle of Java [sic]. Bury, 1778. 8vo. Pirated version by Pidcock.

Nos. 1. 2, 3 and 6 of Perrault's works were published anonymously.

PEYER, J. C. (1653–1712). *Exercitatio anatomico-medica de glandulis intestinorum.* Scaphusiae, 1677. 12mo.
Papers on comp. anat. in *Misc. Med.-Phys. Acad. Nat. Curios.* Dec. II, Ann. I, II and V, 1682, 1683, 1686.
Merycologia. Basileae, 1685. 4to.
PLANCHON, J. E. and G. " Rondelet et ses Disciples." *Montpellier Médical,* 1866, pp. 22 and 43. 8vo.
PLATT, A. " Aristotle on the Heart." Singer's *Studies,* II. Oxford, 1921. 8vo.
PLINY the Elder (23–79). *The Historie of the World.* English trans. by Philemon Holland. 2 vols. London, 1601. Fol.
POLISIUS, G. S. (1636–1700). " De Muscis Polonicis exitiosis [*Sirex*]." *Misc. Med.-Phys. Acad. Nat. Curios.* Dec. II, Ann. IV, 1685. Norimbergae, 1686. 4to.
PORTIUS, L. A. (1639–1723). " De Cancri fluviatilis partibus genitalibus." *Ibid.* Ann. VI, 1687. Pub. 1688.
POUPART, F. (1661–1709). " Letter concerning the insect called Libella." *Phil. Trans.* XXII. London, 1700. 4to.
POWER, H. (1623–1668). *Experimental Philosophy.* London, 1664. 4to.
POWER, JEANETTE. " Osservazioni fisiche sopra il polpo dell' Argonauta Argo." *Atti Accad. Gioen.* XII. Catania, 1837. 4to.
PRENDERGAST, J. S. " The Background of Galen's Life and Activities." *Proc. Roy. Soc. Med.* XXIII. London, 1930. 8vo.
PRESTON, C. " A general idea of the structure of the internal parts of Fish." *Phil. Trans.* XIX. London, 1697. 4to.
RABELAIS, F. (*c.* 1490–1554). *Tiers livre des faictz et dictz heroïques du noble Pantagruel.* Paris, 1546. 8vo.
RACOVITZA, É. G. " Accouplement et fécondation chez l'*Octopus vulgaris,* Lam." *Arch. Zool. Expér.* II. Paris, 1894. 8vo.
RANBY, J. (1703–73). " Some Observations made in an Ostrich." *Phil. Trans.* XXXIII, XXXVI. London, 1725, 1730. 4to.
RATHKE, J. (1769–1855). " Om Dam-Muslingen [*Anodonta anatina*]." *Skr. Naturh. Selsk.* IV. Kjöbenhavn, 1797. 8vo.
RATHKE, M. H. (1793–1860). " Anatomisch-physiologische Untersuchungen über den Athmungsprozess der Insecten." *Schrift. Phys. Oekon. Gesell.* I. Königsberg, 1860. 4to.
RAVEN, C. E. *John Ray Naturalist.* Cambridge, 1942. 8vo.
RAY (Wray), J. (1627–1705). *Catalogus Plantarum circa Cantabrigiam nascentium.* Cantabrigiae, 1660. 12mo.
" An Account of the Dissection of a Porpess." *Phil. Trans.* VI. London, 1671. 4to.
" Some Considerations . . . about the swiming [*sic*] Bladders in Fishes." *Phil. Trans.* X. London, 1675. 4to.
The Wisdom of God Manifested in the Works of the Creation. London, 1691. 8vo.
RÉAUMUR, R.-A. F. DE (1683–1757). *Memoires pour servir a l'histoire des Insectes.* I, V. Paris, 1734, 1740. 4to.
RECCHUS, N. A. See F. Hernandez.
REDI, F. (1626–1697).[1] *Osservazioni intorno alle Vipere [Vipera aspis].* Firenze, 1664. 4to.

---

[1] Other dates given are 1696 and 1698.

ioni intorno agli animali viventi che si trovano negli animali viventi.
e, 1684. 4to.

" De Valvulis In Vena Cava Sub Diaphragmate." *Misc. Med.-*
Acad. Nat. Curios. Dec. I, Ann. IX and X, 1678–9. Vratislaviae &
ae, 1680. 4to.

ᴜꜱᴏɴ, B. W. (1828–96). "Vesalius, and the Birth of Anatomy." *The*
clepiad, II. London, 1885. 8vo.

ᴀʏ, H. (1653–1708). *The Anatomy of the Brain.* London, 1695. 8vo.

ᴇɢᴇʀ, J. C. *Introductio in notitiam rerum naturalium et arte factarum.* 2 voll.
Hagae Comitum, 1742–3. 4to.

Riolan, J. fil (1580–1657). *Osteologia simiae sive ossium hominis et simiae com-*
paratio. Paris, 1614. 8vo. Cf. Tyson, 1699, pp. 60 ff.

Roesel von Rosenhof, A. J. (1705–59). *Der Fluskrebs hiesiges Landes, mit*
seinen merkwürdigen Eigenschafften. Nürnberg, 1755. 4to.

Rolleston, G. (1829–81). *The Harveian Oration, 1873.* London, 1873 8vo.

Rondelet, G. (1507–66). *De Piscibus Marinis.* Lugduni, 1554–5. Fol.

Roth, M. (1839–1914). *Andreas Vesalius Bruxellensis.* Berlin, 1892. 8vo.

Ruini, C. (c. 1530–1598). *Dell' Anotomia, et dell' Infirmita del Cavallo.*
Bologna, 1598. Fol.

La vraye cognoissance dv cheval, . . . avec l'anatomie dv Rvyni. By Jean
Jourdain. Paris, 1647. Fol.

Russell, P. (1726–1805) and E. Home (1756–1832). " Observations on the
Orifices found in certain poisonous Snakes, situated between the Nostril
and the Eye." *Phil. Trans.* XCIV. London, 1804. 4to.

Ruysch, F. (1638–1731). *Dilucidatio valvularum in vasis lymphaticis, et lacteis.*
Hagae-Com., 1665. 8vo.

Museum anatomicum Ruyschianum. Amstelodami, 1691. 4to.

Epistola . . . quarta. De Glandulis . . . &c. Amstelaedami, 1696. 4to

Epistola . . . octava . . . De . . . vasis sanguiferis . . . Cavitatis tympani &
ossiculorum auditus eorumque periostio. Amstelaedami, 1697. 4to.

Thesaurus anatomicus. I-X. Amstelaedami, 1701–15. 4to.

Thesaurus animalium primus. Amstelaedami, 1710. 4to.

Curae posteriores, seu thesaurus anatomicus omnium praecedentium maximus.
Amstelodami, 1724. 4to.

Curae renovatae, seu, thesaurus anatomicus, post curas posteriores, novus.
Amstelodami, 1728. 4to.

Sachs, P. J. (1627–72). " ΓΑΜΜΑΡΟΛΟΓΙΑ." Francofurti & Lipsiae, 1665.
8vo.

Salviani, H. (1514–72). *Aquatilium animalium historiae, Liber primus.* Romae,
1554–7. Fol.

Saunier, J. de, and G. de Saunier (1663–1748). *La parfaite connoissance des*
Chevaux. La Haye, 1734. Fol.

Scaccho da Tagliacozzo, F. *Trattato di Mescalzia.* Rome, 1591. 4to.

Schelhamer, G. C. (1649–1716). Papers on comp. anat. in *Misc. Med.-Phys.*
Acad. Nat. Curios. Dec. II, Ann. I and VI, Dec. III, Ann. V-VI and VII-
VIII, 1682, 1687, 1697–8, 1699–1700.

Anatomes Xiphiae Piscis . . . MDCCIV. Hamburgi, 1707. 4to.

Schenck, J. T. (1619–71). *Epistola ad Authorem Gammarologiae.* Cf. P. J.
Sachs, 1665, p. 935.

SCHRÖCK, L. fil (1646–1730). Papers on comp. anat. in *Misc. Med.-Nat. Curios.* Dec. I, Ann. VIII, Dec. II, Ann. VII, 1677, 1688 *Historia Moschi.* Augustae Vindelicorum, 1682. 4to.

SEGER, G. (1629–78). Papers on comp. anat. in *Misc. Med.-Phys. ⁄ Curios.* Dec. I, Ann. I to V and IX + X, 1670–74 and 1678–9.

SEVERINO, M. A. (1580–1656). *Zootomia Democritaea.* Noribergae, 164 *Vipera Pythia.* Patavii, 1651. 4to. *De piscibus in sicco viventibus.* Neapoli, 1655. Fol. *Antiperipatias.* Neapoli, 1659. Fol.

SINGER, C. " Greek Biology and its Relation to the Rise of Modern Biology.' *Studies,* II. Oxford, 1921. 8vo. *The Evolution of Anatomy.* London, 1925. 8vo.

SLADE, M. [Theod. Aldes] (1628–89). *Dissertatio epistolica contra Gul. Harveum.* Amstelodami, 1667. 8vo. Few copies, with different title, dated 1666. *Observationes naturales in Ovis factae.* Amstelodami, 1673. 12mo.

SLOANE, SIR HANS (1660–1753). " Of Fossile Teeth and Bones of Elephants." *Phil. Trans.* XXXV. London, 1728. 4to.

SMITH, SIR F. (1857–1929). *The Early History of Veterinary Literature.* 4 vols. London, 1919–33. 8vo.

SNAPE, A. (1644–?). *The Anatomy of an Horse.* London, 1683. Fol.

SNODGRASS, R. E. *Anatomy and Physiology of the Honeybee.* New York, 1925. 8vo.

SOEMMERRING, S. T. (1755–1830). *De basi encephali et originibus nervorum cranio egredientium.* Goettingae, 1778. 4to.

SPENCER, W. G. " Vesalius : his delineation of the framework of the human body in the *Fabrica* and *Epitome.*" *British Jour. Surg.* X. Bristol, 1923. 8vo.

SPIGELIUS, A. (1578–1625). *De lumbrico lato liber.* Patavii, 1618. 4to. *De humani corporis fabrica.* Venetiis, 1627. Fol.

STELLUTI, F. (1577–1646). Single sheet illustrating the external morphology of the Honeybee MICROSCOPIO *obseruabat.* Romae, 1625. Fol. *Persio Tradotto in verso sciolto e dichiarato da Francesco Stelluti.* Roma, 1630. 4to.

STENO, N. (1638–86). *De Glandulis Oris.* Inaug. Diss. Lugduni Batavorum, 1661. 4to. *De Musculis & Glandulis observationum specimen.* Hafniae, 1664. 4to. *Elementorum Myologiae Specimen.* Florentiae, 1667. 4to. *Discours . . . Sur L'Anatomie Du Cerveau.* Paris, 1669. 12mo. Papers on comp. anat. in *Acta Med. Hafn.* I, II, 1671–2, 1673. *Nicolai Stenonis Opera Philosophica.* 2 vols. Copenhagen, 1910. 4to.

STUKELEY, W. (1687–1765). *Of the Spleen. . . . Some Anatomical Observations in the Dissection of an Elephant.* London, 1723. Fol.

SWAMMERDAM, J. (1637–80). *Tractatus Physico-Anatomico-Medicus de Respiratione Usuque Pulmonum.* Lugduni Batavorum, 1667. 8vo. *Historia Insectorum Generalis, ofte Algemeene Verhandeling van de Bloedeloose Dierkens.* Utrecht, 1669. 4to. *Miraculum naturae sive uteri muliebris fabrica.* Lugduni Batavorum, 1672. 4to. " Extracts of two Letters . . . concerning some Animals, that having Lungs

are yet found to be without the Arterious Vein [Pulmonary Artery]."
Phil. Trans. VIII. London, 1673. 4to.

Ephemeri Vita. Amsterdam, 1675. 8vo.

Catalogus Van een seer wel gestoffeerde Konstkamer, . . . *Vergaedert deur Johan Jacobsz. Swammerdam.* s.l., 1679. 8vo.

Ephemeri vita: or the Natural History and Anatomy of the Ephemeron. A Fly that Lives but five hours. London, 1681. 4to.

Bybel der Natuure. 2 Dle. Leyden, 1737–8. Fol. Completed *c.* 1675.

The Book of Nature. English trans. by T. Flloyd. London, 1758. Fol. See also under Amsterdammers.

SYLVIUS (Dubois), J. (1478–1555). *In Hippocratis et Galeni physiologiae partem anatomicam isagoge.* Paris, 1555. Fol.

Commentarius in Claudii Galeni de ossibus ad tyronis libellum. Parisiis, 1561. 8vo.

TACHARD, G. (*c.* 1650–1712). *Voyage de Siam des Peres Jesuites. Second voyage du P. Tachard.* Paris, 1686, 1689. 4to.

TENON, J. R. (1724–1816). " Second essai d'étude, par époques, des dents molaires du cheval." *Mém. de l'Inst.* I. Paris, 1797. 4to.

THÉVENOT, M. (*c.* 1620–92). *Recueil des Voyages de M. Thévenot.* Includes " Le Cabinet de Mr. Swammerdam." Paris, 1681. 8vo.

THOMSON, T. *History of The Royal Society.* London, 1812. 4to.

THRUSTON, M. *De Respirationis Usu primario, Diatriba.* Londini, 1670. 8vo.

TODD, R. B. (1809–60). " Nervous System." *Cyclopaedia of Anatomy and Physiology,* III. London, 1844. 8vo.

TODD, R. B., and W. BOWMAN (1816–92). *The Physiological Anatomy and Physiology of Man.* 2 vols. London, 1843–56. 8vo.

TULP, N. (1593–1674). *Observationum medicarum libri tres.* Amstelredami, 1641. 8vo.

TYSON, E. (1651–1708). *Phocaena, or the Anatomy of a Porpess.* London, 1680. 4to.

" Vipera caudi-sona Americana, Or the Anatomy of a Rattle-Snake " : " Lumbricus latus, or . . . the Joynted Worm " : " Lumbricus teres, or some Anatomical Observations on the Round Worm [Ascaris] " : " Tajacu, . . . or the Anatomy of the Mexico Musk-Hog ". *Phil. Trans.* XIII. London, 1683. 4to.

" *Lumbricus hydropicus,* or an essay to prove that the hydatides . . . are a species of worms, or imperfect animals." *Phil. Trans.* XVII. London, 1691. 4to.

" Carigueya, seu Marsupiale Americanum, or, the anatomy of a [female] Opossum." *Phil. Trans.* XX. London, 1698. 4to.

Orang-Outang, sive Homo Sylvestris: or, the Anatomy of a Pygmie. London 1699. 4to.

TYSON, E., and W. COWPER. " Carigueya, . . . or, the anatomy of a male Opossum." *Phil. Trans.* XXIV. London, 1704. 4to.

VALENTINI, M. B. (1657–1729). *Amphitheatrum Zootomicum.* Francof. ad Moenum, 1720. Fol.

VALLISNERI, A. (1661–1730). " Notomia dello Struzzo." Dated 2 April 1712. In *Nuove osservazioni.* Padova, 1726. 4to.

VAROLIUS, C. (1543–75). *Anatomiae, sive de resolutione corporis humani lib. IIII* Francofurti 1591. 8vo.

VAYSSIÈRE, A. " Recherches sur l'organisation des larves des Éphémérines." *Ann. Sci. Nat.* (Zool.), XIII. Paris, 1882. 8vo.

VELSCHIUS (Welsch), G. H. (1624–76). " De anatome Muris alpini." *Misc. Med.-Phys. Acad. Nat. Curios.* Dec. I, Ann. I. Lipsiae, 1670. 4to.

VERANY, G. B. (?–1865). " Mémoire sur six nouvelles espèces de Céphalo- podes." *Torino Mem. Accad.* I. Turin, 1839. 4to.

Céphalopodes de la Méditerranée. Gènes, 1851. 4to.

VÉRANY, J. B., and C. C. VOGT (1817–95). " Mémoire sur les Hectocotyles." *Ann. Sci. Nat.* Sér. III, Zool. XVII. Paris, 1852. 8vo.

VESALIUS, A. (1514–64). *De humani corporis fabrica.* Basileae, 1543. Fol.

Opera Omnia Anatomica & Chirurgica. 2 voll. Lugd. Bat., 1725. Fol.

VESLING, J. (1598–1649). *De viperae partibus. De viperae generatione.* Letters dated 1644 and 1647 printed by Severino in 1651 (*q.v.*).

VICQ-D'AZYR, F. (1748–94). " Mémoires pour servir a l'histoire anatomique des poissons." *Mém. Savans Étrang. Acad. Sci.* VII, Ann. 1773. Paris, 1776. 4to.

" Observations anatomiques sur trois Singes . . . suivies de quelques re- flexions sur plusieurs points d'anatomie comparée." *Hist. Acad. Roy. Sci.* Ann. 1780. Paris, 1784. 4to.

Traité d'anatomie et de physiologie. T. I (all published). Paris, 1786. Fol.

VOLCKAMER, J. G. (1616–93). Papers on comp. anat. in *Misc. Med.-Phys. Acad. Nat. Curios.* Dec. II, Ann. V and VI, Dec. III, Ann. IV, 1686, 1687, 1696.

WALDSCHMID, W. H. " Lampetrae fluviatilis Anatome." *Misc. Med.-Phys. Acad. Nat. Curios.* Dec. III, Ann. VI, 1698. Lipsiae & Francofurti, 1700. 4to.

WALLER, R. (*c.* 1650–1715). " Some observations in the dissection of a Rat." *Phil. Trans.* XVII. London, 1693. 4to.

" A Description of that curious Natural Machine, the Wood-Peckers Tongue, &c." *Ibid.* XXIX, 1716.

WALTHER, J. *Die Kaiserlich Deutsche Akademie der Naturforscher zu Halle.* Leipzig, 1925. 8vo.

WARREN, G. " Observations upon the Dissection of an Ostrich." *Phil. Trans.* XXXIV. London, 1726. 4to.

WEBER, E. H. (1795–1878). " Ueber das Geschmacksorgan der Karpfen und den Ursprung seiner Nerven." Meckel's *Arch. Anat. Phys.* II. Leipzig, 1827. 8vo.

WEDEL, G. W. (1645–1721). " Cygni sterni Anatomiae." *Misc. Med.-Phys. Acad. Nat. Curios.* Dec. I, Ann. II. Jenae, 1671. 4to.

Weekly Memorials for the Ingenious. See Faithorne and Kersey.

WELD, C. R. (1813–69). *A History of the Royal Society.* 2 vols. London, 1848. 8vo.

WEPFER, J. J. (1620–95). " Anatomia aliquot castorum." *Misc. Med.-Phys. Acad. Nat. Curios.* Dec. I, Ann. II. Jenae, 1671. 4to.

" De Ariete Hermaphrodito." *Ibid.* Dec. I, Ann. III, 1672. Lipsiae & Francofurti, 1673. 4to.

Other papers on comp. anat. in Dec. II, Ann. VI, 1687.

WEST, G. S. " On the sensory pit of the Crotalinae." *Quart. Jour. Micr. Sci.* XLIII. London, 1900. 8vo.

WHITLOCK, R. (1616–?). ZΩOTOMI'A, *or, Observations on the present manners of the English.* London, 1654. 8vo.

WILLIS, T. (1621–75). *Cerebri Anatome.* Londini, 1664. 4to. *De Animà Brutorum.* Oxonii, Londini, Amstelodami, 1672. 8vo et 4to. *Pharmaceutice Rationalis.* Pars I. Oxonii, 1673. 4to.

WILLIUS (Wille), J. V. (?–1676). " De aculeo piscis Føsing [*Trachinus draco*]." *Acta Med. Hafn.* III & IV, Ann. 1674–6. Hafniae, 1677. 4to.

WILLUGHBY, F. (1635–72). *De Historia Piscium.* Oxonii, 1686. Fol. Plates dated London, 1685.

WIRSUNG, J. G. (*c.* 1610–43). *Figura ductus cuiusdam cum multiplicibus suis ramulis nouiter in Pancreate in diuersis corporibus humanis obseruati.* Paduae, 1642. Obl. fol.

WOLFSTRIGEL, L. (?–1671). "Anatome Leonum. . . ." "Tigridum Anatome. . . ." *Misc. Med.-Phys. Acad. Nat. Curios.* Dec. I, Ann. II. Jenae, 1671. 4to.

WOTTON, W. (1666–1726). *Reflections upon Ancient and Modern Learning.* London, 1694. 8vo.

ZAMBECCARI, G. " Experimenta circa diversa e variis animalibus viventibus execta viscera." *Misc. Med.-Phys. Acad. Nat. Curios.* Dec. III, Ann. IV, 1696. Lipsiae & Francofurti, 1697. 4to. First published in Italian, 1680.

ZERBIS, G. DE (*c.* 1440–1505). *Liber anathomiae corporis humani.* Venetiis, 1502. Fol.

INDEX

THE END

A CATALOGUE OF SELECTED DOVER BOOKS
IN ALL FIELDS OF INTEREST

A CATALOGUE OF SELECTED DOVER BOOKS
IN ALL FIELDS OF INTEREST

AMERICA'S OLD MASTERS, James T. Flexner. Four men emerged unexpectedly from provincial 18th century America to leadership in European art: Benjamin West, J. S. Copley, C. R. Peale, Gilbert Stuart. Brilliant coverage of lives and contributions. Revised, 1967 edition. 69 plates. 365pp. of text.
21806-6 Paperbound $3.00

FIRST FLOWERS OF OUR WILDERNESS: AMERICAN PAINTING, THE COLONIAL PERIOD, James T. Flexner. Painters, and regional painting traditions from earliest Colonial times up to the emergence of Copley, West and Peale Sr., Foster, Gustavus Hesselius, Feke, John Smibert and many anonymous painters in the primitive manner. Engaging presentation, with 162 illustrations. xxii + 368pp.
22180-6 Paperbound $3.50

THE LIGHT OF DISTANT SKIES: AMERICAN PAINTING, 1760-1835, James T. Flexner. The great generation of early American painters goes to Europe to learn and to teach: West, Copley, Gilbert Stuart and others. Allston, Trumbull, Morse; also contemporary American painters—primitives, derivatives, academics—who remained in America. 102 illustrations. xiii + 306pp.
22179-2 Paperbound $3.50

A HISTORY OF THE RISE AND PROGRESS OF THE ARTS OF DESIGN IN THE UNITED STATES, William Dunlap. Much the richest mine of information on early American painters, sculptors, architects, engravers, miniaturists, etc. The only source of information for scores of artists, the major primary source for many others. Unabridged reprint of rare original 1834 edition, with new introduction by James T. Flexner, and 394 new illustrations. Edited by Rita Weiss. 6⅝ x 9⅝.
21695-0, 21696-9, 21697-7 Three volumes, Paperbound $15.00

EPOCHS OF CHINESE AND JAPANESE ART, Ernest F. Fenollosa. From primitive Chinese art to the 20th century, thorough history, explanation of every important art period and form, including Japanese woodcuts; main stress on China and Japan, but Tibet, Korea also included. Still unexcelled for its detailed, rich coverage of cultural background, aesthetic elements, diffusion studies, particularly of the historical period. 2nd, 1913 edition. 242 illustrations. lii + 439pp. of text.
20364-6, 20365-4 Two volumes, Paperbound $6.00

THE GENTLE ART OF MAKING ENEMIES, James A. M. Whistler. Greatest wit of his day deflates Oscar Wilde, Ruskin, Swinburne; strikes back at inane critics, exhibitions, art journalism; aesthetics of impressionist revolution in most striking form. Highly readable classic by great painter. Reproduction of edition designed by Whistler. Introduction by Alfred Werner. xxxvi + 334pp.
21875-9 Paperbound $3.00

THE RED FAIRY BOOK, Andrew Lang. Lang's color fairy books have long been children's favorites. This volume includes Rapunzel, Jack and the Bean-stalk and 35 other stories, familiar and unfamiliar. 4 plates, 93 illustrations x + 367pp.
21673-X Paperbound $2.50

THE BLUE FAIRY BOOK, Andrew Lang. Lang's tales come from all countries and all times. Here are 37 tales from Grimm, the Arabian Nights, Greek Mythology, and other fascinating sources. 8 plates, 130 illustrations. xi + 390pp.
21437-0 Paperbound $2.75

HOUSEHOLD STORIES BY THE BROTHERS GRIMM. Classic English-language edition of the well-known tales — Rumpelstiltskin, Snow White, Hansel and Gretel, The Twelve Brothers, Faithful John, Rapunzel, Tom Thumb (52 stories in all). Translated into simple, straightforward English by Lucy Crane. Ornamented with head-pieces, vignettes, elaborate decorative initials and a dozen full-page illustrations by Walter Crane. x + 269pp.
21080-4 Paperbound $2.00

THE MERRY ADVENTURES OF ROBIN HOOD, Howard Pyle. The finest modern versions of the traditional ballads and tales about the great English outlaw. Howard Pyle's complete prose version, with every word, every illustration of the first edition. Do not confuse this facsimile of the original (1883) with modern editions that change text or illustrations. 23 plates plus many page decorations. xxii + 296pp.
22043-5 Paperbound $2.75

THE STORY OF KING ARTHUR AND HIS KNIGHTS, Howard Pyle. The finest children's version of the life of King Arthur; brilliantly retold by Pyle, with 48 of his most imaginative illustrations. xviii + 313pp. 6⅛ x 9¼.
21445-1 Paperbound $2.50

THE WONDERFUL WIZARD OF OZ, L. Frank Baum. America's finest children's book in facsimile of first edition with all Denslow illustrations in full color. The edition a child should have. Introduction by Martin Gardner. 23 color plates, scores of drawings. iv + 267pp.
20691-2 Paperbound $3.50

THE MARVELOUS LAND OF OZ, L. Frank Baum. The second Oz book, every bit as imaginative as the Wizard. The hero is a boy named Tip, but the Scarecrow and the Tin Woodman are back, as is the Oz magic. 16 color plates, 120 drawings by John R. Neill. 287pp.
20692-0 Paperbound $2.50

THE MAGICAL MONARCH OF MO, L. Frank Baum. Remarkable adventures in a land even stranger than Oz. The best of Baum's books not in the Oz series. 15 color plates and dozens of drawings by Frank Verbeck. xviii + 237pp.
21892-9 Paperbound $2.25

THE BAD CHILD'S BOOK OF BEASTS, MORE BEASTS FOR WORSE CHILDREN, A MORAL ALPHABET, Hilaire Belloc. Three complete humor classics in one volume. Be kind to the frog, and do not call him names . . . and 28 other whimsical animals. Familiar favorites and some not so well known. Illustrated by Basil Blackwell.
156pp. (USO) 20749-8 Paperbound $1.50

How to Know the Wild Flowers, Mrs. William Starr Dana. This is the classical book of American wildflowers (of the Eastern and Central United States), used by hundreds of thousands. Covers over 500 species, arranged in extremely easy to use color and season groups. Full descriptions, much plant lore. This Dover edition is the fullest ever compiled, with tables of nomenclature changes. 174 full-page plates by M. Satterlee. xii + 418pp. 20332-8 Paperbound $3.00

Our Plant Friends and Foes, William Atherton DuPuy. History, economic importance, essential botanical information and peculiarities of 25 common forms of plant life are provided in this book in an entertaining and charming style. Covers food plants (potatoes, apples, beans, wheat, almonds, bananas, etc.), flowers (lily, tulip, etc.), trees (pine, oak, elm, etc.), weeds, poisonous mushrooms and vines, gourds, citrus fruits, cotton, the cactus family, and much more. 108 illustrations. xiv + 290pp. 22272-1 Paperbound $2.50

How to Know the Ferns, Frances T. Parsons. Classic survey of Eastern and Central ferns, arranged according to clear, simple identification key. Excellent introduction to greatly neglected nature area. 57 illustrations and 42 plates. xvi + 215pp. 20740-4 Paperbound $2.00

Manual of the Trees of North America, Charles S. Sargent. America's foremost dendrologist provides the definitive coverage of North American trees and tree-like shrubs. 717 species fully described and illustrated: exact distribution, down to township; full botanical description; economic importance; description of subspecies and races; habitat, growth data; similar material. Necessary to every serious student of tree-life. Nomenclature revised to present. Over 100 locating keys. 783 illustrations. lii + 934pp. 20277-1, 20278-X Two volumes, Paperbound $7.00

Our Northern Shrubs, Harriet L. Keeler. Fine non-technical reference work identifying more than 225 important shrubs of Eastern and Central United States and Canada. Full text covering botanical description, habitat, plant lore, is paralleled with 205 full-page photographs of flowering or fruiting plants. Nomenclature revised by Edward G. Voss. One of few works concerned with shrubs. 205 plates, 35 drawings. xxviii + 521pp. 21989-5 Paperbound $3.75

The Mushroom Handbook, Louis C. C. Krieger. Still the best popular handbook: full descriptions of 259 species, cross references to another 200. Extremely thorough text enables you to identify, know all about any mushroom you are likely to meet in eastern and central U. S. A.: habitat, luminescence, poisonous qualities, use, folklore, etc. 32 color plates show over 50 mushrooms, also 126 other illustrations. Finding keys. vii + 560pp. 21861-9 Paperbound $4.50

Handbook of Birds of Eastern North America, Frank M. Chapman. Still much the best single-volume guide to the birds of Eastern and Central United States. Very full coverage of 675 species, with descriptions, life habits, distribution, similar data. All descriptions keyed to two-page color chart. With this single volume the average birdwatcher needs no other books. 1931 revised edition. 195 illustrations. xxxvi + 581pp. 21489-3 Paperbound $5.00

AMERICAN FOOD AND GAME FISHES, David S. Jordan and Barton W. Evermann. Definitive source of information, detailed and accurate enough to enable the sportsman and nature lover to identify conclusively some 1,000 species and sub-species of North American fish, sought for food or sport. Coverage of range, physiology, habits, life history, food value. Best methods of capture, interest to the angler, advice on bait, fly-fishing, etc. 338 drawings and photographs. 1 + 574pp. 6⅝ x 9⅜.

22196-2 Paperbound $5.00

THE FROG BOOK, Mary C. Dickerson. Complete with extensive finding keys, over 300 photographs, and an introduction to the general biology of frogs and toads, this is the classic non-technical study of Northeastern and Central species. 58 species; 290 photographs and 16 color plates. xvii + 253pp.

21973-9 Paperbound $4.00

THE MOTH BOOK: A GUIDE TO THE MOTHS OF NORTH AMERICA, William J. Holland. Classical study, eagerly sought after and used for the past 60 years. Clear identification manual to more than 2,000 different moths, largest manual in existence. General information about moths, capturing, mounting, classifying, etc., followed by species by species descriptions. 263 illustrations plus 48 color plates show almost every species, full size. 1968 edition, preface, nomenclature changes by A. E. Brower. xxiv + 479pp. of text. 6½ x 9¼.

21948-8 Paperbound $6.00

THE SEA-BEACH AT EBB-TIDE, Augusta Foote Arnold. Interested amateur can identify hundreds of marine plants and animals on coasts of North America; marine algae; seaweeds; squids; hermit crabs; horse shoe crabs; shrimps; corals; sea anemones; etc. Species descriptions cover: structure; food; reproductive cycle; size; shape; color; habitat; etc. Over 600 drawings. 85 plates. xii + 490pp.

21949-6 Paperbound $4.00

COMMON BIRD SONGS, Donald J. Borror. 33⅓ 12-inch record presents songs of 60 important birds of the eastern United States. A thorough, serious record which provides several examples for each bird, showing different types of song, individual variations, etc. Inestimable identification aid for birdwatcher. 32-page booklet gives text about birds and songs, with illustration for each bird.

21829-5 Record, book, album. Monaural. $3.50

FADS AND FALLACIES IN THE NAME OF SCIENCE, Martin Gardner. Fair, witty appraisal of cranks and quacks of science: Atlantis, Lemuria, hollow earth, flat earth, Velikovsky, orgone energy, Dianetics, flying saucers, Bridey Murphy, food fads, medical fads, perpetual motion, etc. Formerly "In the Name of Science." x + 363pp.

20394-8 Paperbound $3.00

HOAXES, Curtis D. MacDougall. Exhaustive, unbelievably rich account of great hoaxes: Locke's moon hoax, Shakespearean forgeries, sea serpents, Loch Ness monster, Cardiff giant, John Wilkes Booth's mummy, Disumbrationist school of art, dozens more; also journalism, psychology of hoaxing. 54 illustrations. xi + 338pp.

20465-0 Paperbound $3.50

THE PHILOSOPHY OF THE UPANISHADS, Paul Deussen. Clear, detailed statement of upanishadic system of thought, generally considered among best available. History of these works, full exposition of system emergent from them, parallel concepts in the West. Translated by A. S. Geden. xiv + 429pp.
21616-0 Paperbound $3.50

LANGUAGE, TRUTH AND LOGIC, Alfred J. Ayer. Famous, remarkably clear introduction to the Vienna and Cambridge schools of Logical Positivism; function of philosophy, elimination of metaphysical thought, nature of analysis, similar topics. "Wish I had written it myself," Bertrand Russell. 2nd, 1946 edition. 160pp.
20010-8 Paperbound $1.50

THE GUIDE FOR THE PERPLEXED, Moses Maimonides. Great classic of medieval Judaism, major attempt to reconcile revealed religion (Pentateuch, commentaries) and Aristotelian philosophy. Enormously important in all Western thought. Unabridged Friedländer translation. 50-page introduction. lix + 414pp.
(USO) 20351-4 Paperbound $4.50

OCCULT AND SUPERNATURAL PHENOMENA, D. H. Rawcliffe. Full, serious study of the most persistent delusions of mankind: crystal gazing, mediumistic trance, stigmata, lycanthropy, fire walking, dowsing, telepathy, ghosts, ESP, etc., and their relation to common forms of abnormal psychology. Formerly *Illusions and Delusions of the Supernatural and the Occult.* iii + 551pp.   20503-7 Paperbound $4.00

THE EGYPTIAN BOOK OF THE DEAD: THE PAPYRUS OF ANI, E. A. Wallis Budge. Full hieroglyphic text, interlinear transliteration of sounds, word for word translation, then smooth, connected translation; Theban recension. Basic work in Ancient Egyptian civilization; now even more significant than ever for historical importance, dilation of consciousness, etc. clvi + 377pp. 6½ x 9¼.
21866-X Paperbound $4.95

PSYCHOLOGY OF MUSIC, Carl E. Seashore. Basic, thorough survey of everything known about psychology of music up to 1940's; essential reading for psychologists, musicologists. Physical acoustics; auditory apparatus; relationship of physical sound to perceived sound; role of the mind in sorting, altering, suppressing, creating sound sensations; musical learning, testing for ability, absolute pitch, other topics. Records of Caruso, Menuhin analyzed. 88 figures. xix + 408pp.
21851-1 Paperbound $3.50

THE I CHING (THE BOOK OF CHANGES), translated by James Legge. Complete translated text plus appendices by Confucius, of perhaps the most penetrating divination book ever compiled. Indispensable to all study of early Oriental civilizations. 3 plates. xxiii + 448pp.   21062-6 Paperbound $3.50

THE UPANISHADS, translated by Max Müller. Twelve classical upanishads: Chandogya, Kena, Aitareya, Kaushitaki, Isa, Katha, Mundaka, Taittiriyaka, Brhadaranyaka, Svetasvatara, Prasna, Maitriyana. 160-page introduction, analysis by Prof. Müller. Total of 670pp. 20992-X, 20993-8 Two volumes, Paperbound $7.50

MATHEMATICAL PUZZLES FOR BEGINNERS AND ENTHUSIASTS, Geoffrey Mott-Smith. 189 puzzles from easy to difficult—involving arithmetic, logic, algebra, properties of digits, probability, etc.—for enjoyment and mental stimulus. Explanation of mathematical principles behind the puzzles. 135 illustrations. viii + 248pp.
20198-8 Paperbound $2.00

PAPER FOLDING FOR BEGINNERS, William D. Murray and Francis J. Rigney. Easiest book on the market, clearest instructions on making interesting, beautiful origami. Sail boats, cups, roosters, frogs that move legs, bonbon boxes, standing birds, etc. 40 projects; more than 275 diagrams and photographs. 94pp.
20713-7 Paperbound $1.00

TRICKS AND GAMES ON THE POOL TABLE, Fred Herrmann. 79 tricks and games— some solitaires, some for two or more players, some competitive games—to entertain you between formal games. Mystifying shots and throws, unusual caroms, tricks involving such props as cork, coins, a hat, etc. Formerly *Fun on the Pool Table*. 77 figures. 95pp.
21814-7 Paperbound $1.25

HAND SHADOWS TO BE THROWN UPON THE WALL: A SERIES OF NOVEL AND AMUSING FIGURES FORMED BY THE HAND, Henry Bursill. Delightful picturebook from great-grandfather's day shows how to make 18 different hand shadows: a bird that flies, duck that quacks, dog that wags his tail, camel, goose, deer, boy, turtle, etc. Only book of its sort. vi + 33pp. 6½ x 9¼.   21779-5 Paperbound $1.00

WHITTLING AND WOODCARVING, E. J. Tangerman. 18th printing of best book on market. "If you can cut a potato you can carve" toys and puzzles, chains, chessmen, caricatures, masks, frames, woodcut blocks, surface patterns, much more. Information on tools, woods, techniques. Also goes into serious wood sculpture from Middle Ages to present, East and West. 464 photos, figures. x + 293pp.
20965-2 Paperbound $2.50

HISTORY OF PHILOSOPHY, Julián Marias. Possibly the clearest, most easily followed, best planned, most useful one-volume history of philosophy on the market; neither skimpy nor overfull. Full details on system of every major philosopher and dozens of less important thinkers from pre-Socratics up to Existentialism and later. Strong on many European figures usually omitted. Has gone through dozens of editions in Europe. 1966 edition, translated by Stanley Appelbaum and Clarence Strowbridge. xviii + 505pp.   21739-6 Paperbound $3.50

YOGA: A SCIENTIFIC EVALUATION, Kovoor T. Behanan. Scientific but non-technical study of physiological results of yoga exercises; done under auspices of Yale U. Relations to Indian thought, to psychoanalysis, etc. 16 photos. xxiii + 270pp.
20505-3 Paperbound $2.50

Prices subject to change without notice.
Available at your book dealer or write for free catalogue to Dept. GI, Dover Publications, Inc., 180 Varick St., N. Y., N. Y. 10014. Dover publishes more than 150 books each year on science, elementary and advanced mathematics, biology, music, art, literary history, social sciences and other areas.